Energy-Aware Memory Management for Embedded Multimedia Systems

A Computer-Aided Design Approach

T0239650

CHAPMAN & HALL/CRC
COMPUTER and INFORMATION SCIENCE SERIES

Series Editor: Sartaj Sahni

PUBLISHED TITLES

ADVERSARIAL REASONING: COMPUTATIONAL
APPROACHES TO READING THE OPPONENT'S MIND
Alexander Kott and William M. McEneaney

DISTRIBUTED SENSOR NETWORKS
S. Sitharama Iyengar and Richard R. Brooks

DISTRIBUTED SYSTEMS: AN ALGORITHMIC APPROACH
Sukumar Ghosh

ENERGY-AWARE MEMORY MANAGEMENT FOR
EMBEDDED MULTIMEDIA SYSTEMS: A COMPUTER-
AIDED DESIGN APPROACH
Florin Balasa and Dhiraj K. Pradhan

ENERGY EFFICIENT HARDWARE-SOFTWARE
CO-SYNTHESIS USING RECONFIGURABLE HARDWARE
Jingzhao Ou and Viktor K. Prasanna

FUNDAMENTALS OF NATURAL COMPUTING: BASIC
CONCEPTS, ALGORITHMS, AND APPLICATIONS
Leandro Nunes de Castro

HANDBOOK OF ALGORITHMS FOR WIRELESS
NETWORKING AND MOBILE COMPUTING
Azzedine Boukerche

HANDBOOK OF APPROXIMATION ALGORITHMS
AND METAHEURISTICS
Teofilo F. Gonzalez

HANDBOOK OF BIOINSPIRED ALGORITHMS
AND APPLICATIONS
Stephan Olariu and Albert Y. Zomaya

HANDBOOK OF COMPUTATIONAL MOLECULAR BIOLOGY
Srinivas Aluru

HANDBOOK OF DATA STRUCTURES AND APPLICATIONS
Dinesh P. Mehta and Sartaj Sahni

HANDBOOK OF DYNAMIC SYSTEM MODELING
Paul A. Fishwick

HANDBOOK OF ENERGY-AWARE AND GREEN
COMPUTING
Ishfaq Ahmad and Sanjay Ranka

HANDBOOK OF PARALLEL COMPUTING: MODELS,
ALGORITHMS AND APPLICATIONS
Sanguthevar Rajasekaran and John Reif

HANDBOOK OF REAL-TIME AND EMBEDDED SYSTEMS
Insup Lee, Joseph Y-T. Leung, and Sang H. Son

HANDBOOK OF SCHEDULING: ALGORITHMS, MODELS,
AND PERFORMANCE ANALYSIS
Joseph Y.-T. Leung

HIGH PERFORMANCE COMPUTING IN REMOTE SENSING
Antonio J. Plaza and Chein-I Chang

INTRODUCTION TO NETWORK SECURITY
Douglas Jacobson

LOCATION-BASED INFORMATION SYSTEMS:
DEVELOPING REAL-TIME TRACKING APPLICATIONS
Miguel A. Labrador, Alfredo J. Pérez, and
Pedro M. Wightman

METHODS IN ALGORITHMIC ANALYSIS
Vladimir A. Dobrushkin

PERFORMANCE ANALYSIS OF QUEUING AND COMPUTER
NETWORKS
G. R. Dattatreya

THE PRACTICAL HANDBOOK OF INTERNET COMPUTING
Munindar P. Singh

SCALABLE AND SECURE INTERNET SERVICES AND
ARCHITECTURE
Cheng-Zhong Xu

SPECULATIVE EXECUTION IN HIGH PERFORMANCE
COMPUTER ARCHITECTURES
David Kaeli and Pen-Chung Yew

VEHICULAR NETWORKS: FROM THEORY TO PRACTICE
Stephan Olariu and Michele C. Weigle

Energy-Aware Memory Management for Embedded Multimedia Systems

A Computer-Aided Design Approach

Edited by

Florin Balasa
Dhiraj K. Pradhan

CRC Press
Taylor & Francis Group
Boca Raton London New York

CRC Press is an imprint of the
Taylor & Francis Group, an **informa** business

A CHAPMAN & HALL BOOK

CRC Press
Taylor & Francis Group
6000 Broken Sound Parkway NW, Suite 300
Boca Raton, FL 33487-2742

First issued in paperback 2017

© 2012 by Taylor & Francis Group, LLC
CRC Press is an imprint of Taylor & Francis Group, an Informa business

No claim to original U.S. Government works

ISBN 13: 978-1-138-11290-2 (pbk)
ISBN 13: 978-1-4398-1400-0 (hbk)

Library of Congress Cataloging-in-Publication Data

Energy aware memory management for embedded multimedia systems : a
    computer-aided design approach / editors, Florin Balasa, Dhiraj Pradhan.
        p. cm. -- (Chapman & Hall/CRC computer and information science series)
    Includes bibliographical references and index.
    ISBN 978-1-4398-1400-0 (hbk. : alk. paper)
    1. Embedded computer systems--Computer-aided design. 2. Computer storage
devices. 3. Memory management (Computer science) 4. Energy conservation. I. Balasa,
Florin. II. Pradhan, Dhiraj K.

TK7895.E42E69 2012
006.2'2--dc23   2011038430

Visit the Taylor & Francis Web site at
http://www.taylorandfrancis.com

and the CRC Press Web site at
http://www.crcpress.com

Contents

Contents

Contributors

Florin Balasa
American University in Cairo
New Cairo, Egypt

Francky Catthoor
IMEC/Katholieke Unìversìteìt
  Leuven
Leuven, Belgium

Lakshmikantam Chitturi
Zoran Corp.
Sunnyvale, California

Philippe Clauss
Team CAMUS, INRIA
University of Strasbourg
Strasbourg, France

Stefan Cosemans
ESAT-MICAS
K.U. Leuven
Leuven, Belgium

Wim Dahaene
ESAT-MICAS
K.U. Leuven
Leuven, Belgium

Diego Garbervetsky
Departamento de Computación
Universidad de Buenos Aires
Buenos Aires, Argentina

Vincent Loechner
Team CAMUS, INRIA
University of Strasbourg
Strasbourg, France

Ilie I. Luican
Microsoft Corp.
Redmond, Washington

Doru V. Nasui
American Int. Radio, Inc.
Rolling Meadows, Illinois

Preeti Ranjan Panda
Department of Computer Science
  and Engineering
Indian Institute of Technology Delhi
New Delhi, India

Dhiraj K. Pradhan
University of Bristol
Bristol, United Kingdom

Praveen Raghavan
SSET, IMEC
Heverlee, Belgium

Ashoka Sathanur
ULP-DSP, IMEC
Eindhoven, The Netherlands

Guillermo Talavera
Universitat Autònoma de Barcelona
Barcelona, Spain

Ittetsu Taniguchi
Ritsumeikan University
Shiga, Japan

Sven Verdoolaege
Team ALCHEMY
INRIA Saclay
Saclay, France

Doran K. Wilde
Department of Electrical and
    Computer Engineering
Brigham Young University
Provo, Utah

Hongwei Zhu
ARM, Inc.
Sunnyvale, California

Chapter 1

Computer-Aided Design for Energy Optimization in the Memory Architecture of Embedded Systems

Florin Balasa
American University in Cairo, New Cairo, Egypt

Dhiraj K. Pradhan
University of Bristol, Bristol, United Kingdom

Contents

1.1 Introduction

Digital systems had initially only one main design metric: *performance*. Other cost parameters such as *area, energy consumption*, or *testability* were regarded as design constraints. In the early nineties, the role of power consumption changed from that of a design constraint to an actual design metric. This shift occurred due to technological reasons: higher integration and higher frequencies led to a significant increase in power consumption.[1]

The technology used to implement a digital design influences the analysis of the power consumption. For instance, in the complementary metal-oxide-semiconductor (CMOS) device, there are three sources of power dissipation.

---

[1] *Energy consumption* expresses the cumulative power consumption over time.

The most important source is the *switching power* P_{sw}, caused by the charging and discharging of load capacitances. The average switching power of a CMOS gate is given by the formula: $P_{sw} = \frac{1}{2}\alpha C_L V_{dd}^2 f_{clock}$, where C_L is the output load capacitance of the gate, V_{dd} is the supply voltage, f_{clock} is the clock frequency, and α is the switching activity—the probability for a transition to occur at each clock cycle. Low-power design solutions attempt to reduce the load capacitance and/or the switching activity (f_{clock} and V_{dd} being typically design constraints).

The *short-circuit power* P_{sc} is another component of power dissipation. It is caused by the current flow between V_{dd} and ground, originated due to different input and output rise/fall times. This short-circuit current can be kept under control by adequate design of the transition times of devices, so this component accounts for usually less than 5% of the overall power dissipation.

The third component is the *leakage power*. Whereas an ideal CMOS gate does not dissipate any static power, in practice, leakage currents cause power dissipation even in an off-state. In deep-submicron technologies, smaller supply and threshold voltages have caused this component to become more and more relevant.

This book will largely focus on the reduction of the switching power in the memory subsystem of embedded systems.

1.2   Low-Power Design for Embedded Systems

An *embedded system* is a computer system designed to perform one or a few dedicated functions often with real-time computing constraints [1]. It is embedded as part of a complete device often including hardware and mechanical parts. By contrast, a general-purpose computer, such as a personal computer (PC), is designed to be flexible and to meet a wide range of end-user needs. Embedded systems control many devices in common use today. Embedded systems are controlled by one or more main processing cores that are typically either microcontrollers or digital signal processors (DSP). The key characteristic, however, is being dedicated to handle a particular task, which may require very powerful processors.

Embedded systems span all aspects of modern life and there are many examples of their use. Telecommunication systems employ numerous embedded systems from telephone switches for the network to mobile phones at the end user. Computer networking uses dedicated routers and network bridges to route data. Consumer electronics include personal digital assistants (PDAs), mp3 players, mobile phones, video-game consoles, digital cameras, DVD players, GPS receivers, and printers. Many household appliances, such as microwave ovens, washing machines, and dishwashers, are including embedded systems to provide flexibility, efficiency, and features.

The design trade-off between performance and energy efficiency has become a major point of concern in embedded systems [2,3]. This is mainly due to the fact that such systems appear in electronic products available on the market that are portable and battery-operated (such as cellular phones and laptop computers). This implies that their functionality be fulfilled with energy delivered by a battery of minimum weight and size [4].

In the hardware platform of an embedded system, three types of operations are responsible for energy consumption: (1) data processing, (2) data transfers, and (3) data storage [2–4]. Since the software component of an embedded system does not have a physical realization, suitable models for estimating the software impact on the hardware energy consumption have been proposed (e.g., [5]). The choice of the software implementation may affect the energy consumption of the three operations performed by the hardware platform. For instance, software compilation affects the instructions used by the computing elements (processors, DSPs), each one with a specific energy cost, consequently, having an impact on the energy consumption of data processing.

1.3 The Role of On-Chip Memories

The key system design challenge is searching for the best-suited architecture for implementing the functional specification [4]. Designing from scratch a system architecture perfectly tailored to a given application is too time consuming. The typical system-level design strategy is to start from a generic system architecture (sometimes called *architectural template*) and customize it for the target application. This paradigm is known as *platform-based design* [6], and it was adopted by many ASIC developers and electronic design automation (EDA) tools. The function of the system is specified at a high level of abstraction (typically, in a high-level programming language) and the design becomes a top-down process of tailoring the abstract hardware template to the specification. EDA tools play an important part in this mapping process, since they facilitate the exploration of design alternatives, and the optimization of different components and design metrics.

In practice, most designs are derived from *processor-based templates*, where one or more processors and memory units are connected through an on-chip communication network (often a standardized bus), together with various peripherals, I/O controllers, and coprocessors (application-specific computational units), as shown in Figure 1.1.

Between processors and memories there is a substantial amount of traffic of information. Programmable processors are more demanding—compared to the application-specific data processors—in terms of memory bandwidth (i.e., the rate at which data can be read from or stored into a memory by a processor) because instructions must be fetched during the program execution. Providing

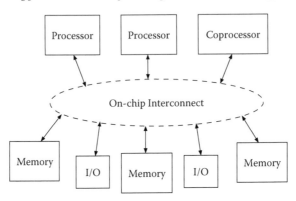

Figure 1.1   Processor-based template [4]. © 2002. With kind permission from Springer Science and Business Media.

sufficient bandwidth to sustain fast and energy-efficient program execution is a challenge for system designers. External memory chips can be interfaced through *printed circuit board* (PCB) interconnections to the embedded system, providing large storage space at a very low cost per bit. Unfortunately, due to data transfer from/to external memories, and the relatively high energy consumption per memory access, a fully off-chip storage would yield a poor performance and a large amount of dissipated energy. In response to this challenge, *on-chip* (also called *embedded*) memories took the center stage and started being exploited effectively [7].

1.4   Optimization of the Energy Consumption of the Memory Subsystem

In the earlier days of digital system design, memory was expensive; so, researchers focused on memory size optimization. Nowadays, the cost per memory bit is very low due to the progress of semiconductor technology and the consequent increase in the level of integration. Gradually, the memory size optimization decreased in importance, while performance and power consumption became the key challenges.

Memory latency (i.e., the time it takes to access a particular location in storage) and energy consumption per access increase with the memory size. Hence, memory may become a bottleneck—both in terms of energy and performance—for applications with large storage requirements [2,4,7], especially when *flat* memory architecture—that is, when data is stored in a single, off-chip memory—is adopted. Therefore, reducing the memory requirements of the target applications continues to be used as a first-phase design flow strategy for reducing the storage power budget and increasing performance. During this phase, the designer attempts to improve the temporal locality of data (i.e., the results of a computation should be used as soon as possible by next computations in order to reduce the need for temporary storage) by performing code (especially, loop) transformations on the behavioral specifications [8–11]. Data compression is another technique for reducing storage requirements, which targets finding efficient representations of data (e.g., [12]). However, data memory size has steadily increased over time due to the fact that system applications grew more complex.

Further efforts to tackle the memory bottleneck focused on (1) energy-efficient technologies and circuit design [13,14] and (2) hierarchical memory architectures. This book addresses the latter direction targeting low-power memory subsystems.

1.4.1 Hierarchical memory architectures

In general, lower levels in the hierarchy are made of small memories, close and tightly coupled to computation units; higher hierarchy levels are made of increasingly large memories, far from the computation units. The terms "close" and "far" imply here the effort needed to fetch or store a given amount of data from/to the memory. This effort can be expressed in units of time or units of energy, depending on the cost function.

Figure 1.2 shows a memory hierarchy with four levels [15]. Three levels of cache are on the same chip with the execution units. The cache size increases with the level of hierarchy. Every location contained into a low-level cache is contained in all higher-level caches (i.e., caches are said to be *fully inclusive*). The last level of the memory hierarchy is an off-chip dynamic random access memory (DRAM), organized in banks. The DRAM access is routed through a memory control and buffer that generates, for example, addresses row/column address strobes. To allow nonblocking external memory access, data from DRAM are routed through a data-path control buffer that signals the processor when the data is available. Average energy for accessing the level-0 cache is 1.5 nJ, the energy for level-1 cache access is 3 nJ, and for level-2 cache is 7 nJ; the average energy for accessing the external DRAM is 127 nJ—almost two orders of magnitude larger than the energy for accessing the lowest level of hierarchy [15].

The main objective of energy-aware memory design is to minimize the overall energy cost for accessing memory, within performance and memory size constraints [4].

Hierarchical memory organizations—like the one in Figure 1.2—reduce energy consumption by exploiting the nonuniformities of memory accesses; most applications access a relatively small amount of data with high frequency, while the rest of the data are accessed only a few times [16]. In a hierarchical memory organization, the reduction in power consumption can be achieved by assigning the frequently accessed data in low hierarchy levels.

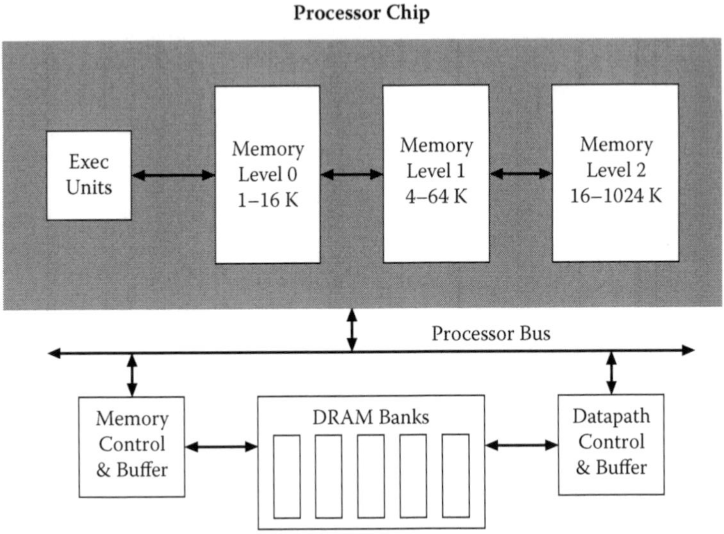

Processor Chip

Figure 1.2   Memory hierarchy example [15]. © 2000 IEEE.

The memory optimization targeted by this book starts from a given high-level behavioral specification (therefore, the memory access pattern is fixed) and aims to produce an energy-efficient customized memory hierarchy. Notice that the memory optimization problem could be addressed in different ways: for instance, one may assume a fixed memory hierarchy and modify the access pattern of the target application to optimally match the given hierarchy.

When comparing time and energy per access in a memory hierarchy, it may be observed that these two metrics often have similar behavior; namely, they both increase as we move from low to high hierarchy levels. While it sometimes happens that a low-latency memory architecture is also a low-power one, optimizing memory performance does not imply power optimization,[2] or vice-versa [17,18]. Shiue and Chakrabarti gave an example of a two-level memory hierarchy—an on-chip cache and an off-chip main memory, showing how energy and performance can have a contrasting effect [19]. There are two basic reasons for this: first, power consumption and performance do not increase in the same way with memory size and hierarchy level; second, performance is a worst-case quantity, while power is an average-case quantity: for instance, the removal of a critical computation that improves performance may be harmful in terms of power consumption.

As on-chip storage, the scratch-pad memories (SPMs)—compiler-controlled static random access memories (SRAMs), more energy-efficient than the hardware-managed caches—are widely used in embedded systems, where

---

[2]It is true though that some architectural solutions originally devised for performance optimization are also beneficial in terms of energy.

caches incur a significant penalty in aspects like area cost, energy consumption, and hit latency. A detailed study [20] comparing the trade-offs of caches as compared to SPMs found in their experiments that the latter exhibit 34% smaller area and 40% lower power consumption than a cache of the same capacity. Even more surprisingly, the runtime measured in cycles was 18% better with an SPM using a simple static knapsack-based allocation algorithm. As a general conclusion, the authors of the study found absolutely no advantage in using caches [21], even in high-end embedded systems in which performance is important.[3]

SPMs are used to *statically* store a portion of the off-chip memory (or a memory farther in the hierarchy). This is in contrast with caches that *dynamically* map a set of noncontiguous addresses from a slower, larger memory. Different from caches, the SPM occupies a distinct part of the virtual address space, with the rest of the address space occupied by the main memory. The consequence is that there is no need to check for the availability of the data in the SPM. Hence, the SPM does not possess a comparator and the miss/hit acknowledging circuitry [21]. This contributes to a significant energy (as well as area) reduction. Another consequence is that in cache memory systems, the mapping of data to the cache is done during the code execution, whereas in SPM-based systems this can be done at compilation time, using a suitable algorithm. SPMs are particularly effective in application-specific systems, running data-intensive applications, whose memory profile can be determined a priori, thus providing intuitive candidates for the data to be assigned into the SPMs.

The energy consumption caused by data storage has two components: the *dynamic energy* consumption—caused by memory accesses, and the *static energy* consumption—caused by leakage currents. As already explained, savings of dynamic energy can be potentially obtained by accessing frequently used data from smaller on-chip memories rather than from the large off-chip main memory, the problem being how to optimally assign the data to the memory layers. Note that this problem is basically different from caching for performance [22,23], where the question is to find how to fill the cache such that the data needed have been loaded in advance from the main memory.

With the scaling of the technology below 0.1 µm, the static energy due to leakage currents has become increasingly important. While leakage is a problem for any transistor, it is even more critical for memories: their high density of integration translates into a higher power density that increases temperature, which in turn increases leakage currents significantly. As technology scales, the importance of static energy consumption increases even when memories are idle (not accessed). To reduce the static energy, proper schemes to put a memory block into a dormant (*sleep*) state with negligible energy spending

---

[3]Caches have been a big success for desktops though, where the usual approach to adding SRAM is to configure it as a cache.

are required. These schemes normally imply a timing overhead: transitioning a memory block into and, especially, out of the dormant state consumes energy and time. Putting a memory block into the dormant state should be done only if the cost in extra energy and decrease in performance can be amortized. For dealing with dynamic energy, we are interested only in the total number of accesses, and not in their distribution over time. Introducing the *time* dimension makes the problem of energy reduction much more complex.

The storage allocation in a hierarchical organization (on- and off-chip) [7] and the energy-aware partitioning of the storage blocks into several banks [4] must be complemented with a comprehensive solution for mapping the multidimensional arrays from the code of the application into the physical memories.

1.4.2   A practical example

This section will give a brief overview of a commercial System-on-a-Chip (SoC)—the MPEG4 video codec SoC [24]—emphasizing the memory-related aspects.

The MPEG4 standard is one of the video coding methods used in wireless telephony, especially in the so-called *third-generation* mobile telephony— supporting transmission of data streams, as well as speech and data. Instead of a chip-set with multiple integrated circuits (ICs) that would have implied high-bandwidth I/O interfaces among the various ICs conducive to a significant power consumption, Takahashi et al. opted for a SoC solution that integrates on the same silicon substrate most of the digital baseband functionality [24]. This SoC implements a video codec, a speech codec, or an audio decoder, multiplexing and de-multiplexing between multiple video and speech/audio streams.

The overall block diagram of the SoC is shown in Figure 1.3. The chip contains a 16 Mb embedded DRAM and three signal processing cores: a video core, a speech/audio core, and a multiplexing core. Several peripheral interfaces (camera, display, audio, external CPU host for configuration) are also implemented on-chip.

Each of the major signal processing cores contains a 16-bit RISC processor and dedicated hardware accelerators. Data transfers among the three processors are performed via the DRAM. A virtual FIFO is configured on the DRAM for each processor pair, as well as shared control register used as *read* and *write* pointers.

In addition to the RISC processor, the video processing core contains a 4 KB instruction cache (I$) and an 8 KB data cache (D$). The cache sizing was determined by profiling video decoding applications. The video processor also includes several custom coprocessors: two discrete cosine transform (DCT) coprocessors, a motion compensation block (MCB), two (fine and coarse) motion

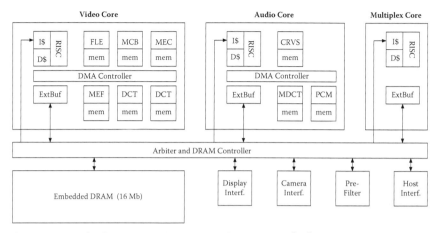

Figure 1.3 Architecture of the MPEG4 decoder [24].

estimation blocks (MEF and MEC), and a filter block (FLE). All these hardware accelerators have local SRAM buffers (5.3 KB in total) for speeding up local operations, thus limiting the number of accesses to the shared DRAM.

The audio core has a similar organization as the video core. It also contains an RISC processor with caches, but it includes different coprocessors. The multiplexing core contains an RISC processor and a network interface block, but no hardware accelerators.

This SoC was designed in a 0.25 μm variable-threshold CMOS technology.[4] The chip contains 20.5 million transistors; the chip area is 0.84×10.84 mm^2. The 16 Mb embedded DRAM occupies about 40% of the chip, and the various caches and SRAM buffers occupy roughly 20% of the area. Thus, more than 60% of the active chip area is dedicated to embedded memories. The chip consumes 240 mW at 60 MHz, and the memory access power (both to embedded SRAM and DRAM memories) is the dominant contributor to the overall chip power budget.

1.5 The Goal and Organization of the Book

According to the International Technology Roadmap for Semiconductors (ITRS) published by the Semiconductor Research Corporation, power management is one of the near-term grand challenges for the semiconductor industry: "Power management challenges need to be addressed across multiple

---

[4]In active mode, the threshold voltage of the transistors is 0.55 V; in standby mode, it is raised to 0.65 V in order to reduce leakage [24].

levels, especially system, design, and process technology.... The implementation challenges expand upward into system design requirements, the continuous improvement in computer-aided design (CAD), design tools for power optimization, and downward into leakage and performance requirements of new device architectures" (ITRS 2007, p.14). In the same document, the management of leakage power consumption is considered a long-term grand challenge. The goal of this book is in line with the strategic document mentioned above.

Memory system design for multiprocessor [25,26] and embedded systems has always been a crucial problem because system-level performance depends strongly on the memory organization. The storage organization in data-intensive (in the sense that their cost-related aspects, namely power consumption and footprint are heavily influenced, if not dominated, by the data access and storage aspects) multidimensional signal processing applications, particularly in the multimedia and telecommunication domains, has an important impact on the system-level energy budget. Such embedded systems are often designed under stringent energy consumption constraints, to limit heat generation and battery size.

The design of embedded systems warrants a new perspective because of the following two reasons. First, slow and energy-inefficient memory hierarchies have already become the bottleneck of the embedded systems (documented in the literature as the memory wall problem). Second, the software running on the contemporary embedded devices is becoming increasingly complex.

Contemporary system design focuses on the trade-off between performance and energy consumption in processing and storage units, as well as in their interconnections; moreover, on how to balance these two major cost parameters in the design of the memory subsystem. The heterogeneity of components and structures within embedded systems and the possibility of using application-specific storage architectures has added a new dimension to memory system design.

The high-level goal of this book is to present more recent CAD ideas for addressing memory management tasks, in particular, the optimization of energy consumption in the memory subsystem and, also, CAD solutions (i.e., theoretical methodologies, novel algorithms, and their efficient implementation). Wherever possible, we tried to adopt an algorithmic style that will help electronic design automation (EDA) researchers and tool developers to create prototype software tools for the system-level exploration in order to obtain an optimized architectural solution of the memory subsystem.

As the design of the memory subsystem (or any memory management task) is nowadays compiler-oriented (since the input is the high-level specification of the application), an introductory sections on data-dependence analysis techniques will be necessary. This chapter reviews and illustrates classic algorithms for solving Diophantine linear systems, bringing a matrix to a Hermite reduced form, the Fourier–Motzkin elimination, counting the number of points in **Z**-polytopes (bounded and closed polyhedra restricted to the points having integer coordinates) and the more recent advance Barvinok polynomial-time algorithm, computing the vertices of a polytope from its hyperplane representation, elimination of redundant inequalities.

The energy-aware optimization of the foreground memory organization will be addressed in Chapter 5.

Historically, the research for the evaluation of the data storage requirements can be classified in five categories: (1) memory size estimation methods for procedural, nonparametric, specifications—where the loop structure and sequence of instructions induce the (fixed) execution ordering; (2) memory size estimation methods for non-procedural, non-parametric, specifications—where the execution ordering is still not (completely) fixed; (3) memory size estimation for specifications with dynamically allocated data structures; (4) exact computation of the minimum data memory size for procedural, nonparametric, specifications; (5) memory size estimation methods for procedural, parametric, specifications. Two chapters of the book present works focusing on the more recent algorithms from the last two categories: (a) the exact computation of the minimum data storage for procedural specifications based on the decomposition of the array references from the algorithmic specification into disjoint linearly bounded lattices and (b) the memory size estimation approach for parametric specifications—technique based on the Bernstein expansion of polynomials that is used to compute their upper bounds over parametric convex domains.

While the evaluation of the data storage and the memory-aware loop transformations play an important role in the system-level exploration stage, signal-to-memory mapping models and algorithms are crucial to an energy-aware memory management methodology, as it will become apparent from the book. This task, solving the memory allocation and assignment problem, has the following goals: (a) to map the data structures from the behavioral specification into an amount of data storage as small as possible; (b) to compute this amount of storage (after mapping) and be able to determine the memory location of any scalar signal from the specification; (c) to use mapping functions simple enough in order to ensure an address generation hardware of a reasonable complexity; (d) to ascertain that any distinct scalar signals simultaneously alive are mapped to distinct storage locations. Chapter 6 will present a polyhedral framework allowing an efficient implementation of canonical-linearization and bounding-window mapping models, together with metrics for the quality assessment of the memory allocation solutions. Extensions of the mapping approaches into hierarchical memory organizations, as well as a recent model exploiting the possibility of memory sharing between elements of distinct arrays, will be finally presented.

While the memory management tasks addressed in Chapters 3 and 4 seemed to be mainly aimed to the reduction in the chip area (e.g., by means of reducing the data memory size), other sections focus explicitly on the reduction in the energy consumption in the memory subsystem. This latter goal (minimization of the energy consumption) is not orthogonal to the former one (minimization of the area); it only adds a new dimension to the optimization problem. Chapter 6 will address the problem of reducing the dynamic energy in the memory subsystem, that is, the energy which expands only when memory accesses occur. Savings of dynamic energy can be obtained by accessing frequently used data from smaller on-chip memories rather than from large background (off-chip) memories. As on-chip storage, the scratch-pad memories (SPMs) are widely used in embedded systems. Whereas in cache memory systems, the mapping of data to the cache is done during run time, in SPM-based systems this can be done either manually by the designer, or automatically, using a suitable mapping algorithm. One of the problems to be addressed is the optimal assignment of the signals to the memory layers (off-chip and scratch-pad). A formal approach operating with lattices, allowing to identify with accuracy those parts of multidimensional arrays intensely accessed by read/write memory operations is presented. Storing on-chip these highly accessed parts of arrays will yield the highest reduction in the dynamic energy consumption in a hierarchical memory subsystem. The reduction in the dynamic energy is not only achieved by partitioning the data between the memory layers, but also by splitting the virtual address space into independently accessed smaller memory blocks and partitioning the data between these blocks. The memory banking approaches have the advantage of reducing the static power consumption as well. For the reduction of the static power consumption (Chapters 7 and 8), we are also interested in the distribution of memory accesses over time.

The increased bandwith offered by multiport memories makes them attractive architectural candidates in performance-sensitive system designs, since they can lead to significantly shorter schedules. Chapter 9 will discuss techniques for assigning memory accesses to the ports of multiport memories.

Chapter 10 will provide a survey of method and techniques that optimize the address generation task for embedded systems, explaining current research trends and needs for the future. Since more and more sophisticated architectures with a high number of computational resources running in parallel are emerging, the access to data is becoming the main bottleneck that limits the available parallelism. To alleviate this problem, in current embedded architectures, a special unit, the address generator, works in parallel with the main computing resources to ensure efficient feed and storage of the data. Future architectures will have to deal with enormous memory bandwidths in distributed memories and the development of address generator units will be crucial since global trade-offs between reaction-time, bandwidth, area, and energy must be achieved. Address calculations often involve linear and polynomial arithmetic expressions, which must be evaluated during program execution under strict timing constraints. Memory address computation can significantly degrade the performance and increase power consumption. This chapter focuses on architectures for the address generator units and on compilation techniques to optimize the address generation process for scratch-pad memories subject to the power restrictions of the embedded systems. The emphasis is on address generation for multimedia signal processing, where the applications are data-intensive.

References

1. S. Heath, *Embedded Systems Design*, EDN series for design engineers (2nd ed.), 2003.

2. F. Catthoor, S. Wuytack, E. De Greef, F. Balasa, L. Nachtergaele, and A. Vandecappelle, *Custom Memory Management Methodology: Exploration of Memory Organization for Embedded Multimedia System Design*, Boston: Kluwer Academic Publishers, 1998.

3. P.R. Panda, F. Catthoor, N. Dutt, K. Dankaert, E. Brockmeyer, C. Kulkarni, and P.G. Kjeldsberg, "Data and memory optimization techniques for embedded systems," *ACM Trans. Design Automation of Electronic Syst.*, vol. 6, no. 2, pp. 149–206, April 2001.

4. A. Macii, L. Benni, and M. Poncino, *Memory Design Techniques for Low Energy Embedded Systems*, Boston: Kluwer Academic Publishers, 2002.

5. T.C. Lee, S. Malik, V. Tiwari, and M. Fujita, "Power analysis and minimization techniques for embedded DSP software," *IEEE Trans. VLSI Systems*, vol. 5, no. 1, pp. 123–135, March 1997.

6. K. Keutzer, A. Newton, J. Rabaey, and A. Sangiovanni-Vincentelli, "System-level design: Orthogonalization of concerns and platform-based design," *IEEE Trans. Computer-Aided Design on IC's and Systems*, vol. 19, no. 12, pp. 1523–1543, Dec. 2000.

7. P.R. Panda, N.D. Dutt, and A. Nicolau, "On-chip vs. off-chip memory: the data partitioning problem in embedded processor-based systems," *ACM Trans. Design Automation for Electronic Systems*, vol. 5, no. 3, pp. 682–704, 2000.

8. M.E. Wolf and M.S. Lam, "A data locality optimization algorithm," *Proc. ACM SIGPLAN'91 Conf.*, Toronto, Canada, pp. 30–44, June 1991.

9. A. Darte, "On the complexity of loop fusion," *Parallel Computing*, vol. 26, no. 9, pp. 1175–1193, 2000.

10. Q. Hu, A. Vandecapelle, M. Palkovic, P.G. Kjeldsberg, E. Brockmeyer, and F. Catthoor, "Hierarchical memory size estimation for loop fusion and loop shifting in data-dominated applications," *Proc. Asia & South-Pacific Design Automation Conf.*, Yokohama, Japan, pp. 606–611, Jan. 2006.

11. F. Balasa, P.G. Kjeldsberg, M. Palkovic, A. Vandecappelle, and F. Catthoor, "Loop transformation methodologies for array-oriented memory management," *Proc. IEEE Int. Conf. on Application-Specific Systems, Architectures, and Processors*, Steamboat Springs, CO, pp. 205–212, Sept. 2006.

12. G. Chen, M. Kandemir, N. Vijaykrishnan, M.J. Irwin, and W. Wolf, "Energy savings through compression in embedded Java environments," *Proc. Int. Conf. Hardware-Software Codesign and System Synthesis*, 2002.

13. K. Itoh, K. Sasaki, and Y. Nakagome, "Trends in low-power RAM circuit technologies," *Proc. of the IEEE*, vol. 83, no. 4, pp. 524–543, April 1995.

14. S. Bhattacharjee and D.K. Pradhan, "LPRAM: A novel low-power RAM design with testability," *IEEE Trans. Computer-Aided Design on IC's and Systems*, vol. 23, no. 5, pp. 637–651, May 2004.

15. U. Ko, P.T. Balsara, and A.K. Nanda, "Energy optimization of multi-level cache architectures for RISC and CISC processors," *IEEE Trans. on VLSI Syst.*, vol. 6, no. 2, pp. 299–308, June 1998.

16. F. Balasa, I.I. Luican, H. Zhu, and D.V. Nasui, "Automatic generation of maps of memory accesses for energy-aware memory management," *Proc. IEEE Int. Conf. on Acoustics, Speech, and Signal Processing*, pp. 629–632, Taipei, Taiwan, April 2009.

17. C.L. Su and A.M. Despain, "Cache design trade-offs for power and performance optimization: A case study," *Proc. ACM/IEEE Int. Symposium on Low-Power Design*, pp. 63–68, Dana Point CA, April 1995.

18. R.I. Bahar, G. Albera, and S. Manne, "Power and performance trade-offs using cache strategies," *Proc. ACM/IEEE Int. Symposium on Low-Power Design*, pp. 64–69, Monterey CA, Aug. 1998.

19. W. Shiue and C. Chakrabarti, "Memory exploration for low power embedded systems," *Proc. 36th ACM/IEEE Design Automation Conf.*, pp. 140–145, New Orleans LA, June 1999.

20. R. Banakar, S. Steinke, B.-S. Lee, M. Balakrishnan, and P. Marwedel, "Comparison of cache- and scratch-pad based memory systems with respect to performance, area and energy consumption," *Technical Report #762*, University of Dortmund, Sept. 2001.

21. R. Banakar, S. Steinke, B.-S. Lee, M. Balakrishnan, and P. Marwedel, "Scratchpad memory: A design alternative for cache on-chip memory in embedded systems," *Proc. 10th Int. Workshop on Hardware/Software Codesign*, Estes Park, CO, May 2002.

22. J.Z. Fang and M. Lu, "An iteration partition approach for cache or local memory thrashing on parallel processing," *IEEE Trans. Computers*, vol. 42, no. 5, pp. 529–546, 1993.

23. N. Manjikian and T. Abdelrahman, "Reduction of cache conflicts in loop nests," *Technical Report CSRI-318*, Computer Systems Research Institute, Univ. Toronto, 1995.

24. M. Takahashi, T. Nishikawa, M. Hamada, T. Takayanagi, H. Arakida, N. Machida, H. Yamamoto, T. Fujiyoshi, Y. Ohashi, O. Yamagishi, T. Samata, A. Asano, T. Terazawa, K. Ohmori, Y. Watanabe, H. Nakamura, S. Minami, T. Kuroda, and T. Furuyama, "A 60 MHz 240 mW MPEG-4 videophone LSI with 16 Mb embedded DRAM," *IEEE J. Solid-State Circuits*, vol. 35, no. 11, pp. 1713–1721, Nov. 2000.

25. M. Verma and P. Marwedel, *Advanced Memory Optimization Techniques for Low-Power Embedded Processors*, Springer, Dordrecht, Netherlands, 2007.

26. F. Catthoor, K. Danckaert, C. Kulkarni, E. Brockmeyer, P.G. Kjeldsberg, T.V. Achteren, and T. Omnes, *Data Access and Storage Management for Embedded Programmable Processors*, Kluwer Academic Publishers, Boston, 2010.

Chapter 2

The Power of Polyhedra

Doran K. Wilde
Brigham Young University
Provo, Utah

Contents

2.1 Introduction

Polyhedra have been studied in several unrelated fields: from the geometric point of view by computational geometrists [1], from the algebraic point of view by the operations research and linear programming communities [2], and from the structural/lattice point of view by the combinatorics community [3]. Each community takes a different view of polyhedra, and they sometimes use different notation and terminology; however, they all share a common need to be able to represent and perform computations with polyhedra.

In computer science, we use polyhedra as part of the abstract model for representing the index domains over which variables are defined and the index domains over which mathematical equations are evaluated. Operations on polyhedra are needed for doing static analysis of programs involving these equations to help in debugging, performing code optimizations, and specifying program transformations. The study and analysis of an important class of static programs became feasible in the early 1990s when we started modeling index domains as the lattice of integer points in a polyhedron or union of polyhedra. This is the reason that the understanding of polyhedra is so important in this book.

2.1.1   Equational languages

When thinking about algorithms such as those used in signal processing or numerical analysis applications, a person naturally thinks in terms of mathematical equations. Mathematical notation obeys certain fundamental rules: 1. Given a function and an input, the same output must be produced each time the function is evaluated. If the function is time varying, then time must be a parameter to the function. The term *static* is given to describe this property. 2. Consistency in the use of names: a variable stands for the same value throughout its scope. This is called *referential transparency*. An immediate consequence of referential transparency is that equality is substitutive—equal expressions are always and everywhere interchangeable. These properties are what gives mathematical notation its deductive power [4].

Using a language that shares the properties of mathematical notation eases the task of representing a mathematical algorithm as a program. An equation specifies an assertion on a variable which must always and everywhere be true. Reasoning about programs can thus be done in the context of the program itself, and relies essentially on the fact that equational programs respect the **substitution principle**. This principle states that an equation $X = Expression$ specifies a total synonymy between the variable on the left-hand side of the equation and the expression on the right-hand side of the equation. Thus any instance of a variable found on the left-hand side of any equation may be replaced with the corresponding expression on the right-hand side. Likewise, any subexpression may be replaced with a variable identifier, provided that an equation exists, or one is introduced, in which that variable is defined to be equal to that subexpression.

A system of recurrence equations is a natural formalism for expressing an equational language and it shares both of the two key properties of mathematical notation: it is static and it is referentially transparent. Systems of recurrence equations have the following important properties:

- Recurrence equations are written unordered. They are executed on a demand-driven basis, independent of the order they are written.

- Recurrence equations are single assignment. Each variable element can only ever hold a single value, which is computed once and then does not change.

- A system of recurrence equations is a *static program*, meaning that its execution behavior can be analyzed at compile time.

- Recurrence equations do not support any notion of global variables. The execution of a system of recurrence equations only affects the outputs of the system—there are never any side effects.

- Recurrence equations are strongly typed. Each variable is either explicitly or implicitly predeclared as a domain of index points that map to values of a specific data type.

Systems of recurrence equations adopt the classic principles of functional languages that are structured and strongly typed. A system of recurrence equations defines (possibly recurrent) functions that map system input variables to system output variables. The notion of a functional language is embedded in the definition of a system of recurrence equations.

2.1.2 Systems of recurrence equations

In this section we define the basic concepts of systems of recurrence equations. Recurrence equations are able to formally represent algorithms that have a high degree of regularity and parallelism because recurrence equations are not necessarily executed in any fixed or sequential order, but rather can be executed using the simple demand-driven semantics that respect the causality requirements of data dependences.

The formalism of recurrence equations has often been used in various forms by several authors [5–7], all of which were based on the introduction of uniform recurrence equations by Karp et al. [8]. These definitions given as follows are taken primarily from the work of Rajopadhye and Fujimoto [5], Yaacoby and Cappello [9], Delosme and Ipsen [10], and Quinton and Van Dongen [11].

In describing systems of recurrence equations, there are two equally valid points of view that can be taken. The first is a purely functional point of view in which every variable identifier is a *function*. A recurrence equation defines a function on the left hand side in terms of the functions on the right hand side. Alternately, each identifier can be thought of as a *single assignment variable* and equations equate the variable on the left hand side to a function of variables on the right.

Definition 2.1 (Recurrence Equation) *A **Recurrence Equation** over a domain $\mathcal{D}$ is defined to be an equation of the form*

$$f(z) = g(f_1(I_1(z)), f_2(I_2(z)), \dots, f_k(I_k(z)))$$

where

- *$f(z)$ is an array variable indexed by z, an n-dimensional vector of integers. It can also be thought of as a function mapping index points z in its domain to values in its datatype.*

- *$z \in \mathcal{D}$, where $\mathcal{D}$ is the (possibly parameterized) index domain of variable f.*

- *$f_1, \dots, f_k$ are any number of array variables found on the right-hand side of the equation. They may include the variable f itself, or multiple instances of any variable.*

- *$I_1, \dots, I_k$ are index mapping functions (also called dependency mapping functions) that map index point $z \in \mathcal{D}$ to index point $I_i(z) \in \mathcal{D}_i$,* where $\mathcal{D}, \mathcal{D}_1, \ldots, \mathcal{D}_k$ are the (possibly parameterized) index domains of variables $f, f_1, \ldots, f_k$, respectively.

- g is a strict single-valued function whose complexity is $\mathcal{O}(1)$ (can be executed in a fixed number of clock cycles) defining the right hand side of the equation.

Example 2.1 Let's look at the definition of a recurrence equation using an example.

$$x(i) = \frac{b(i) - \sum_{j=i+1}^{n} u(i,j) \cdot x(j)}{u(i,i)}$$

is a recurrence equation used to solve an $n \times n$ upper triangular linear system $Ux = b$ [12].

Variables x and b are defined over index domains $\{i \mid 1 \leq i \leq n\}$ and variable u is defined over index domain $\{i, j \mid 1 \leq i \leq n; i \leq j \leq n\}$. Notice that only the nonzero upper triangular matrix values of u are stored. In all three variables, all index points in the variable domains are mapped to real values, or in other words, all three variables are of floating point type.

The index mapping functions for variables on the right-hand side of the equation are:

$b(i) : I(i) = i$, a one-dimensional identity function
$u(i,j) : I(i,j) = (i,j)$, a two-dimensional identity function
$x(j) : I(i,j) = (j)$, an affine projection function
$u(i,i) : I(i) = (i,i)$, an affine function.

The strict-valued function g is composed of multiplication and summation operations, followed by a subtraction and division operation.

A variation of an equation allows f to be defined by a finite number of disjoint "cases" consisting of subdomains of the index space whose union is f's entire index domain:

$$f(z) = \begin{cases} z \in \mathcal{D}_1 & \Rightarrow & g_1(\ldots f_1(I_1(z)) \cdots) \\ z \in \mathcal{D}_2 & \Rightarrow & g_2(\ldots f_2(I_2(z)) \cdots) \\ \vdots & & \end{cases} \tag{2.1}$$

where the index domain of variable f is $\mathcal{D} = \bigcup_i \mathcal{D}_i$ and $(i \neq j) \rightarrow (\mathcal{D}_i \cap \mathcal{D}_j = \emptyset)$.

Example 2.2

$$fib(i) = \begin{cases} (i) \in \{i \mid 0 \leq i \leq 1\} \Rightarrow 0 \\ (i) \in \{i \mid i \geq 2\} \quad \Rightarrow fib(i-1) + fib(i-2) \end{cases}$$

is a recurrence equation to compute the Fibonacci sequence. There are two disjoint cases that partition the computation into two distinct subdomains whose union is $\{i \mid i \geq 0\}$, the entire domain of the variable *fib*.

Definition 2.2 (Dependency) *For a system of recurrences, we say that an array variable f_i at index point $p \in D_i$ (directly)* **depends on** *array variable f_j at index point q, (denoted by $p_i \mapsto q_j$), whenever $f_j(q)$ occurs on the right-hand side of the equation defining $f_i(p)$.*

The transitive closure of this relation is called the **dependency relation,** *denoted by $p_i \rightarrow q_j$.*

A dependence is called a **uniform dependence** *if it is of the form $z \mapsto z + b$, where b is a constant vector. It is called an* **affine dependence** *if it is of the form $z \mapsto Az + b$, where A is a constant matrix and b is a constant vector. Uniform dependences are also affine dependences with A being the identity matrix.*

Example 2.3

$(i) \mapsto (i)$	uniform dependence	$I(z) = z + (0)$
$(i) \mapsto (i-1)$	uniform dependence	$I(z) = z + (-1)$
$(i) \mapsto (i+1, i+2)$	affine dependence	$I(z) = \begin{bmatrix} 1 \\ 1 \end{bmatrix} \cdot z + \begin{bmatrix} 1 \\ 2 \end{bmatrix}$
$(i) \mapsto (i^2)$	neither uniform nor affine	
$(i, j) \mapsto (j, i)$	affine dependence	$I(z) = \begin{bmatrix} 0 & 1 \\ 1 & 0 \end{bmatrix} \cdot z + \begin{bmatrix} 0 \\ 0 \end{bmatrix}$
$(i, j) \mapsto (i-3, j+2)$	uniform dependence	$I(z) = z + \begin{bmatrix} -3 \\ 2 \end{bmatrix}$
$(i, j) \mapsto (j)$	affine dependence	$I(z) = \begin{bmatrix} 0 & 1 \end{bmatrix} \cdot z + \begin{bmatrix} 0 \end{bmatrix}$

Definition 2.3 (Uniform and Affine Recurrence Equations) *A recurrence equation of the form defined above is called a* **uniform recurrence equation** *(URE) if all of the dependency functions I_i are of the form $I(z) = z + b$, where b is a constant vector. It is called an* **affine recurrence equation** *(ARE) if $I(z) = Az + b$, where A is a constant matrix and b is a constant vector.*

Example 2.4 Affine and Uniform Equations
An example of an affine recurrence equation is:

$$x(i) = \sum_{j=1}^{N} y(i, j)$$

The variable N is a problem-size parameter. This equation has the affine dependence $(i) \mapsto (i, j)$ that is many-to-one. The summation along with the many-to-one dependence makes this type of equation a *reduction*. This reduction can be eliminated by a technique called *serialization*. After serialization, the following recurrence equations are produced that perform the same computation as the reduction above.

$$s(i,j) = \begin{cases} (i,j) \in \{i,j \mid j = 0\} & \Rightarrow 0 \\ (i,j) \in \{i,j \mid 1 \leq j \leq N\} \Rightarrow s(i,j-1) + y(i,j) \end{cases}$$

$$x(i) = s(i,N)$$

The first equation is uniform, having uniform dependences (s to s) $(i,j) \mapsto (i,j-1)$ and (s to y) $(i,j) \mapsto (i,j)$. The second equation is an affine recurrence having the affine dependence (x to s) $(i) \mapsto (i,N)$ that is one-to-one for a given parameter value N.

Definition 2.4 (System of Affine Recurrence Equations, or SARE)
A **system** *of recurrence equations is a set of m affine recurrence equations defining the array functions $f_1 \ldots f_m$ over index domains $\mathcal{D}_1 \cdots \mathcal{D}_m$, respectively. The equations may be (mutually) recursive. Array variables are designated as either input, output, or local variables of the system. Each variable (that is not a system input) appears on the left-hand side of an equation once and only once. Variables may appear on the right-hand sides of equations as often as needed.*

Example 2.5

Back-Substitution Function

```
Declarations:
parameter { N | N>0 }
input float { A[i,j] | 1<=i<=N; 1<=j<=i }
input float { B[i] | 1<=i<=N }
output float { X[i] | 1<=i<=N }
local float { b[i,j] | 1<=i<=N; 0<=j<i }

System of recurrence equations:
b[i,j] = (i,j) in { j=0 } => 0;
         (i,j) in { j>0 } => A[i,j] * X[j] + b[i,j-1];
X[i]   = ( B[i] - b[i,i-1] )/A[i,i];
```

Such equations serve as a purely functional definition of a computation and are in the form of a **static program**—a program whose dependency graph can be statically determined and analyzed (for any given instance of the parameters). Static programs require that all g_i be strict functions and that any conditional expressions be limited to linear inequalities involving the indices of the left-hand side variable and not array element values. By convention, it is assumed that input values are all specified whenever needed for any function evaluation. In spite of the limitations of systems of affine recurrence equations, they are very expressive.

2.1.3   Using polyhedra to model recurrence equations

Systems of affine recurrence equations are most often based on array variables with polyhedral-shaped domains. The domain of an equation or expression is the index domain that limits the range of indices over which a computation is performed and is also called the control domain or context domain. Index domains consist of the integer index points that lie inside a polyhedron, or union of polyhedra. The semantics of recurrence equations guarantee that a control domain can always be statically derived for every computation; that is, by analyzing a system of affine recurrence equations, you can always know in advance the set of index points for which any expression needs to be computed. By designating certain indices as time and other indices as space, all temporal and spatial control information that determines when and where a computation is performed is also contained in these context domains. Formal program transformations can be done by manipulating control domains and reformulating recurrence equations using polyhedral operations until the control domains and equations are expressed in a desired form. Basic transformations can be mathematically proven to be correct, and thus a program that is derived by starting from a known correct program and incrementally modifying it by applying a chain of these basic transformations; the resulting program is necessarily correct by construction.

In a related area, static nested loop programs can also be analyzed using polyhedra since nested loops with affine lower and upper bounds can be described in terms of polyhedral index spaces. The spatial point of view of a loop nest goes back to the work of Kuck [13] who showed that the domain of nested loops with affine lower and upper bounds can be described in terms of a polyhedron. Loop reindexing can then be done using transformations based on polyhedral operations. Program transformations such as change of index basis, localization of communication, and space-time mapping, etc. all require that polyhedral operations be performed as part of the transformation.

2.1.4   Domain-based array variables

Domain-based array variables are an abstraction of the standard array variables used in modern programming languages. They are a powerful generalization of the standard variable that has fewer shape restrictions and more representational power. For instance, in C, array variables can only be rectangularly shaped. A triangular-shaped variable like the variable u in Example 2.1 has to be embedded in a rectangular-shaped variable, which wastes storage.

As noted in the definition of a recurrence equation, all variables must have an associated index domain that maps to a given datatype. This information can be explicitly given as part of the variable declaration, or implicitly computed by analysis. Each variable is a function over a domain of the integer lattice points $\mathcal{Z}^n$ that lie inside a polyhedron, or union of polyhedra. When specifying recurrence equations, unions of polyhedra are used to describe the domains of computation of system variables.

Definition 2.5 *A* **variable** X *of type* "datatype" *declared over an index domain $\mathcal{D}$ is defined as*

$$X = \{X[z] \mid X[z] \in \text{datatype}, z \in \mathcal{D}\} \tag{2.2}$$

where $X[z]$ is the element of X corresponding to the index point vector z in index domain $\mathcal{D}$ and "datatype" *is a number representation such as integer, boolean, or real.*

2.1.5 Polyhedra and operations on polyhedra

In this section, we briefly and informally introduce a polyhedron with its dual algebraic and geometric definitions. A more formal definition will be given later in Section 2.2.

Definition 2.6 *A* **polyhedron** *can be defined two ways:*
    Geometric definition:
A **polyhedron**, $\mathcal{P}$ *is a convex subspace of $\mathcal{Q}^n$ (rational or real space) bounded by a finite number of hyperplanes.*
    Algebraic definition:
A **polyhedron**, $\mathcal{P}$ *is the intersection of a finite family of closed linear half-spaces $\{x \mid ax \geq c\}$ where a is a nonzero row vector and c is a scalar constant.*

From the algebraic definition, it follows that the set of solution points that satisfy a mixed system of linear constraints form a polyhedron $\mathcal{P}$ and serve as the *implicit definition* of the polyhedron

$$\mathcal{P} = \{x \mid Ax = b, Cx \geq d\} \tag{2.3}$$

given in terms of equations (rows of A, b) and inequalities (rows of C, d), where A, C are matrices, and b, d and x are vectors.

In representing index spaces of variables, we use the term *polyhedral index domain* that is based on the polyhedron but is not actually a polyhedron. A polyhedron is a region containing an infinite number of rational (or real) points, whereas a *polyhedral domain*, as the term is used in this chapter, refers

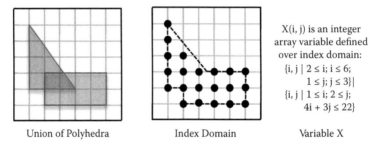

Union of Polyhedra   Index Domain   Variable X

Figure 2.1 Comparison of polyhedra, domain, and variable.

to the lattice of integral points $\mathcal{Z}^n$ that are inside a polyhedron (or union of polyhedra). Figure 2.1 illustrates this difference.

Definition 2.7 *A polyhedral* **index domain** *of dimension n is defined as*

$$\mathcal{D} : \{z \mid z \in \mathcal{Z}^n, z \in \mathcal{P}_1 \cup \mathcal{P}_2 \cup \cdots \cup \mathcal{P}_m\} = \mathcal{Z}^n \cap (\mathcal{P}_1 \cup \mathcal{P}_2 \cup \cdots \cup \mathcal{P}_m) \quad (2.4)$$

where $\mathcal{P}_1 \cup \mathcal{P}_2 \cup \cdots \cup \mathcal{P}_m$ *is a union of finitely many polyhedra of dimension n.*

In a system of recurrence equations, every variable is declared over an index domain as just described. Elements of a variable are in a one-to-one correspondence with index point vectors in an index domain as illustrated in Figure 2.2 and defined in Definition 2.5.

The syntax we will use for representing a single polyhedral domain is as follows:

```
{ <index-list> | <constraint-list> }
```

More complicated domains can be built up using the three domain operators: union, intersection, and difference. Union is written by combining two domains with a vertical bar: { ... } | { ... }, intersection is written with an ampersand: { ... } & { ... }, and difference with an ampersand-tilde: { ... } &~{ ... }

Example 2.6 Some examples of index domains are given below:

```
{ x,y,z | -5 <= x-y <= 5; -5 <= x+y <= 5; z = 2x - 3y }
{ i,j | 0 <= i <= N-1; N <= j <= 2N-1 }    -- domain parameterized by N
{ i,j,k | k = 5 }                          --   a plane in 3 space
{ i,j | 1>=0 }                   -- 2 dimensional universe domain
{ i | 1 = 0 }                    -- 1 dimensional    empty domain
{ i,j | i=1; 0<=j<=2} | { i,j | i=3; 1<=j<=5 }   -- union of domains
```

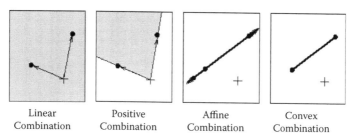

| Linear | Positive | Affine | Convex |
| Combination | Combination | Combination | Combination |

Figure 2.2 Geometric interpretations of the combinations of two points (origin="+").

2.1.6 Summary of this chapter

The remainder of this chapter is organized as follows. Section 2.2 is background information and a review of the fundamental definitions relating to polyhedra. Section 2.3 discusses issues relating to how polyhedra are represented in memory. Section 2.4 defines and discusses the fundamental polyhedral operations. Section 2.5 describes how variables that are defined over a polyhedral domain can be serialized in terms of imperative loops that scan the domain. A method for synthesizing loop nests to scan all index points in a domain is presented. Section 2.6 presents the program transformations used to localize a recurrence equation program, that is, for transforming a system of affine recurrence equations to a system of uniform recurrence equations. This is needed for implementing a program in hardware.

2.2 Polyhedra

This section is a quick review of fundamental definitions relating to polyhedra and cones. The major part of this summary is taken from the works of Grunbaum, *Convex Polytopes* [1], and of Schrijver, *Theory of Linear and Integer Programming* [2], and of Edelsbrunner, *Algorithms in Combinatorial Geometry* [3].

In this chapter, polyhedra are restricted to being in the n-dimensional rational Cartesian space, represented by the symbol $\mathcal{Q}^n$. All matrices, vectors, and scalars are thus assumed to be rational unless otherwise specified.

Given a vector x and a scalar coefficient vector λ, the following different combinations are defined:

> A *linear combination* $\sum \lambda_i x_i$
> A *positive*[1] *combination* $\sum \lambda_i x_i$ where all $\lambda_i \geq 0$

---

[1] Also called *nonnegative* or *conic* combination.

An *affine combination* $\sum \lambda_i x_i$ where $\sum \lambda_i = 1$

A *convex combination* $\sum \lambda_i x_i$ where $\sum \lambda_i = 1$ and all $\lambda_i \geq 0$.

Figure 2.2 shows the geometries generated by the different combinations of two points in 2-space.

2.2.1   The two representations of polyhedra

Every polyhedron $\mathcal{P}$ has two ways it can be represented: an implicit and a parametric representation. The set of solution points that satisfy a mixed system of constraints form a polyhedron $\mathcal{P}$ and serve as the *implicit definition* of the polyhedron

$$\mathcal{P} = \{x \mid Ax = b, Cx \geq d\} \tag{2.5}$$

given in terms of equations (rows of A, b) and inequalities (rows of C, d), where A, C are matrices, and b, d, and x are vectors. This form corresponds to the definition of a polyhedron as the intersection of a finite family of closed linear halfspaces, defined by the inequalities: $Ax \geq b$, $Ax \leq b$, and $Cx \geq d$.

$\mathcal{P}$ has an equivalent dual *parametric representation* (also called the *Minkowski characterization* after Minkowski–1896 [2, p. 87]):

$$\mathcal{P} = \left\{ x \mid x = L\lambda + R\mu + V\nu, \quad \mu, \nu \geq 0, \quad \sum \nu = 1 \right\} \tag{2.6}$$

in terms of a linear combination of lines (columns of matrix L), a convex combination of vertices (columns of matrix V), and a positive combination of extreme rays (columns of matrix R). The parametric representation shows that a polyhedron can be generated from a set of lines, rays, and vertices.

A *vertex* of a polyhedron $\mathcal{P}$ is any point in $\mathcal{P}$ that cannot be expressed as a convex combination of any other distinct points in $\mathcal{P}$. A *ray* of $\mathcal{P}$ is a vector r, such that $x \in \mathcal{P}$ implies $(x + \mu r) \in \mathcal{P}$ for all $\mu \geq 0$ and is thus a direction in which $\mathcal{P}$ is infinite. A ray of $\mathcal{P}$ is an *extreme ray* if and only if it cannot be expressed as a positive combination of any other two distinct rays of $\mathcal{P}$. A *line* of $\mathcal{P}$ is a vector l, such that $x \in \mathcal{P}$ implies $(x + \mu l) \in \mathcal{P}$ for all μ. Allowing μ to have both positive and negative values creates a bidirectional ray in the direction of l and $-l$.

Procedures exist to compute the dual representations of $\mathcal{P}$, that is, given A, b, C, d, compute L, V, R, and vice versa. Such a procedure is an important part of computing polyhedral operations and will be described later in Section 2.4.1.

Polyhedral *cones* are a special case of polyhedra that have only a single vertex. (Without loss of generality, the vertex is at the origin.) A cone $\mathcal{C}$ is defined parametrically as

$$\mathcal{C} = \{x \mid x = L\lambda + R\mu, \ \mu \geq 0\} \tag{2.7}$$

where L and R are matrices whose columns are the lines and extreme rays, respectively. If L is empty, then the cone is pointed.

Since the origin is always a solution point in Equation 2.5, the implicit description of a cone has the following form

$$\mathcal{C} = \{x \mid Ax = 0, \; Cx \geq 0\} \tag{2.8}$$

the solution of a mixed system of *homogeneous* inequalities and equations.

A set $\mathcal{K}$ is called a *linear (sub)space* if it has the property: $x, y \in \mathcal{K}$ implies all linear combinations of x, y are in $\mathcal{K}$. The *dimension* of a space is the rank of a set of lines that span the space. A linear space of dimension m is called an *m-space*.

The *lineality space* of a polyhedron is the dimensionally largest linear subspace contained in the polyhedron. A lineality space of dimension m is represented by a fundamental set of m lines that form a linearly independent basis of the subspace. The lineality space of a polyhedron is unique, although it may be represented using any appropriate basis that is set of m linearly independent lines.

A set $\mathcal{K}$ is called a *flat* if it has the property: $x, y \in \mathcal{K}$ implies all affine combinations of x, y are also in $\mathcal{K}$. The *dimension* of a flat is the rank of a set of lines that span the flat. A flat of dimension d is called an *d-flat*. A 0-flat, 1-flat, and 2-flat are called respectively a *point*, *line*, and *plane*. Flats are containers that contain polyhedra.

2.2.2 Decomposition

In 1936, Motzkin gave the decomposition theorem for polyhedra. Any polyhedron $\mathcal{P}$ can be uniquely decomposed into a polytope $\mathcal{V} = $ conv.hull$\{v_1, \ldots, v_m\}$ generated by convex combination of the vertices of $\mathcal{P}$, and a cone $\mathcal{C} = $ char.cone$\mathcal{P}$ as follows:[2]

$$\mathcal{P} = \mathcal{V} + \mathcal{C} \, . \tag{2.9}$$

A nonpointed convex cone can in turn be partitioned into two parts,

$$\mathcal{C} = \mathcal{L} + \mathcal{R} \tag{2.10}$$

the combination of its lineality space $\mathcal{L}$ generated by a linear combination of the lines (bidirectional-rays) of $\mathcal{P}$, and a pointed cone $\mathcal{R}$ generated by positive combination of the extreme rays of $\mathcal{P}$. Combining equations Equations 2.9 and 2.10, a polyhedron may be fully decomposed into

$$\mathcal{P} = \mathcal{V} + \mathcal{R} + \mathcal{L} \, . \tag{2.11}$$

---

[2]The symbol '+' in the Equations 2.9 through 2.11, is called the Minkowski sum and is defined: $R + S = \{r + s : r \in R, s \in S\}$.

This decomposition implies that any polyhedron may be decomposed into its vertices, rays (unidirectional rays), and lines (bidirectional rays). This can be clearly seen in the parametric description in Equation 2.6.

A decomposition that has a practical application is the decomposition of a polyhedron into its *lineality space* and its *ray space*. This division separates lines (bidirectional rays) from vertices and rays (unidirectional rays). In the alternate conic form of a polyhedron, developed in Section 2.3.1, both rays and vertices are representable as unidirectional rays in the cone. In a cone, this decomposition simply separates lines and rays (Equation 2.10).

2.3   Representation of Polyhedra in a Computer

In this section, we want to motivate a way of representing polyhedra in a computer that is reasonably space efficient and in a form that makes operations on polyhedra as time efficient as possible.

2.3.1   Equivalence of homogeneous and inhomogeneous systems

We want to be able to represent a *mixed inhomogeneous system* of equations as given in Equations 2.5 and 2.6. This is the most general type of constraint system. A memory representation of an n-dimensional-mixed inhomogeneous system consisting of j equalities and k inequalities would require the storage of the following arrays: $A(j \times n), b(j \times 1), C(k \times n), d(k \times 1)$. The dual representation would require the storage of R, V, and L, the arrays representing the rays, vertices, and lines. The representation in memory can be simplified, however, with a transformation $x \to \begin{pmatrix} \xi x \\ \xi \end{pmatrix}, \xi \geq 0$ that changes an inhomogeneous system $\mathcal{P}$ of dimension n into a homogeneous system $\mathcal{C}$ of dimension $n + 1$, as shown here:

$$\mathcal{P} = \{x \mid Ax = b, Cx \geq d\}$$

$$= \{x \mid Ax - b = 0, Cx - d \geq 0\}$$

$$\mathcal{C} = \left\{ \begin{pmatrix} \xi x \\ \xi \end{pmatrix} \mid \xi Ax - \xi b = 0, \xi Cx - \xi d \geq 0, \xi \geq 0 \right\}$$

$$= \left\{ \begin{pmatrix} \xi x \\ \xi \end{pmatrix} \mid (A \mid -b) \begin{pmatrix} \xi x \\ \xi \end{pmatrix} = 0, \left(\begin{array}{c|c} C & -d \\ \hline 0 & 1 \end{array} \right) \begin{pmatrix} \xi x \\ \xi \end{pmatrix} \geq 0 \right\}$$

$$= \{\hat{x} \mid \hat{A}\hat{x} = 0, \hat{C}\hat{x} \geq 0\}$$

The transformed system $\mathcal{C}$ is now an $(n+1)$ dimensional cone that contains the same information as the original n-dimensional polyhedron. Goldman showed that the mapping $x \to \begin{pmatrix} \xi x \\ \xi \end{pmatrix}$ is one-to-one and inclusion preserving, and thus the two are combinatorially equivalent [14]. The original polyhedron $\mathcal{P}$ is in fact the intersection of the cone $\mathcal{C}$ with the hyperplane defined by the equality $\xi = 1$. Given any $\mathcal{P}$ as defined in Equation 2.5, a unique homogeneous cone form exists defined as follows:

$$\mathcal{C} = \{\hat{x} \,|\, \hat{A}\hat{x} = 0, \hat{C}\hat{x} \geq 0\}$$
$$= \text{homogeneous.cone } \mathcal{P},$$

$$\text{where } \hat{x} = \begin{pmatrix} \xi x \\ \xi \end{pmatrix}, \hat{A} = \left(A\,|\,-b \right), \hat{C} = \left(\begin{array}{c|c} C & -d \\ \hline 0 & 1 \end{array} \right) \qquad (2.12)$$

The homogeneous system requires the storage of arrays: $\hat{A}(j \times (n+1))$, $\hat{C}((k+1) \times (n+1))$. Thus the total amount of memory needed for the original system is $(j + k + 1)(n + 1)$ words for the cone versus $(j + k)(n + 1)$ words for the polyhedron. The amount of memory for a cone is slightly larger; however, the cone representation is structurally simpler (two matrices versus two matrices and two vectors). Likewise, the parametric (geometric) representation of the cone is also simpler. The decomposition of a cone is $\mathcal{R} + \mathcal{L}$, and thus only rays and lines have to be represented. During the transformation process from a polyhedron to a cone, vertices of $\mathcal{P}$ are transformed into rays of $\mathcal{C}$. The vertices and rays of an inhomogeneous polyhedron $\mathcal{P}$ have a unified and homogeneous representation as rays in a polyhedral cone $\mathcal{C}$. Thus the rays of the cone represent both the vertices and rays of the original polyhedron. As before, the amount of memory needed to store the dual representation is nearly the same; however the structure of the representation is simpler (two matrices versus three matrices). Table 2.1 shows the equivalent forms of inhomogeneous and homogeneous systems, polyhedra and cones, along with their dual implicit and parametric representations. The table highlights the fundamental relationships between the polyhedron and cone.

Using the homogeneous cone form simplifies not only the data structure used to represent the polyhedron but also computation. From practical experience with the implementation of polyhedral operations, it is known that fewer array references have to be done and fewer "end cases" have to be handled when computing with the homogeneous form. This results in smaller and more efficient procedures.

2.3.2 The dual representation of a polyhedron in memory

A polyhedron may be fully described as either a system of constraints or by its dual form, a collection of rays and lines. Given either form, the other may be computed. However, since the duality computation is an expensive operation (see Section 2.4.1) and since both forms are needed for computation of

Table 2.1   Duality between polyhedra and cones.

	Inhomogeneous system	Homogeneous system
Structure	**Polyhedron** $\mathcal{P}$**, dimension** d	**Cone** $\mathcal{C}$**, dimension** $d+1$
Implicit representation using equations and inequalities	$\mathcal{P} = \{x \mid Ax = b, Cx \geq d\}$	$\mathcal{C} = \{\hat{x} \mid \hat{A}\hat{x} = 0, \hat{C}\hat{x} \geq 0\}$ $\hat{x} = \begin{pmatrix} \xi x \\ \xi \end{pmatrix}$ $\hat{A} = \begin{pmatrix} A & \mid & -b \end{pmatrix}$ $\hat{C} = \left(\begin{array}{c\|c} C & -d \\ \hline 0 & 1 \end{array} \right)$
Parametric representation using vertices, rays, and lines	$\mathcal{P} = \{x \mid x = L\lambda + R\mu + V\nu,$ $\mu, \nu \geq 0, \sum \nu = 1\}$	$\mathcal{C} = \{\hat{x} \mid \hat{x} = \hat{L}\lambda + \hat{R}\mu,$ $\mu \geq 0\}$
Vertices	$v = \begin{pmatrix} v_1 \\ v_2 \\ \vdots \\ v_d \end{pmatrix}, \quad v \in V$	$\hat{r}_v = \begin{pmatrix} \xi v_1 \\ \xi v_2 \\ \vdots \\ \xi v_d \\ \xi \end{pmatrix}, \ \xi > 0, \ \hat{r}_v \in \hat{R}$
Rays	$r = \begin{pmatrix} r_1 \\ r_2 \\ \vdots \\ r_d \end{pmatrix}, \quad r \in R$	$\hat{r}_r = \begin{pmatrix} r_1 \\ r_2 \\ \vdots \\ r_d \\ 0 \end{pmatrix}, \quad \hat{r}_r \in \hat{R}$
Lines	$l = \begin{pmatrix} l_1 \\ l_2 \\ \vdots \\ l_d \end{pmatrix}, \quad l \in L$	$\hat{l} = \begin{pmatrix} l_1 \\ l_2 \\ \vdots \\ l_d \\ 0 \end{pmatrix}, \quad \hat{l} \in \hat{L}$

different operations (see Section 2.4), it is advantageous to represent polyhedra redundantly using both forms.

Even though this way of representing polyhedra is redundant and takes more memory, keeping both forms in memory reduces the number of duality computations that have to be made and improves the efficiency of polyhedral operations. This is a basic memory/execution time trade-off made in favor of execution time.

A second advantage is that the two dual forms can be used to reduce each other to a minimal size. It will be shown below how the rays and lines can be used to remove redundant inequalities, and how the constraints can be used to remove redundant (nonextremal) rays. The process of reducing the representation of polyhedra to a minimal normal form is greatly simplified by keeping and maintaining the two dual forms of polyhedra in memory.

2.3.3 Saturation and the incidence matrix

After being transformed to a homogeneous coordinate system, a polyhedron is represented as a cone (Equation 2.12). The dual representations of the cone are:

$$\mathcal{C} = \{x \mid Ax = 0, Cx \geq 0\} \qquad \text{(implicit form)}$$

$$= \{x \mid x = L\lambda + R\mu, \mu \geq 0\} \quad \text{(parametric form)}$$

Substituting the equation for x in the parametric form into the equations involving x in the implicit form, we obtain:

$$\forall (\mu \geq 0, \lambda) : \begin{cases} AL\lambda + AR\mu = 0 \\ CL\lambda + CR\mu \geq 0 \end{cases} \implies \begin{cases} AL = 0, AR = 0 \\ CL = 0, CR \geq 0 \end{cases} \tag{2.13}$$

where rows of A and C are equalities and inequalities, respectively, and where columns of L and R are lines and rays, respectively.

An incidence matrix is a memory-efficient bookkeeping structure that is useful in reducing the dual representation of a polyhedron to a minimal and normal form by helping to identify and remove redundant rays and inequalities. Before discussing the incidence matrix, the notion of saturation needs to be defined.

Definition 2.8 *A ray r is said to* **saturate** *an inequality $a^T x \geq 0$ when $a^T r = 0$, it* **verifies** *the inequality when $a^T r > 0$, and it* **does not verify** *the inequality when $a^T r < 0$. Likewise, a ray r is said to* **saturate** *an equality $a^T x = 0$ when $a^T r = 0$, and it* **does not verify** *the equality when $a^T r \neq 0$. Equalities and inequalities are collectively called* **constraints**. *A constraint is* **satisfied** *by a ray if the ray saturates or verifies the constraint.*

The incidence matrix S is a boolean matrix that has a row for every constraint (rows of A and C) and a column for every line or ray (columns of L and R). Each element s_{ij} in S is defined as follows:

$$s_{ij} = \begin{cases} 0, & \text{if constraint } c_i \text{ is saturated by ray(line) } r_j, \text{ i.e. } c_i^T r_j = 0 \\ 1, & \text{if constraint } c_i \text{ is verified by ray(line) } r_j, \text{ i.e. } c_i^T r_j > 0 \end{cases}$$

From the demonstrations in Equation 2.13, we know that all rows of the S matrix associated with equations (A) are 0, and all columns of the S matrix associated with lines (L) are also 0. Only entries associated with inequalities (C) and rays (R) can have 1s as well as 0s. This is illustrated in the following diagram representing the saturation matrix S.

S	L	R
A	(0)	(0)
C	(0)	(0 or 1)

2.3.4   Expanding the model to unions of polyhedra

Polyhedra are closed under intersection, convex union (convex.hull($A \cup B$), and affine transformation. However, they are not closed under (simple) union since the union of any two polyhedra is not necessarily convex. Likewise, polyhedra are not closed under the difference operation. To obtain closure of these two operations (union and difference), it is necessary to expand the model from a simple polyhedron to a finite union of polyhedra. By supporting an extended model of a union of polyhedra, the operations of intersection, union, and difference are all closed.

2.3.5   Validity rules

These are rules used to keep the dual form representation of polyhedra in memory in a minimal and normal form. While the representation of a polyhedron is not necessarily unique, all polyhedra (including empty and universe polyhedra) satisfy three general rules. In this section, the consistency rules that govern the dual form representation of polyhedra are described.

Given a polyhedron $\mathcal{P} = \mathcal{L} + \mathcal{R} + \mathcal{V}$, the following meanings of the term *dimension* are defined:

1. The **dimension** of a lineality space $\mathcal{L}$ is m ($\mathcal{L}$ is an m-space) when the number of linearly independent lines in a basis for $\mathcal{L}$ is m (m is the rank of a set of lines which span $\mathcal{L}$).

2. The **dimension** of the ray space $\mathcal{R} + \mathcal{V}$ is d (affine.hull($\mathcal{R} + \mathcal{V}$) is a d-flat) when the dimension of the smallest lineality space that contains $\mathcal{R} + \mathcal{V}$ is d.

3. The **dimension** of the polyhedron $\mathcal{P} = \mathcal{L} + \mathcal{R} + \mathcal{V}$ is $p = m + d$ or in other words, affine.hull $\mathcal{P}$ is a p-flat. The dimensions of the lineality space and ray space are unique and separable since no irredundant ray is equal to a linear combination of lines (else the ray is redundant) and no line is a linear combination of rays (else the basis of ray space is redundant). Thus, the lineality space and ray space of a polyhedron are dimensionally distinct and the sum of their dimensions is the dimension of the polyhedron.

4. The **dimension** of the system that contains the polyhedron is n, where n is the number of indices in the system.

5. The **dimension** of the space in a system that cannot contain a polyhedron because of equality constraints is j, the number of linearly independent equalities. Each irredundant equality restricts the flat that contains the polyhedron by one dimension.

Property 2.1 (**Dimensionality Rule**)
The dimension **n** of the system is equal to the dimension **d** of the ray space plus the number of linearly independent lines **m** plus the number of linearly independent equalities **j**.

The dimension d of the ray space is an important number, and it is used to locate and remove redundant rays and inequalities. It is the key number used in the saturation rule (Prop. 2.2) given below. It is computed according to the dimensionality rule (Prop. 2.1) as $d = n - m - j$.

Example 2.7 $\{x, y \mid x = 2;\ y = 3\}$
The polyhedron consisting of a single point (2,3) is of dimension 0. The dimension of the system is 2 since there are two indices (x and y), there are two equalities, and the dimension of the lineality space is 0; thus the dimension of the ray space is 2-2-0=0 (Prop. 2.1).

Property 2.2 **(Saturation Rule)**
In a d-dimensional ray space,

   a. Every inequality must be saturated by at least d vertices/rays.
   b. Every vertex must saturate at least d inequalities, and a ray must saturate at least $d - 1$ inequalities plus the positivity constraint.
   c. Every equation must be saturated by all lines and vertices/rays.
   d. Every line must saturate all equalities and inequalities.

The independence rule is an invariant of the dual form representation of polyhedra in which only a (nonunique) minimal representation of a polyhedron is stored.

Property 2.3 **(Independence Rule)**

   a. No inequality is a positive combination of any other two inequalities or equalities.
   b. No ray is a linear combination of any other two rays or lines.
   c. The set of j equalities must be linearly independent.
   d. The set of m lines must be linearly independent.

Definition 2.9 (Redundancy) *Inequalities that don't satisfy Prop. 2.2.a or Prop. 2.3.a are* **redundant**.
Vertices/rays that don't satisfy Prop. 2.2.b or Prop. 2.3.b are **redundant**.

Example 2.8 $\{x, y \mid 1 \le x \le 3; 2 \le y \le 4\}$

	x>=1	x<=3	y>=2	y<=4
vertex(1,2)	sat		sat	
vertex(1,4)	sat			sat
vertex(3,2)		sat	sat	
vertex(3,4)		sat		sat

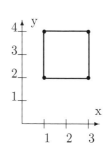

The dimension of the ray space is 2. Every constraint saturates two vertices and every vertex saturates two inequalities. This is a perfectly nonredundant system.

2.3.5.1   The positivity constraint

In the language of algebraists, the trivial constraint $1 \geq 0$ is called the "positivity[3] constraint." When true, you know that positive numbers are positive (a nice thing to know). It was generated as a side effect of converting from an inhomogeneous polyhedron to a homogeneous cone representation as can be seen in Equation 2.12. As stated earlier, rays may be thought of as points at infinity. In this vein of thought, the positivity constraint generates the face that connects those points, creating a face at infinity that "closes" unbounded polyhedra. The following property gives the reasoning behind this.

Property 2.4 All rays are saturated by the positivity constraint, and no vertex is saturated by the positivity constraint.

As surprising as it may seem, the positivity constraint is not always redundant, as is shown in Examples 2.9 and 2.10. The following property gives a rule for when the positivity constraint will be needed.

Property 2.5 The positivity constraint will be irredundant if the size of the set of rays is $\geq d$, the dimension of the ray space, and the rank of the ray set is d.

Positivity constraints are included so polyhedra can have valid dual form representations (according to Prop. 2.1 and Prop. 2.2). The positivity constraint is kept in polyhedra where it is needed (according to Prop. 2.5) so that all polyhedra have valid representations according to Prop. 2.1 and Prop. 2.2. Since humans don't want to see the constraint $1 \geq 0$ printed when displaying a polyhedron, the positivity constraint will have to be filtered out by the pretty printer when displaying the constraints.

Example 2.9 $\{x, y \,|\, x \geq 1; y \geq 2\}$

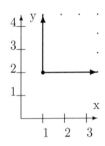

	x>=1	y>=2	1>=0
vertex(1,2)	sat	sat	
ray(1,0)		sat	sat
ray(0,1)	sat		sat

---

[3]Also called the nonnegativity constraint. Here the term *positive* is used in a nonstrict way to include zero.

Here, every constraint saturates two vertices/rays and every vertex/ray saturates two inequalities. This is also a nonredundant system. However the positivity constraint is also irredundant ... it is needed to support the presence of the two rays. Without it, the two rays are not supported and appear mistakenly to be redundant.

Example 2.10 $\{x \mid x \geq 1\}$

	x>=1	1>=0
line(0,1)	sat	sat
vertex(1,2)	sat	
ray(1,0)		sat

A halfplane.

2.3.5.2 Empty polyhedra

Example 2.11 An empty polyhedron: $\{x, y \mid 1 = 0\}$

```
Empty Polyhedron, Dimension 2
Constraints  ( 3 equalities, 0 inequalities )
   x = 0
   y = 0
   1 = 0
Lines/Rays ( 0 lines, 0 rays )
   -none-

dim(ray space) = dimension - numlines - numequalities
  = 2    - 0          - 3
         = -1
dim(lineality space) = numlines
         = 0
```

An empty domain is a polyhedron that includes no points. It is caused by over-constraining a system such that no point can satisfy all of the constraints. Empty polyhedra have the following properties:

Property 2.6 In an empty polyhedron

    a. the dimension of the lineality space is 0,
    b. the dimension of the ray space is -1, and
    c. there are no rays (and no vertices, to be more specific).

2.3.5.3　Universe polyhedron

Example 2.12　Universe Polyhedron: $\{x, y \mid 1 \geq 0\}$

```
Universe Polyhedron, Dimension 2
Constraints  ( 0 equalities, 1 inequality )
   1 >= 0
Lines/Rays ( 2 lines, 0 rays)
   line    (1,0) (x-axis)
   line    (0,1) (y-axis)
   vertex (0,0) (origin)

dim(ray space) = dimension - num_lines - num_equalities
        = 2    - 2          - 0
        = 0
dim(lineality space) = num_lines
        = 2
```

A universe polyhedron is one that encompasses all points within a certain dimensional linear subspace. It is therefore unbounded in all directions. It is created by not constraining a system at all (except with the positivity constraint). A universe polyhedron has the following properties:

Property 2.7 In a universe polyhedron

    a. the dimension of the lineality space is the dimension of the polyhedron,
    b. the dimension of the ray space is 0,
    c. there are no constraints, other than the positivity constraint.

2.3.6　Summary

The polyhedra can be represented by either one of two dual forms:

    1. a set of constraints—inequalities and equalities, and
    2. a set of geometric features—lines, rays, and vertices.

However, there is significant advantage in keeping both representations in the computer. Polyhedra can be created starting from either a list of constraints or a list of geometric features. Given one representation form, the dual form can be computed as will be shown in Section 2.4.1.

It is convenient to represent polyhedra in their pointed cone form, which results from the transformation $x \rightarrow \begin{pmatrix} \xi x \\ \xi \end{pmatrix}$. This transformation maps inhomogeneous constraints to homogeneous constraints and maps both the vertices and rays of a polyhedron to rays in a pointed cone.

2.4 Polyhedral Operations

Domains are made up of a finite union of polyhedra. All of the important operations are closed when operating on domains. This section builds on Section 2.3 and describes the algorithms used to operate on domains.

2.4.1 Computation of dual forms

An important problem in computing with polyhedral domains is being able to convert from a domain described implicitly in terms of linear constraints (equalities and inequalities, Equation 2.5), to a parametric description (Equation 2.6) given in terms of the geometric features of the polyhedron (lines, rays, and vertices). The computation of the dual form is one of the rare and interesting algorithms, which is its own inverse. The same algorithm that computes the lines, rays, and vertices from a list of constraints can also compute the constraints, given the lines, rays, and vertices of the polyhedron. An equivalent problem is called the *convex hull problem*, which computes the facets of the convex hull surrounding a given set of points.

The algorithms to solve this problem are categorized into one of two general classes: the pivoting and non-pivoting methods [15]. The pivoting methods are derivatives of the simplex method, which finds new vertices located adjacent to known vertices using simplex pivot operations.

The nonpivoting methods are based on an algorithm called *the double description method* invented by Motzkin et al. in 1953 [16]. Motzkin described a general algorithm that iteratively solves the dual computation problem for a cone. (Since polyhedra may be converted to cones, it works for all polyhedra.) In each iteration, one new constraint is added to the current cone. Rays in the cone are divided into three groups: R^+ the rays that verify the constraint, R^0 the rays that saturate the constraint, and R^- the rays that do not verify the constraint. A new cone is then constructed from the ray sets R^+, R^0, plus the convex combinations of pairs of rays, one each from sets R^+ and R^-. The main problem with the nonpivoting methods is that they can generate a nonminimal set of rays by creating nonextreme or redundant rays. If allowed to stay, the number of rays would grow exponentially and would seriously test the memory capacity of the hardware, as well as degrade the performance of the procedure. Motzkin proposed a simple and rather elegant test to solve this problem. He showed that a convex combination of a pair of rays $(r^- \in R^-, r^+ \in R^+)$ will result in an extreme ray in the new cone if and only if the minimum face that contains them both: (1) is dimension one greater than r^- and r^+, and (2) only contains the two rays r^- and r^+. This test inhibits the production of unwanted rays and keeps the solution in a minimal form.

Chernikova [17,18] described a similar algorithm to solve the restricted case of the mixed constraint problem with the additional constraint that variables are all nonnegative ($x \geq 0$). Chernikova's method was similar to Motzkin's method, except that she used a slightly smaller and improved tableau. Fernández and Quinton [19] extended the Chernikova method by removing the restriction that $x \geq 0$ and adding a heuristic to improve speed by ordering the constraints. A large portion of the computation time is spent doing the adjacency test. Le Verge [20] improved the speed of the redundancy checking procedure used in [19], which is the most time consuming part of the algorithm. Seidel described an algorithm for the equivalent convex hull problem [21], which executes in $\mathcal{O}(n^{\lfloor \frac{d}{2} \rfloor})$ expected running time where n is the number of points and d is the dimension. This is provably the best one can do, since the output of the procedure is of the same order. He solves the adjacent ray problem (the adjacent facet problem in his case) by creating and maintaining a facet graph in which facets are vertices and adjacent facets are connected by edges. It takes extra code and memory to maintain the graph, but then he does not need to do the Motzkin adjacency test on all pairs of vertices (facets).

2.4.2   Reducing the dual form representation to a minimal form

After computing the dual of a set of constraints, the set of rays produced is guaranteed to be nonredundant by virtue of the adjacency test which is done when each ray was produced. However, the constraints are still possibly redundant. There remain a number of simplifications that can still be done on the resulting polyhedron, among which are:

1. Detection of implicit lines such as line (1,2) given that there exist rays (1,2) and (−1,−2).

2. Finding a reduced basis for the lines (using Gaussian elimination).

3. Removing the positivity constraints $1 \geq 0$ if it is redundant.

4. Detection of redundant inequalities such as $y \geq 4$ given $y \geq 3$, or $x \geq 2$ given $x = 1$, or the less obvious, $x + y \geq 5$ given $x \geq 3$ and $y \geq 2$.

5. Reducing (or solving) the system of equalities using Gaussian elimination.

The algorithm to do all of these reductions is sketched out below. In the procedure, each constraint and each ray needs a small amount of memory to store status counts.

Algorithm to minimize the dual form representation

Step 0 Count the number of vertices among the rays while initializing the ray status counts to 0. If no vertices are found, quit the procedure and return an empty polyhedron as the result. (Every nonempty polyhedron must have at least one vertex.)

Step 1 Compute status counts for both rays and inequalities. For each constraint, count the number of vertices/rays saturated by that constraint, and put the result in the status word. At the same time, for each vertex/ray, count the number of constraints saturated by it.

Delete any positivity constraints you find, but give rays credit in their status counts for saturating the positivity constraint.

Step 2 Sort equalities out from among the constraints, leaving only inequalities. Equalities are constraints that saturate all of the rays. Status count == number of rays

Step 3 Perform Gaussian elimination on the list of equalities. Obtain a minimal basis by solving for as many variables as possible. Use this solution to reduce the inequalities by eliminating as many variables as possible. Set j to the rank of the system of equalities.

Step 4 Sort lines out from among the rays, leaving only unidirectional rays. Lines are rays that saturate all of the constraints. Status count = number of constraints + 1(for the positivity constraint).

Step 5 Perform Gaussian elimination on the lineality space to obtain a minimal basis of lines. Use this basis to reduce the representation of the unidirectional rays. Set m to the rank of the system of lines.

Step 6 Filter the inequalities and identify the equalities.

New positivity constraints may have been created by step 3. Check for and eliminate them.

Compute $d = n - j - m$

if (Status==0) or (Status< d) Constraint is redundant.

else if (Status== the number of vertices and rays) Constraint is an equality.

else Constraint is an irredundant inequality.

Step 7 Filter the rays and identify the lines.

if (Status< d) Ray is redundant.

else if (Status== number of Constraints+1) Ray is a line.

else Ray is an irredundant unidirectional ray.

Step 8 Create the polyhedron.

2.4.3 Intersection

Intersection is performed by concatenating the lists of constraints from two (or more) polyhedra into one list and finding the polyhedron that satisfies all

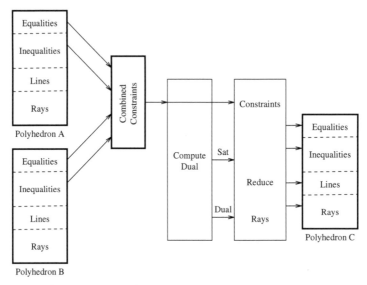

Figure 2.3   Computation of intersection.

of the combined constraints. This is done by finding the extremal rays that satisfy the combined constraints, (finding the dual of the list of constraints) and then reducing both the constraints and rays into one polyhedron. This procedure is illustrated in Figure 2.3.

To intersect two domains, A and B, which are unions of polyhedra, $A = \cup_i A_i$ and $B = \cup_j B_j$, the pairwise intersection of the component polyhedra from A and B must be computed, and the union of the results is the resulting domain of intersection, as shown below:

$$
\begin{aligned}
A \cap B \quad &= \quad (\cup_i A_i) \cap (\cup_j B_j) \\
&= \quad \cup_{i,j}(A_i \cap B_j)
\end{aligned}
$$

2.4.4   Union

The domain (non-convex) union operation simply combines two domains into one. The lists of polyhedra associated with the domains are combined into a single list. However, combining the two lists blindly may create non-minimal representations. For instance, if, in forming the union of domains $A = \{i \,|\, i \geq 1\}$ and $B = \{i \,|\, i \geq 2\}$, the fact that $A \supset B$ is taken into consideration, then the union can be reduced to simply A. The algorithm to compute union should perform this kind of simplification during the union operation to control the amount of memory used. Thus, before adding any new polyhedron to an existing list of polyhedra, first check to see if that polyhedron is covered by some polyhedron already in the domain. If it *is* covered, then the new polyhedron is not added to the domain. Likewise, polyhedra in the existing list may be deleted if they are covered by the new polyhedron. In the new combined list, no polyhedron should be a subset of any other polyhedron. However, polyhedra are not necessarily disjoint, and polyhedra in the list may intersect each other.

The test for when a polyhedron $p1$ covers or includes another polyhedron $p2$ is straightforward, using the dual representation of polyhedra: $p1 \supseteq p2$ if all of the rays of $p2$ *satisfy* (see Definition 2.8) all of the constraints of $p1$. This is an example of when a difficult computation is done efficiently using the dual representation. The constraint representation of $p1$ and the dual ray representation of $p2$ are used to determine $p1 \supseteq p2$. Since the dual form representations of both $p1$ and $p2$ are kept in memory, the dual does not need to be (re)computed in order to do this test.

2.4.5   Difference

Domain difference $A - B$ computes the domain that is part of A but not part of B. It is equivalent to $A \cap \sim B$, where $\sim B$ is the *complement* domain of B. If B is the intersection of a set of hyperplanes (representing the equalities) and closed halfspaces (representing the inequalities), then the complement of B is computed as follows:

$$\sim B = \sim (\cap_i H_i)$$
$$= \cup_i (\sim H_i)$$

where

$$\sim H_i = \begin{cases} \{x \mid a^T x < 0\} & \text{when} \quad H_i = \{x \mid a^T x \geq 0\} \\ \{x \mid (a^T x < 0 \cup a^T x > 0)\} & \text{when} \quad H_i = \{x \mid a^T x = 0\} \end{cases}$$

Since the set of $\sim H_i$ does not consist of *closed* halfspaces, the complement of a polyhedron is not itself a union of polyhedra. Thus, unions of rational polyhedra are not closed under the difference operation. However, in many applications we are only interested in the lattice of integer points contained in a polyhedron. For this case, we can define a difference operation that works for integer lattices. Normalizing for the integer case, the following closed complements contain the same integer points as the previous nonclosed complements:

$$\sim H_i = \begin{cases} \{x \mid -a^T x + 1 \geq 0\} & \text{when} \quad H_i = \{x \mid a^T x \geq 0\} \\ \{x \mid (-a^T x + 1 \geq 0 \cup a^T x - 1 \geq 0)\} & \text{when} \quad H_i = \{x \mid a^T x = 0\} \end{cases}.$$

The computation of difference is the same as the computation of intersection after taking the integer complement of B. Since the integer complement of B is a union of polyhedra, the difference of two polyhedra is a union of polyhedra.

Figure 2.4 $C = \mathrm{Simplify}(A, B)$, $B = context$

2.4.6   Simplification in context

The operation *simplify* is defined as follows (refer to Figure 2.4):

> Given domains A and B, $\mathrm{Simplify}(A, B) = C$ when $C \cap B = A \cap B$, $C \supseteq A$ and there does not exist any other domain $C' \supset C$ such that $C' \cap B = A \cap B$. If $A \cap B =$ the null set, then C is defined to be the null set.

The domain B is called the *context*. The simplify operation therefore finds the largest domain set (or smallest list of constraints) that, when intersected with the context B, is equal to $A \cap B$. In the example in Figure 2.4, domain A is simplified (resulting in domain C) by eliminating the two constraints that are redundant with context domain B.

The simplify operation is done by computing the intersection $A \cap B$, and, while checking for redundant constraints, by recording which constraints of A are "redundant" with the intersection. The result of the simplify operation is then the domain A with the "redundant" constraints removed.

2.4.7   Convex union

Convex union is performed by concatenating the lists of rays and lines of the two (or more) polyhedra in a domain into one combined list and finding the set of constraints that tightly bound all of those objects. This is done by finding the dual of the list of rays and lines, and then reducing both the constraints and rays into one polyhedron. This procedure is illustrated in Figure 2.5.

This procedure is very similar to the intersection procedure, already described in Section 2.4.3. Convex union finds the polyhedron generated from the union of the lines and rays of the two input polyhedra. Intersection finds the polyhedron generated from the union of the equalities and inequalities of the inputs.

One of the useful applications of convex union is for computing projections. If you compute the convex union of a polyhedron with a line, the result is a cylindrical polyhedron whose cross section is the projection of the original polyhedron in the direction along the axis of the line.

2.4.8   Image

The function *image* transforms a domain $\mathcal{D}$ into another domain $\mathcal{D}'$ according to a given affine mapping function, $Tx + t$. (Refer to Figure 2.6.) The resulting

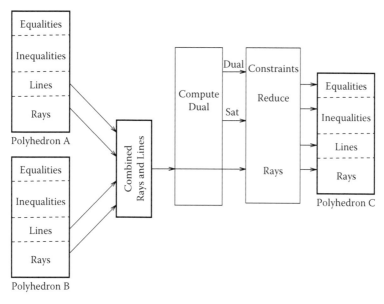

Figure 2.5 Computation of convex union.

domain $\mathcal{D}'$ is defined as:

$$\mathcal{D}' = \{x' \,|\, x' = Tx + t, x \in \mathcal{D}\}$$

The image of a polyhdron is in general not a polyhedron, but a *linear bounded lattice* [22]. However, under the restriction that T is unimodular, $Image(\mathcal{P}, T)$ is a polyhedron. To force closure for non-unimodular *Image* functions, the operation can compute the convex hull of the image.

In homogeneous terms, the transformation is expressed as

$$C' = \left\{ \begin{pmatrix} \xi x' \\ \xi \end{pmatrix} \,\Big|\, \begin{pmatrix} \xi x' \\ \xi \end{pmatrix} = \left(\begin{array}{c|c} T & t \\ \hline 0 & 1 \end{array} \right) \begin{pmatrix} \xi x \\ \xi \end{pmatrix}, \begin{pmatrix} \xi x \\ \xi \end{pmatrix} \in C \right\}$$

Thus in the homogeneous representation, an affine transfer function becomes a linear transfer function (no constant added in). In the analysis that follows, we will treat the transfer function from the linear point of view. The transformation function $\left(\begin{array}{c|c} T & t \\ \hline 0 & 1 \end{array} \right)$ is a matrix dimensioned by $(n+1) \times (m+1)$, where n and m are the dimensions of x and x', respectively. This transformation matrix is passed as a parameter to the image procedure. If $n = m$, x and x' are the same dimension. If $n \neq m$, the transformed space is of a larger (or smaller) dimension. The transformation does not have to be one-to-one, and therefore may not be invertible. Also, if $\det T \neq 1$, then the volume of the domain (the number of points in the domain) will be scaled by the magnitude of the determinant. To compute an image of D, given the full

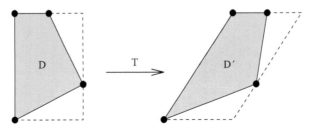

Figure 2.6   Affine transformation of $\mathcal{D}$ to $\mathcal{D}'$.

redundant representation:

$$D = \{x \mid Ax \geq 0, x = R\mu, AR \geq 0, \mu \geq 0\}$$

and given the transformation $x' = Tx$, the resulting D' is

$$D' = \{x' \mid A'x' \geq 0, x' = TR\mu, A'TR \geq 0, \mu \geq 0\}$$
$$= \{x' \mid A'x' \geq 0, x' = R'\mu, A'R' \geq 0, \mu \geq 0\}$$

A' can be computed as the dual of R'. Thus, $R' = TR$ and $A' = \text{dual}(R')$. This computation is illustrated in Figure 2.7. The image of a domain is simply the union of the images of the component polyhedra contained in the domain, as follows:

$$T.\mathcal{D} = T.(\cup_i \mathcal{P}_i)$$
$$= \cup_i (T.\mathcal{P}_i)$$

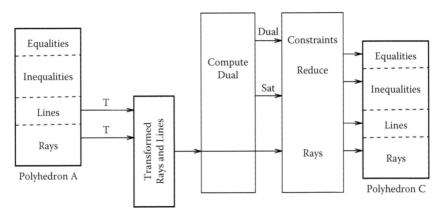

Figure 2.7   Computation of image.

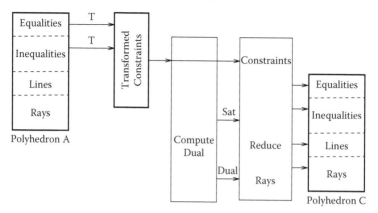

Figure 2.8 Computation of preimage.

2.4.9 Preimage

Preimage is the inverse operation of image. That is, given a domain D' defined as

$$D' = \{x' \mid A'x' \geq 0, x' = R'\mu, A'R' \geq 0, \mu \geq 0\}$$

and a transformation T, find the domain D which when transformed by T gives D'. The relation $x' = Tx$ still holds. (Refer again to Figure 2.6.) The result D is

$$D = \{x \mid A'Tx \geq 0, x = R\mu, A'TR \geq 0, \mu \geq 0\}$$

$$= \{x \mid Ax \geq 0, x = R\mu, AR \geq 0, \mu \geq 0\}$$

In the result, $A = A'T$ and $R = \text{dual}(A)$. This procedure is illustrated in Figure 2.8. The preimage of a domain is simply the union of the preimages of the component polyhedra contained in the domain, as follows:

$$T^{-1}.\mathcal{D} = T^{-1}.(\cup_i \mathcal{P}_i)$$

$$= \cup_i (T^{-1}.\mathcal{P}_i)$$

T^{-1} here is simply a notation for the preimage operation and does not mean to imply that T is invertible.

2.4.10 Practical considerations

Experience has shown that two practical problems have to be considered when implementing polyhedral operations. The first is a numeric overflow problem. If operations are performed using exact rational computation, and numbers are stored using 32-bit integer numerators and denominators, then numeric overflow can occur. If two rational numbers are multiplied and there is no cancellation, then the storage requirement for the result is the sum of the storage for the two operands (measured in number of bits). The solution to this problem is to use a *multi-precision arithmetic package* in which numeric storage grows to meet demand.

The second problem is a memory overflow problem. Given a d-dimensional polyhedron with n constraints, as many as $n^{\lfloor \frac{d}{2} \rfloor}$ vertices might be required in the dual representation in the worse case. This effectively limits computation to small dimensional polyhedra in the worst case. This also makes it difficult to statically allocate a fixed amount of work space to perform a computation. A dynamic work space would be better, in light of this problem. Fortunately, the worst case is very rarely seen.

A library called polylib was written in 1994 and 1995 in the C–language that provided the capability for doing geometric operations on unions of polyhedra. The operations union, intersection, difference, simplification in context, convex union, image, and preimage were all supported. The library was written while the author was working as a researcher at the IRISA laboratory in Rennes, France. Early parts of the library were done in collaboration with the late Hervé Le Verge who died in a tragic accident before the library was completed. Since then, the library has been maintained and extended by Vincent Loechner in collaboration with Sven Verdoolaege.

2.5   Loop Nest Synthesis Using Polyhedral Operations

In a system of affine recurrence equations, equations are defined over a region of index space called a domain. To convert a system of affine recurrence equations programs to imperative code, equations that are defined over a domain must be elaborated in terms of imperative loops that scan the domain. The synthesis of loop nests to scan a domain is thus fundamental in the generation of imperative code from a system of affine recurrence equations.

2.5.1   Introduction

The spatial point of view of a loop nest goes back to the work of Kuck [13] who showed that the domain of nested loops with affine lower and upper bounds can be described in terms of a polyhedron (Figure 2.9). Loop nest synthesis grew out of the earlier loop transformation theory [23,24] where it was shown that all loop transformations could be performed by doing a reindexing of the underlying index domain, followed by a rescanning (perhaps in a different order) of the domain (Figure 2.10). Loop nest synthesis is based on the polyhedral scanning problem that poses the problem of finding a set of nested do–loops that visit each integral point in a polyhedron. In this section,

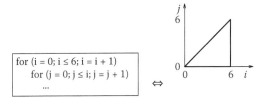

Figure 2.9 The spatial interpretation of a loop nest.

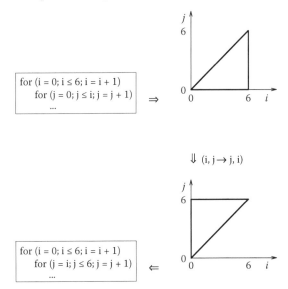

Figure 2.10 Loop transformations.

I present a method published by Le Verge, et al. [25] to scan parameterized polyhedra using polyhedral operations.

2.5.2 The polyhedron scanning problem

2.5.2.1 Introduction to parameterized polyhedra

This section quickly introduces the concept of parameterized polyhedra to help in the understanding of the polyhedron scanning problem. A polyhedron is defined to be the set of points bounded by a set of hyperplanes. Each hyperplane is associated with an inequality $(ax \geq b)$ that divides space into two halfspaces: a closed halfspace that satisfies the inequality and an open halfspace that does not. A system of such inequalities induces a polyhedron $\mathcal{D} = \{x : Ax \geq b\}$ where A and b are a constant matrix and vector, respectively.

Often, one is interested in describing families of polyhedra $\mathcal{D}(p)$, one polyhedron per instance of the parameters p. This can be done by replacing vector b above with an affine combination of a set of parameters p. By so doing, one obtains a parameterized polyhedron:

$$\mathcal{D}(p) = \{x : Ax \geq Bp + b\}$$

where A and B are constant matrices and b is a constant vector. This parameterized polyhedron can be rewritten in the form of a canonical projection of a non-parameterized polyhedron $\mathcal{D}'$ in the combined index and parameter space as shown by:

$$\mathcal{D}(p) = \left\{ x : (A \ -B) \begin{pmatrix} x \\ p \end{pmatrix} \geq b \right\}$$

$$\mathcal{D}' = \left\{ \begin{pmatrix} x \\ p \end{pmatrix} : A' \begin{pmatrix} x \\ p \end{pmatrix} \geq b \right\}$$

2.5.2.2 The polyhedron scanning problem

To generate sequential code for operations and variables declared over polyhedra, a loop nest that scans the given polyhedral region must be generated. The *polyhedron scanning problem* is formally stated as:

Given a parameterized polyhedral domain $\mathcal{D}(p)$ in terms of a parameter vector p and a set of k constraints:

$$\mathcal{D}(p) = \{x : Ax \geq Bp + b\}$$

where A and B are constant matrices of size $k \times n$ and $k \times m$, respectively, and b is a constant k-vector, produce the set of loop bound expressions $L_1, U_1, \ldots L_n, U_n$ such that loop nest:

$$
\begin{aligned}
&\texttt{DO} \quad x_1 = L_1, U_1 \\
&\qquad \vdots \\
&\qquad \texttt{DO} \qquad\qquad x_n = L_n, U_n \\
&\qquad\qquad\qquad\qquad\quad \text{body} \\
&\qquad \texttt{END} \\
&\texttt{END}
\end{aligned}
$$

will visit once and only once each and every integer point in the domain $\mathcal{D}(p)$ in lexicographic order of the elements of $x = (x_1, \ldots, x_n)$.

When talking about a particular loop variable x_i, the term *outer loops* refers to loops that enclose the x_i–loop, that is, the loops of indices $x_j, j < i$. *Inner loops* refers to the loops contained in the x_i–loop, that is, the loops of indices $x_j, j > i$.

The problem of finding loop bounds is related to the linear programming problem and shares its complexity. Fortunately, these problems tend to be relatively small (in terms of the dimension and number of constraints) due to the fact that loops are not deeply nested, and exact solutions for typical problems can be found in reasonable time.

Example 2.13 Given the parameterized domain defined as :

```
{i,j,k | i>=0; -i+M>=0; j>=0; -j+N>=0; k>=0; i+j-k>=0}:S
```

and the context domain `{N,M | N>0; M>0}` describing what is known to be true a priori, the following four different loop nests (in a system of affine recurrence equations syntax) were generated by the method described in this chapter. Each loop nest scans the domain in a different order.

```
{i | 0<=i<=M} ::
    {j,i | 0<=j<=N} ::
        {k,j,i | 0<=k<=i+j} :: S
```

a. The loop nest in {i, j, k} scan order.

```
{j | 0<=j<=N} ::
    {k,j | 0<=k<=j+M} ::
        {i,k,j | 0<=i<=M; i>=k-j} :: S
```

b. The loop nest in {j, k, i} scan order.

```
{k | 0<=k<=N+M>=0} ::
    {j,k | 0<=j<=N; j>=k-M} ::
        {i,j,k | 0<=i<=M; i>=k-j} :: S
```

c. The loop nest in {k, j, i} scan order.

```
{i | 0<=i<=M} ::
    {k,i | 0<=k<=i+N} ::
        {j,k,i | 0<=j<=N; j>=k-i} :: S
```

d. The loop nest in {i, k, j} scan order.

2.5.3 Description of the method

In this section, I solve the polyhedron scanning problem in terms of polyhedral operations and show how it can be solved using the *Simplify in Context* (see Section 2.4.6) and the *Convex Union* (see Section 2.4.7) operations previously described in this chapter. The method projects the polyhedron in the canonical direction of inner loop variables in order to eliminate dependencies on them. Thus bounds on a loop are found independent of inner loop indices. This method relies on the fact that computational domains tend to be relatively small polyhedra (in terms of the dimension and number of constraints) due to the fact that loops are not deeply nested.

The polyhedral operation *convex union* (see Section 2.4.7) joins one polyhedron to another and produces a resulting polyhedron with all redundant constraints and geometries eliminated. It is used in this section to eliminate (or project out) the inner loop indices from the bound expressions of outer

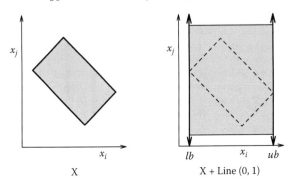

X

X + Line (0, 1)

Figure 2.11   Adding a line to project out an index.

loops. This is illustrated in Figure 2.11 where x_i is an outer loop variable and x_j is an inner loop variable. To compute the loop bounds for the x_i–loop as a function of parameters and outer loop variables, the inner loop variables $x_j, j > i$ must be removed from the domain. This is done by projecting the domain in the direction of the inner loop variables onto the x_i-axis, giving lb and ub as the lower and upper bounds of x_i, respectively. This can be accomplished using convex union by adding a polyhedron consisting of a set of lines $\{l_{i+1}, l_{i+2}, \cdots\}$ in the directions of all of the inner loop variables $\{x_{i+1}, x_{i+2}, \cdots\}$ to the polyhedron being scanned. The resulting polyhedron is a cylinder open in the direction of inner loop variables (as shown in the figure), which has no constraints in terms of the inner loop variables.

This polyhedron can be further simplified by considering the context of the surrounding loops. Once in the code body of the surrounding loops, the constraints imposed by those loops have already been enforced and can be assumed true. Thus, retesting for those constraints in the body of the loop is pointless and inefficient. The polyhedron resulting from the projection described above includes constraints from the outer loops. To remove these constraints, we can use the simplification in context operation described in Section 2.4.6.

Before giving the algorithm, the polyhedral scanning problem is restated in terms of polyhedral loop domains instead of loops.

> Given a d-dimensional polyhedron $\mathcal{D}$ to be scanned, with an initial context domain $\mathcal{D}_0$ (constraints that are known to be true at this point in the program and that do not need to be reverified), find a sequence of loop domains $\mathcal{D}_1, \mathcal{D}_2, \cdots, \mathcal{D}_d$ such that
>
> $$\mathcal{D} = \mathcal{D}_0 \cap \mathcal{D}_1 \cap \mathcal{D}_2 \cap \cdots \cap \mathcal{D}_d$$
>
> where each loop domain $\mathcal{D}_i$ is only a function of outer loop variables $x_j, 1 \leq j \leq i$.

Using the above two polyhedral operations, a function can be written that takes a specified domain $\mathcal{D}$ and separates (or factors) it into an intersection of the initial context domain $\mathcal{D}_0$ and a sequence of loop domains $\mathcal{D}_1, \mathcal{D}_2, \dots, \mathcal{D}_d$ (so that $\mathcal{D} = \mathcal{D}_0 \cap \mathcal{D}_1 \cap \cdots \cap \mathcal{D}_d$) where each loop domain is not a function of inner loop variables.

The loop domain $\mathcal{D}_i, 1 \leq i \leq d$ is computed in two steps. First, the inner loop variables are projected out of the original domain $\mathcal{D}$ by adding lines $\{l_{i+1}, l_{i+2}, \dots, l_d\}$ in the directions of the inner loop dimensions using the *ConvexUnion* operation. The resulting domain is not a function of inner loop variables. This domain is then simplified in context of the initial context domain $\mathcal{D}_0$ and all of the outer loop domains $\mathcal{D}_1, \mathcal{D}_2, \dots, \mathcal{D}_{i-1}$.

Accordingly, the loop domain $\mathcal{D}_i, i \geq 1$ can be recursively computed as:

$$\mathcal{D}_i = SimplifyInContext(ConvexUnion(\mathcal{D}, \{l_{i+1}, \cdots, l_d\}), \mathcal{D}_0 \cap \cdots \cap \mathcal{D}_{i-1});$$

After finding the domain $\mathcal{D}_i$, the loop bounds of the ith loop can be easily extracted from the domain. Polyhedral operations can thus be used to efficiently compute loop nests to scan a polyhedral domain, which is an essential step in generating sequential imperative programs to execute recurrence equations.

2.6   Localizing Affine Dependences

The purpose of this section is to show how polyhedral operations can be used to create localizing transformations for systems of affine recurrence equations.

Algorithms expressed as uniform recurrence equations (URE) have been shown to map well on to regular architectures known as *regular array processors*, or *systolic arrays*. These architectures take advantage of the small grain parallelism in regular algorithms. Interconnection between processors is *local*, meaning only nearest neighbors communicate. The modularity of the processor array and locality of communication makes these architectures completely scalable.

Systems of affine recurrence equations (ARE) are not intrinsically local, and therefore, do not map directly to regular array architectures. A transformation process called *localization* must be applied to *affine* recurrence equations to rewrite them as a functionally equivalent system of *uniform* recurrence equations that only uses local interconnect. For many algorithms, this process is easily accomplished. For others, the localization of dependences can be quite complex.

The localization of affine dependences is a major problem in the methodical derivation of systolic arrays from systems of affine recurrence equations. A paper by Roychowdhury, et al. [26] describes a mathematical approach to the localization problem that elegantly handles both the pipelining of broadcasts and the serialization of reductions with a unified approach. This section summarizes their method and then discusses an algorithm that was developed to implement this method using polyhedral operations.

2.6.1 Dependences

As was discussed in Section 2.1.2, a **recurrence equation** equates an array variable on the left hand side of the equation to a function of other array variables on the right hand side. An array variable can be thought of as a function mapping points i in the index domain of a variable to values in the variable's data type set (such as integer, real, or boolean) as illustrated graphically in Figure 2.12a.

In a system of recurrences, we say that a variable X at a point $p \in \mathcal{D}_x$ (directly) **depends on** variable Y at a point $q \in \mathcal{D}_y$, (denoted by $X[p] \mapsto Y[q]$), whenever $Y[q]$ occurs on the right hand side of an equation defining $X[p]$. This is shown graphically in Figure 2.12b. In general, each dependence has the form:

$$\forall i \in \mathcal{D} : X[Pi + p] \mapsto Y[Qi + q] \qquad (2.14)$$

and is read as, "for all i in index domain $\mathcal{D}$, X at $Pi + p$ depends on Y at $Qi + q$." The domain $\mathcal{D}$ is the common iteration space. For each point i in $\mathcal{D}$ there is a single data dependence between X and Y. The image of the domain $\mathcal{D}$ by the affine function $Pi + p$ must be a subset of the index domain $\mathcal{D}_x$ of X, and image of $\mathcal{D}$ by $Qi + q$ must be a subset of the index domain $\mathcal{D}_y$ of Y. A general dependence of this type is called an **affine dependence**. A simpler

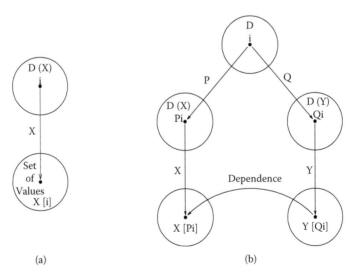

(a)   (b)

Figure 2.12   Graphical representation of variable and dependences.

dependence of the form $\forall i \in \mathcal{D} : X[i] \mapsto Y[i+b]$, where b is a constant vector, is called a **uniform dependence**.

In Figure 2.12b, circles represent index domains and value sets, and the arrows represent mapping functions between them. Functions P and Q map points in $\mathcal{D}$ to the index domain of variables X and Y, respectively, and X and Y map points in their respective index domains to the actual values. The dependence is represented by the arrow between $X[Pi]$ and $Y[Qi]$. (For this analysis, the additive constants p and q are not significant.) The arrow head points in the direction of data flow, opposite the direction of the dependence. Thus we have a graphical representation of the dependence $X[Pi] \mapsto Y[Qi]$.

Dependences may also be represented on a *dependence graph*, where variables are represented by nodes, and dependences are represented as directed arcs with the tail at the node where data originates and the head at the node where the data is used. (Note that an arc on a dependence graph goes in the direction of dataflow, which is exactly opposite the direction of dependence.) Figure 2.13 shows dependence graphs for the different forms of dependences to be considered in this chapter. Local dependences (Figure 2.13a.) and non-local dependences (Figure 2.13b.) are a one-to-one relationship between the data source and destination. A broadcast (Figure 2.13c.) occurs when data generated at one node is needed in computations at several different nodes and is a one-to-many data communication. A reduction (Figure 2.13d.) is a many-to-one communication that occurs when data from several different nodes are combined with an associative and commutative operator. In this chapter, we are concerned with the pipelining and localization of broadcasts and reductions.

2.6.2 Method to classify dependences

The basic mathematical representation of a dependence given in Equation 2.14 can represent all four types of dependences mentioned above. To determine the type of a dependence, we must look at the *null spaces* of the functions P and Q. We digress here for a short discussion of null spaces, and then we will illustrate how null spaces in P and Q are used to differentiate the different types of dependences.

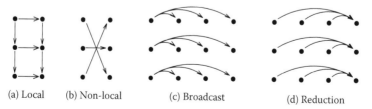

| (a) Local | (b) Non-local | (c) Broadcast | (d) Reduction |

Figure 2.13   Simple dependence graphs illustrating types of dependences.

2.6.2.1   Null spaces

Null space   Consider an affine mapping function $Az + b$ that maps one set of points to another set of points. Consider two distinct points z_1 and z_2 that are mapped to the same point z'. Then,

$$Az_1 + b = z' \quad \text{and} \quad Az_2 + b = z' \Rightarrow A(z_2 - z_1) = 0$$
$$Av = 0, \quad v = z_2 - z_1$$

The vector v that separates z_1 and z_2 is said to be in the *null space* or *kernel* of A. The **null space** of A is defined as:

$$null(A) = \{x : Ax = 0\} = \{x : x = B\lambda\} = lin(B)$$

The null space of A is a linear space that is spanned by a set of basis lines (columns of matrix B). In the equation above, $B\lambda$ is a linear combination of the basis lines (λ is any vector). The null space of A is unique; however, the matrix B is not unique since the basis of a linear space is not unique. The rank of matrix B is called the *dimension* of the null space. An implication of the definition of a null space is that $AB = 0$.

We also need to refer to the inverse of the *null* function. If the matrix B is a basis for the null space of A, then we can say $null^{-1}(lin(B)) = A$.

Null space in context   Consider an affine mapping function $Az + b$ that maps all points in a polyhedral domain $\mathcal{D}$ to another domain $\mathcal{D}' = $ Image $(\mathcal{D}, (z \rightarrow Az + b))$. Consider two points z_1 and z_2 in $\mathcal{D}$ that are mapped to the same point z' in $\mathcal{D}'$. If $\mathcal{D}$ is defined as $\mathcal{D} = \{z : Cz = d; Ez \geq f\}$, then any distinct two points z_1 and z_2 in $\mathcal{D}$ must both satisfy the equality $Cz = d$.

$$Cz_1 = d \quad \text{and} \quad Cz_2 = d \Rightarrow C(z_2 - z_1) = 0$$
$$Cv = 0, \quad v = z_2 - z_1$$

and thus $v \in null(C)$. The **affine hull** of $\mathcal{D}$ is the smallest dimensional flat that contains all points in $\mathcal{D}$. If $\mathcal{D}$ is defined above, then the affine hull of $\mathcal{D}$ is defined as:

$$\text{aff}\mathcal{D} = \{z : Cz = d\}.$$

The null space of A in the context of a domain $\mathcal{D}$ whose affine hull $= \{z : Cz = d\}$ is:

$$\{x : Ax = 0; Cx = 0\} = null(A) \cap null(C)$$

2.6.2.2   Null spaces in P and Q

Roychowdhury, et al. [26] describe how to interpret the presence of nonempty null spaces in P and Q and how this enables us to detect broadcasts and reductions, or a combination of both. In this section, we summarize their work using the domain figure we have developed.

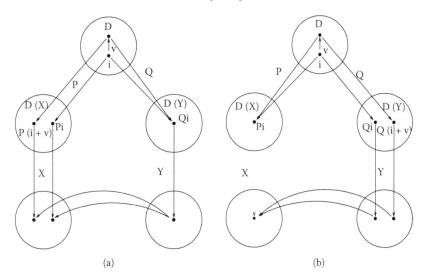

Figure 2.14   Broadcast and reduction.

Broadcasts   Broadcasts are created when there exists a vector v in $\text{Null}(Q)$, which is not in $\text{Null}(P)$. Refer to Figure 2.14a. If v is the vector separating two points in $\mathcal{D}$ and v is in the null space of Q, then the two points in $\mathcal{D}$ will map to the same point in $\mathcal{D}(Y)$. If v is not in the null space of P, then these two points will map to two distinct points in $\mathcal{D}(X)$. This means that two or more values in X are dependent on a single value in Y and this is a broadcast.

Reductions   Refer to Figure 2.14b. If v is in the null space of P and not in the null space of Q, then the two points in $\mathcal{D}$ will map to a single point in $\mathcal{D}(X)$ and two distinct points in $\mathcal{D}(Y)$. This creates a dependence of $X[Pi]$ on two points or more points in Y and this is a reduction.

Broadcast-reductions   It is also possible that both P and Q have nonempty and nonintersecting null spaces ($\text{Null}(P) \cap \text{Null}(Q) = \emptyset$). In this case the dependence is both a broadcast and a reduction. This case is illustrated in Figure 2.15a.

In the figure, v is in the null space of P and u is in the null space of Q and $u \neq v$. Any point separated by one of these vectors will map to a single point in one domain and to two points in the other. The result, as can be seen in the figure, is that each point in the left domain is dependent on both points in the right domain. This type of dependence may be separated into a reduction followed by a broadcast. The details of this transformation will be covered later in this section.

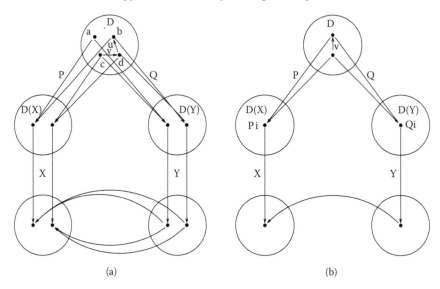

Figure 2.15 Broadcast-reduction and redundant iteration space.

Redundant iteration space   It is possible that the domain $\mathcal{D}$ is larger than it needs to be to complete the desired calculation. This happens when the null spaces of P and Q have a nonempty intersection: $\text{Null}(P) \cap \text{Null}(Q) \neq \emptyset$. A subset of this case, $\text{Null}(P) = \text{Null}(Q)$, is represented in Figure 2.15b. We will look at this simple case here, and abstract the idea later.

In the figure, $\text{Null}(P) = \text{Null}(Q) = v$. Since v is in the null space of both P and Q, any points in $\mathcal{D}$ separated by this vector will map to a single point in both $\mathcal{D}(X)$ and $\mathcal{D}(Y)$. Thus, the two points pictured map to a single point in both domains. Both points in $\mathcal{D}$ lead to a single one-to-one dependence between $X[Pi]$ and $Y[Qi]$.

2.6.3   Description of the algorithm

2.6.3.1   Computing null spaces using Hermite normal forms

First, we need to introduce and define the two Hermite normal forms. The definitions are similar and so differences are noted.

<u>**Right**</u> **Hermite normal form**   If A is an integer matrix of full <u>column</u> rank, then there exist integer matrices U and H such that $UA = \begin{bmatrix} H \\ 0 \end{bmatrix}$, where U is unimodular, and H is a nonnegative[4] <u>upper triangular</u> matrix

---

[4]A different definition of Hermite normal form defines only the diagonal elements positive with all other elements in the matrix zero or negative [27, p. 190]. Here, we use the definition given by Schrijver [2, p. 45].

with positive diagonal elements. Further, each off-diagonal element is strictly less than the diagonal element in its same <u>column</u>. The right Hermite normal form is computed using row operations.

<u>Left</u> Hermite normal form If A is an integer matrix of full <u>row</u> rank, then there exist integer matrices U and H such that $AU = [\ H\ \ \ 0\ ]$, where U is unimodular and H is a nonnegative <u>lower triangular</u> matrix with positive diagonal elements. Further, each off-diagonal element is strictly less than the diagonal element in its same <u>row</u>. The left Hermite normal form is computed using column operations.

We can find a basis set for the null space of a full row matrix A using the Hermite normal form. Let H be the left Hermite normal form of matrix A, then

$$AU = [H\quad 0]$$

where U is a unimodular matrix. Let the matrix U be written as the column catenation of two matrices U_1 and U_2, such that the number of columns of U_1 matches the number of columns of H.

$$AU = A[U_1\quad U_2] = [H\quad 0]$$

$$AU_1 = H$$

$$AU_2 = 0$$

The columns of the matrix U_2 form a minimal basis for the null set of A:

$$null(A) = lin(U_2)\quad \text{and}\quad null^{-1}(lin(U_2)) = A$$

The columns of U_2 are guaranteed not to be linear combinations of each other, and thus it is a minimum-sized basis, and the dimension of the null space of A is the number of columns in U_2.

The columns of U_1 form a minimum basis of another linear space, which we will call the *nonnull space* defined as follows:

$$nonnull(A) = \{v : Av \neq 0\}$$

And thus,

$$nonnull(A) = lin(U_1)$$

The following relationships can be shown to be true:

$$null(A) \cap null(B) = null\left(\begin{bmatrix} A \\ B \end{bmatrix}\right)$$

$$null(A) - null(B) = null(A) \cap nonnull(B)$$

$$nonnull(A) \cup nonnull(B) = nonnull\left(\begin{bmatrix} A \\ B \end{bmatrix}\right)$$

$$null(A) = lin(B) \Rightarrow null(B^T) = lin(A^T)$$

$$null^{-1}(lin(A)) = B \Rightarrow null(A^T) = lin(B^T)$$

The last of these relationships provides a basis for computing $null^{-1}$.

It is possible to do a chain of transformations on the recurrence equations while still preserving the computational meaning of the recurrences. The transformations we are interested in doing are those that will help us to localize the dependences in our calculation. There are three main transformations we want to make: reducing the redundant iteration space; introducing an intermediate variable to separate dependences with both broadcasts and reductions; and introducing pipelining vectors to localize broadcasts and reductions. We pose the algorithm in terms of transformations of dependences for simplicity. In the larger sense, it is really the recurrence equations which are transformed, and the dependences are changed as a result.

2.6.3.2   Step 1: Reducing redundant iteration spaces

As was shown previously, when $null(P) = null(Q)$, the index space is redundant. This can be extended to the general case when $null(P) \cap null(Q) \neq \emptyset$.

Referring to Figure 2.16a, both u and v are in the null space of Q. Thus all four points a, b, c, and d map to a single point in $\mathcal{D}(Y)$. Then, there would be only one element in Y, and the dependence would be a broadcast. One may see that there are redundancies in the iteration space: the two iteration points a and b in $\mathcal{D}$ lead to the same dependence and computation. Likewise, the points c and d perform the same computation.

The redundant iteration space $\mathcal{D}$ can be reduced to $\mathcal{D}'$ as shown in Figure 2.16b, in which the same computation is accomplished as shown in Figure 2.16a, only without the redundancy. To reduce the system, common components of the null spaces of P and Q (in this case v) are removed. When the iteration space is changed to $\mathcal{D}'$, the mapping functions P and Q must also change in order to preserve the proper mapping.

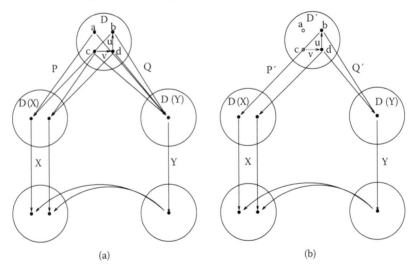

(a)   (b)

Figure 2.16   Redundant iteration space before and after simplification.

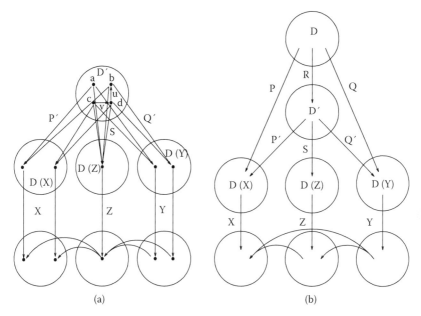

Figure 2.17 a. Introduction of intermediate variable Z to separate a broadcast-reduction dependence. b. All domains and mappings used in the localization algorithm.

To transform $\mathcal{D}$ to $\mathcal{D}'$, we simply "compress" the domain $\mathcal{D}$ so that vectors in the common null space are no longer in the domain. As shown in Figure 2.17b, the new domain $\mathcal{D}'$ is computed by mapping (projecting) all points in $\mathcal{D}$ by a function R onto the image domain $\mathcal{D}'$ where $null(R) = null(P) \cap null(Q)$. In our example, $R = null^{-1}(lin(v))$. The image function maps all points in $\mathcal{D}$ separated by the vector v to a single point in $\mathcal{D}'$, and thus the redundant points in the domain are removed. New mapping functions P' and Q' that map from the new domain $\mathcal{D}'$ to the index domains of X and Y, respectively, must also be computed.

Using the techniques just discussed, it is possible to reduce the calculation of interest to its minimum iteration space. The end effect of these reductions is to reduce the hardware of the final systolic array, and/or reduce the time of the computation, both of which are desirable as they lead to a more efficient implementation of the algorithm.

1. Let $null(R) = null(P) \cap null(Q)$.
   If $null(R) \neq \emptyset$ Then
       a. $\mathcal{D}' = Image(\mathcal{D}, R)$ (Project out the redundancy of $\mathcal{D}$)
       b. Compute P' such that $null(P') = null(P) - null(R)$
       c. Compute Q' such that $null(Q') = null(Q) - null(R)$
      Else $\mathcal{D}' = \mathcal{D}$; $P' = P$; $Q' = Q$

After reducing the iteration space to $\mathcal{D}'$, $null(P') \cap null(Q') = \emptyset$ and thus dependences are not redundant.

2.6.3.3   Step 2: Separating broadcast-reductions

Some nonlocal dependences may be broadcasts, others may be reductions, and yet others may have both a broadcast and a reduction component in them. This last type of dependence requires a step in the algorithm that introduces an intermediate variable to separate a combined reduce-broadcast into a reduce-operation followed by a broadcast.

The introduction of an intermediate variable Z is illustrated in Figure 2.17a. A vector v is not in $Null(Q')$ but is in $Null(S)$, creating a reduction from Y to Z. Also, a vector u is in $Null(S)$, but not in $Null(P')$, creating a broadcast from Z to X. The intermediate variable Z is found by defining a new mapping function S such that both u and v are in the null space of S. The function S maps all points in $\mathcal{D}'$ that are separated by vectors in the null space of either P or Q to the same point in the new domain $\mathcal{D}(Z)$. The resulting intermediate variable Z is dependent on variable Y and this dependence is a reduction. The variable X is dependent on intermediate variable Z and this dependence is a broadcast.

2. If $null(P') \neq \emptyset$ and $null(Q') \neq \emptyset$ Then
    Compute S: If $null(P') = lin(V)$ and $null(Q') = lin(U)$ then
        $S = null^{-1}(lin([V\ U]))$
        ($null(S) = lin(V) + lin(U)$)
    Create new intermediate variable Z with domain $\mathcal{D}(z) = image(\mathcal{D}', S)$
    Split recurrence equation with dependence:
        $\forall i \in \mathcal{D}' : X[P'i] \mapsto Y[Q'i]$
    into two recurrences so that the new dependences are:
        $\forall i \in \mathcal{D}' : Z[Si] \mapsto Y[Q'i]$ (reduction)
        $\forall i \in \mathcal{D}' : X[P'i] \mapsto Z[Si]$ (broadcast)

After the recurrence is separated, the reduction and the broadcast can be individually localized and simplified in the next step.

2.6.3.4   Step 3: Computing the pipelining vectors

At this point in the algorithm, all broadcasts and reductions have been separated and recurrences can only have dependences of three forms. The first is a one-to-one dependence. This is the case if both $null(P')$ and $null(Q')$ are empty. The second case is a broadcast. This comes about if $null(Q')$ is not empty and $null(P')$ is empty. We localize this dependence by introducing pipelining vectors chosen from the basis of the null space of Q'. The third case is a reduction, which is similar in form to the broadcast. It arises when the $null(P')$ is not empty and the $null(Q')$ is empty. Pipelining vectors are then chosen from a basis of the null space of P'. Through use of pipelining, data is passed from node to node using only local communication. By pipelining, all broadcasts and reductions can be broken up into local dependences. To initialize a pipeline, a bottom (or beginning) point must be found. Pipelining originates from this point and flows in the direction of the pipelining vectors to the rest of the nodes that need the data. The domain of bottom points will be designated as $\mathcal{D}_\perp$. The dependence would then take on the following forms.

3. If $null(P') = \emptyset$ and $null(Q') \neq \emptyset$ Then (broadcast)
  Choose a basis vector ρ from $null(Q')$ to be the pipelining vector
  Compute $\mathcal{D}_\perp$
  Transform the recurrence so that the following dependences exist:
  $$\forall i \in \mathcal{D}_\perp : X[P'i] \mapsto Y[Q'i] \text{ (pipeline initialization)}$$
  $$\forall i \in (\mathcal{D} - \mathcal{D}_\perp) : X[P'i] \mapsto X[P'(i - \rho)] \text{ (pipeline broadcast)}$$
  If $null(P') \neq \emptyset$ and $null(Q') = \emptyset$ Then (reduction)
  Choose a basis vector ρ from $null(P')$ to be the pipelining vector
  Compute $\mathcal{D}_\perp$
  Transform the recurrence so that the following dependences exist:
  $$\forall i \in \mathcal{D}_\perp : X[P'i] \mapsto Y[Q'i] \text{ (pipeline initialization)}$$
  $$\forall i \in (\mathcal{D} - \mathcal{D}_\perp) : Y[Q'i] \mapsto Y[Q'(i - \rho)] \text{ (pipeline reduction)}$$

In some cases it is necessary to use two or more pipelining vectors. In general, the number of pipelining vectors needed is equal to the dimension of the null space of Q or P for broadcast or reduction operations, respectively. If there are multiple dimensions to be pipelined, then you must choose a primary pipelining vector, a secondary pipelining vector, and etc. The data is then pipelined from the bottom point in the direction of the primary pipelining vector. Then it is sent from these points in the direction of the secondary vectors. This process is repeated until the full dimensionality of the null space is covered. The basis vectors of a null space are not unique. Any basis of the null space may be used as the pipelining vectors. In most cases, it will be obvious which set of vectors will be most advantageous.

2.6.4 Summary

An algorithm has been presented that uses polyhedral operations to localize broadcasts and reductions. The resulting transformations are performed on the system of recurrence equations, and at each step, non-local dependences are eliminated and the system becomes more and more localized. Figure 2.17b shows all of the transformation steps with their domains together on one graph.

By following these steps, affine dependencies can be localized. At the completion of the process, all one-to-many, many-to-many, and many-to-one dependences will have been replaced with one-to-one dependences. The majority of these dependences will be local. Non-local one-to-one dependences may also possibly arise in connection with the initialization of pipelines. The localization of these dependencies is a different discussion.

As has been shown, polyhedral operations are pervasively used in the specification, analysis, and transformation of recurrence equations.

References

1. B. Grunbaum. *Convex Polytopes*, volume 16 of *Pure and Applied Mathematics*. John Wiley & Sons, London, England, 1967.

2. A. Schrijver. *Theory of Linear and Integer Programming*. John Wiley and Sons, NY, 1986.

3. H. Edelsbrunner. *Algorithms in Combinatorial Geometry*, volume 10 of *Monographs on Theoretical Computer Science*. Springer-Verlag, Berlin, 1987.

4. D.A. Turner. Recursion equations as a programming language. In J. P. Darlinton, D.A. Henderson, and Turner, editors, *Functional Programming and Its Applications: An Advanced Course, 1981*, pp. 1–28. Cambridge University Press, New York, 1982.

5. S.V. Rajopadhye and R.M. Fujimoto. Synthesizing systolic arrays from recurrence equations. *Parallel Computing 14*, 14:163–189, 1990.

6. C. Guerra. A unifying framework for systolic design. In F. Makedon, T. Melhorn, T. Papatheodorou, and P. Spirakis, editors, *VLSI Algorithms and Architectures: Aegean Workshop on Computing*, Loutraki, Greece, Springer-Verlag, 1986.

7. P. Quinton. Automatic synthesis of systolic arrays from uniform recurrence equations. *Proceedings 11th Annual International Symposium on Computer Architecture, Ann Arbor*, pp. 208–214, June 1984.

8. R.M. Karp, R.E. Miller, and S. Winograd. The organization of computations for uniform recurrence equations. *JACM*, 14(3):563–590, July 1967.

9. Yoav Yaacoby and Peter R. Cappello. Scheduling a system of nonsingular affine recurrence equations onto a processor array. *Journal of VLSI Signal Processing*, 1(2):115–125, 1989.

10. J.M. Delosme and I.C.F. Ipsen. Systolic array synthesis: Computability and time cones. In M. Cosnard et al., editors, *Parallel Algorithms and Architectures*, pp. 295–312. Elsevier Science Publishers B. V. (North-Holland), 1986.

11. P. Quinton and V. Van Dongen. The mapping of linear recurrence equations on regular arrays. *Journal of VLSI Signal Processing*, 1(2):95–113, 1989. Also appeared as *IRISA Technical Report 485*, 1989.

12. Gene H. Golub and Charles F. Van Loan. *Matrix Computations, Third Edition.* The John Hopkins University Press, Baltimore, MD, 1996.

13. D.J. Kuck. *The Structure of Computers and Computations.* J. Wiley and Sons, NY, 1978.

14. A.J. Goldman. Resolution and separation theorems for polyhedral convex sets. In H.W. Kuhn and A.W. Tucker, editors, *Linear Inequalities and Related Systems*, number 38 in *Annals of Mathematics Studies*. Princeton University, Princeton, NJ, 1956.

15. T.H. Mattheiss and D. Rubin. A survey and comparison of methods for finding all vertices of convex polyhedral sets. *Mathematics of Operations Research*, 5(2):167–185, May 1980.

16. T.S. Motzkin, H. Raiffa, G.L. Thompson, and R.M. Thrall. The double description method. *Theodore S. Motzkin: Selected Papers*, 1953.

17. N.V. Chernikova. Algorithm for finding a general formula for the nonnegative solutions of a system of linear inequalities. *U.S.S.R. Computational Mathematics and Mathematical Physics*, 5(2):228–133, 1965.

18. D. Rubin. Vertex generation and cardinality constrained linear programs. *Operations Research*, 23(3):555–565, May 1975.

19. F. Fernandez and P. Quinton. Extension of Chernikova's algorithm for solving general mixed linear programming problems. *Technical Report 437*, IRISA, Rennes, France, Oct. 1988.

20. H. Le Verge. A note on Chernikova's algorithm. *Technical Report PI 635*, IRISA, Feb. 1992.

21. R. Seidel. Small-dimensional linear programming and convex hulls made easy. *Discrete & Computational Geometry*, Vol. 6, num. 1, pp. 423–434, Springer-Verlag, New York, 6:423–434, 1991.

22. Hervé Le Verge. Recurrences on lattice polyhedra and their applications. *Technical Report*, based on an unpublished manuscript written by Hervé Le Verge just before his untimely death, Rennes, France, 1995.

23. F. Irigoin. Code generation for the hyperplane method and for loop interchange. *Technical Report ENSMP-CAI-88-E102/CAI/I*, Ecole Nationale Superieure des Mines de Paris, Oct. 1988.

24. M.E. Wolf and M. Lam. Loop transformation theory and an algorithm to maximize parallelism. *IEEE Transactions on Parallel and Distributed Systems*, 2(4):452–471, Oct. 1991.

25. H. Le Verge, V. Van Dongen, and D. Wilde. La synthèse de nids de boucles avec la bibliothèque polyédrique. *RenPar'6*, June 1994. English version "Loop Nest Synthesis Using the Polyhedral Library" in IRISA *Technical Report 830*, May 1994.

26. V.P. Roychowdhury, L. Thiele, S.K. Rao, and T. Kailath. On the localization of algorithms for VLSI processor arrays. In *VLSI Signal Processing III*, pp. 459–470. IEEE Press, New York, 1989.

27. G. Nemhauser and L. Wolsey. *Integer and Combinatorial Optimization.* John Wiley and Sons, NY, 1988.

Chapter 3

Computation of Data Storage Requirements for Affine Algorithmic Specifications

Florin Balasa
American University in Cairo, New Cairo, Egypt

Hongwei Zhu
ARM, Inc., Sunnyvale, California

Ilie I. Luican
Microsoft Corp., Redmond, Washington

Contents

3.1   Introduction

In the earlier days of digital system design, memory was expensive, so researchers focused on memory size optimization. Nowadays, the cost per memory bit is very low due to the progress of the semiconductor technology and the consequent increase of the level of integration. Gradually, the memory size optimization decreased in direct importance, while performance and power consumption became the key challenges. Memory may become a bottleneck—both in terms of energy and performance—for data-intensive embedded applications in areas like multimedia and multidimensional signal processing due to the impact of data storage and transfer versus data processing [1–3].

However, memory latency and energy consumption per access increase with the memory size. Therefore, reducing the memory requirements of the target applications continues to be used in the early phase of the design flow, the ultimate goals being reducing the storage power budget and increasing performance. During this phase, the designer attempts to improve the temporal locality of data (i.e., the results of a computation should be used as soon as possible by next computations in order to reduce the need for temporary storage) by performing code transformations on the behavioral specifications [4–6]. Loop transformations are important system-level design techniques, used to enhance the locality of data and the regularity of data accesses. The overall reduction of the lifetimes[1] of scalar signals (array elements) increases the possibility of memory sharing, since data with nonoverlapping lifetimes can be mapped to the same physical location. This leads to the overall reduction in the data storage requirements and, hence, of the chip area [7]. Data compression is another technique for reducing the storage requirements, which targets finding efficient representations of data [8]. In spite of applying such techniques in the early phase of the design flow, data memory size has steadily increased over time due to the fact that system applications grew more complex.

In deriving an optimized memory architecture, memory size computation continues to be an important evaluation tool in the early phase of the design, the system-level exploration. The problem is to determine the *minimum* amount of memory locations necessary to store the signals of a given multimedia algorithm during its execution, or, equivalently, the *maximum* storage occupancy assuming any scalar signal needs to be stored only during its lifetime. The total number of scalars in each of the three equivalent codes in Figure 3.1 is 4,096. In the first code, the lifetimes of any two array elements are

---

[1]The lifetime of a scalar signal is the time interval between the clock cycles when the scalar is *produced* or written, and when it is read for the last time, i.e., *consumed*, during the code execution. Two scalars are simultaneously alive if their lifetimes do overlap. Obviously, in such a case, they must occupy different memory locations; otherwise, they can share the same location.

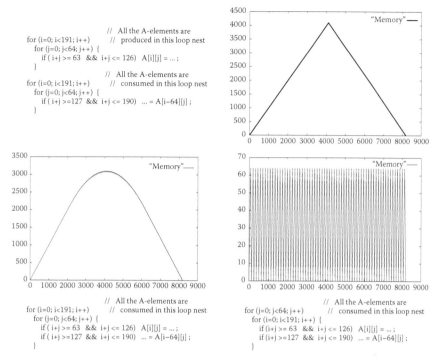

```
for (i=0; i<191; i++)          // All the A-elements are
  for (j=0; j<64; j++) {       //   produced in this loop nest
    if ( i+j >= 63  && i+j <= 126)  A[i][j] = ... ;
  }
for (i=0; i<191; i++)          // All the A-elements are
  for (j=0; j<64; j++) {       //   consumed in this loop nest
    if ( i+j >=127  && i+j <= 190)  ... = A[i-64][j] ;
  }
```

```
for (i=0; i<191; i++)          // All the A-elements are
  for (j=0; j<64; j++) {       //   consumed in this loop nest
    if ( i+j >= 63  && i+j <= 126)  A[i][j] = ... ;
    if ( i+j >=127  && i+j <= 190)  ... = A[i-64][j] ;
  }
```

```
for (j=0; j<64; j++)           // All the A-elements are
  for (i=0; i<191; i++) {      //   consumed in this loop nest
    if (i+j >= 63  && i+j <= 126)  A[i][j] = ... ;
    if (i+j >=127  && i+j <= 190)  ... = A[i-64][j] ;
  }
```

Figure 3.1 Memory traces showing the effect of loop fusion and loop interchange on the storage requirement. The storage requirement of the initial code is 4096 memory locations. After loop fusion, the needed data storage decreases to 3104 locations; additionally, after loop interchange, it becomes only 64 locations (about 1.5% of the initial value). (Reprinted from [56] with permission from IOS Press.)

overlapping (since all the *A*-elements are produced before any is consumed), so each array element needs a distinct memory location. However, only 64 locations are necessary for the third code (as confirmed by our tool presented in Section 3.6): due to the fact that scalars having disjoint lifetimes can share the same memory location, the amount of storage (64) can be much smaller than the total number of scalar signals (4096).

One of the possible applications of the exact memory size computation is the evaluation of the impact of different code (and, in particular, loop) transformations on the data storage, during the early design phase of system-level exploration. For instance, the minimum memory size needed by the array *A* in the exemplifying code with two loop nests from Figure 3.1 is 4096 locations. The variation of the storage requirement has a simple pattern: in the first loop nest, it increases uniformly due to the production of the *A*-elements; in the second loop nest, it decreases uniformly due to the consumption of the same elements, as shown in the first graph. After the fusion of the nested loops, the storage requirement decreases to 3104 locations, the new trace of the memory variation being displayed in the second graph of the figure. Interchanging the loops drastically decreases the storage requirements (with 98%) to only 64 locations, the final trace being the third graph shown in Figure 3.1.

Different variants of code of a same application can be compared one against another in storage point of view, without the need of performing a proper memory allocation for each variant—a significantly more expensive solution.

Another possible application is the evaluation of signal-to-memory mapping techniques,[2] which are used to compute the physical addresses in the data memory for the array elements in the application code. These mapping models actually trade-off an excess of storage for a less complex address generation hardware. The mapping techniques known nowadays do not provide consistent information on *how good their models are*, that is, how large is the oversize of their resulting storage amount after mapping in comparison to the minimum data storage. The effectiveness of their mapping model is assessed only *relatively*; that is, they compare the storage resulted after applying their mapping model either to the total number of array elements in the code or to the storage results when applying a different mapping model. The exact computation of the memory size and, also, the minimum storage windows for each array allow to better evaluate the performance of *any* signal-to-memory mapping model [9].

In order to solve *exactly* the memory size computation problem, the simulated execution of the code may seem, at a first glance, an appealing strategy. In principle, it could indeed solve the problem by brute-force computation. However, the simulated execution exhibits a poor scalability: while being fast for small and even medium-size applications, the computation times increase steeply, becoming ineffectual when processing examples with millions of scalars (array elements) and deep loop nests with iterators having large ranges, like many image and video processing applications. Enumerative techniques or scalar approaches for register-transfer level (RTL) programs (see Section 3.2) are too computationally expensive in such cases, often prohibitive to use.

The algorithm presented in this chapter is a non-scalar technique for computing the maximum number of array elements simultaneously alive. This number represents the minimum data storage necessary for the code execution because the simultaneously alive scalar signals must be stored in distinct memory locations. The basic reasons of performance of this approach are: (a) an efficient decomposition of the array references of the multidimensional signals into disjoint *linearly bounded lattices* (LBLs) [10], and (b) an efficient mechanism of pruning the code of the algorithmic specification, concentrating the analysis in the parts of the code where the larger storage requirements are likely to happen.

The rest of the chapter is organized as follows. Section 3.2 gives a brief overview of the previous storage estimation techniques. Section 3.3 explains

---

[2]An overview of signal-to-memory mapping models is given in Chapter 6.

the memory size computation problem for affine algorithmic specifications, introducing terminology and basic concepts. Section 3.4 details some more significant algorithms used by this polyhedral framework, while Section 3.5 describes the global flow of the storage computation algorithm. Section 3.6 presents basic implementation aspects and discusses the experimental results. Section 3.7 summarizes the main conclusions of this work.

3.2 The Memory Size Computation Problem: A Brief Overview

This problem has been tackled in the past both for register-transfer level (RTL) programs at scalar level (since the earlier digital systems were simpler) and for behavioral specifications at non-scalar level.

The problem of optimizing the register allocation/assignment in programs at RTL was initially formulated in the field of software compilers [11], aiming at a high-quality code generation phase. The problem of deciding which values in a program should reside in registers (allocation) and in which register each value should reside (assignment) has been solved by a graph coloring approach. As in the code generation, the register structure of the targeted computer is known, the k-coloring of a register-interference graph (where k is the number of computer registers) can be systematically employed for assigning values to registers and managing register spills.[3] Although the problem of determining whether a graph is k-colorable is NP-complete, effective heuristic techniques were proposed.

In the field of synthesis of digital systems, starting from a behavioral specification, the register allocation was first modeled as a clique partitioning problem [12]. The register allocation/assignment problem has been optimally solved for nonrepetitive schedules, when the life time of all scalars is fully determined [13]. The similarity with the problem of routing channels without vertical constraints [14] has been exploited in order to determine the minimum register requirements (similar to the number of tracks in a channel) and to optimally assign the scalars to registers (similar to the assignment of one-segment wires to tracks) in polynomial time by using the *left-edge* algorithm [15]. A nonoptimal extension for repetitive and conditional schedules has been proposed in [16]. A lower bound on the register cost can be found at any stage of the scheduling process using force-directed scheduling [17]. Integer linear programming (ILP) techniques are used in [18] to find the optimal

---

[3]When a register is needed for a computation but all available registers are in use, the content of one of the used registers must be stored (spilled) into a memory location in order to free a register.

number of memory locations during a simultaneous scheduling and allocation of functional units, registers, and busses. Employing circular graphs, [19] and [20] proposed optimal register allocation/assignment solutions for repetitive schedules. A lower bound for the register count is found in [21] without fixing the schedule, using ASAP and ALAP constraints on the operations. Good overviews of these techniques can be found in [22].

Common to all the scalar-based techniques is that they break down when used by flattening large multidimensional arrays, each array element being considered a separate scalar signal. The nowadays data-intensive signal processing applications are described by high-level, loop-organized, algorithmic specifications whose main data structures are typically multidimensional arrays. Flattening the arrays from the specification of a video or image processing application would typically result in millions of scalars.

To overcome the shortcomings of the scalar estimation techniques for high-level algorithmic specifications where the code has an organization based on loop nests and multidimensional arrays are present, several research teams proposed various techniques exploiting the fact that, due to the loop structure of the code, large parts of an array can be produced or consumed within a single array reference. These estimation approaches can be basically split into two categories: those requiring a fully fixed execution ordering, and those assuming non-procedural specification where the execution ordering is still not (completely) fixed. The techniques falling in the first category will be addressed first.

Verbauwhede et al. consider a production axis for each array to model the relative production and consumption time (or date) of the individual array accesses [23]. The difference between these two dates equals the number of array elements produced between them, while the maximum difference gives the storage requirement for the considered array. The total storage requirement is the sum of the requirements for each array. An ILP approach is used to find the date differences. Since each array is treated separately, only the internal in-place mapping of an array (*intra-array in-place*) is considered; the possibility of mapping arrays in-place of each other (*inter-array in-place*) is not exploited. (When data with nonoverlapping lifetimes are mapped to the same physical memory locations, this is sometimes referred to as *in-place mapping*.)

In [24], the data-dependency relations between the array references in the code are used to find the number of array elements produced or consumed by each assignment. The storage requirement at the end of a loop equals the storage requirement at the beginning of the loop, plus the number of elements produced within the loop, minus the number of elements consumed within the loop. The upper bound for the occupied memory size within a loop is computed by producing as many array elements as possible before any elements are consumed. The lower bound is found with the opposite reasoning. From this, a memory trace of bounding rectangles as a function of time is found. The total storage requirement equals the peak bounding rectangle. If the difference between the upper and lower bounds for this critical rectangle is too large, better estimates can be achieved by splitting the corresponding loop into two loops and rerunning the estimation. In the worst-case situation, a full loop unrolling is necessary to achieve a satisfactory estimate.

Zhao and Malik developed a methodology for so-called exact memory size estimation for array computation [25]. It is based on live variable analysis and integer point counting for intersection/union of mappings of parameterized polytopes. In this context, a polytope is the intersection of a finite set of half-spaces and may be specified as the set of solutions to a system of linear inequalities. It is shown that it is only necessary to find the number of live variables for one statement in each innermost loop nest to get the minimum memory size estimate. The live variable analysis is performed for each iteration of the loops, however, which makes it computationally hard for large multidimensional loop nests.

Ramanujam et al. use for each array a reference window containing at any moment during execution the array elements alive (that have already been referenced and will also be referenced in the future) [26]. The maximal window size gives the storage requirement for the corresponding array. If multiple arrays exist, the maximum reference window size equals the sum of the windows for individual arrays. Treating the arrays separately, the technique does not consider the possibility of inter-array in-place mapping.

All the techniques above estimate the memory size assuming a single memory. A hierarchical memory size estimation is performed in [6], taking data reuse and memory hierarchy allocation into account.

In contrast to the array-based methods described so far in this section, the storage requirement estimation technique presented in [27] assumes a non-procedural execution of the application code. It traverses a dependency graph based on an extended data-dependency analysis, resulting in a number of nonoverlapping array sections (so called basic sets) and the dependencies between them. The basic set sizes and the sizes of the dependencies are found using an efficient lattice point counting technique [28]. The maximal combined size of simultaneously alive basic sets found through a greedy graph traversal gives an estimation of the storage requirement.

The estimation technique described in [29] assumes a *partially fixed* execution ordering. The authors employ a data dependence analysis similar to [27], their major improvement being to add the capability of taking into account available execution ordering information (based mainly on loop interchanges), thus avoiding the possible overestimates due to the total ordering freedom (less the data dependence constraints).

Good overviews of some of these techniques can be found in [30] and [31].

It must be noticed that specifications of multimedia algorithms are often parametrized. These specifications may contain more than one parameter, and the array indexes—although linear functions of the loop iterators—may be nonlinear functions of the parameters. A lot of research involving parametric polytopes has been done in the compiler community like, for instance, affine loop nest analysis and transformation [32,33], the improvement of data locality of nested loops [34], counting lattice points, their images, their projections [35], computing the number of distinct memory locations accessed by a loop nest [36], and parametrized integer programming [37]. Since the values of the parameters are known anyway before the implementation of data-intensive embedded applications, the memory size computation problem addressed in this chapter considers the values of the parameters to be known.[4] Chapter 4 will present a memory size evaluation approach for parametric specifications.

3.3   Computation of the Minimum Data Storage for Affine Specifications

The algorithms for telecom and (real-time) multimedia applications are typically specified in a high-level programming language, where the code is organized in sequences of loop nests having as boundaries linear functions of the outer loop iterators, conditional instructions where the conditions may be both data-dependent or data-independent (relational and/or logical operators of linear functions of loop iterators), and multidimensional signals whose array references have (possibly complex) linear indices (see the codes in Figures 3.1 and 3.2). This class of specifications is often referred to as *affine* [1]. Sometimes, in image and video processing, there may be also indices containing modulo operators, but these situations can be brought into the affine specification class [38].

These algorithms describe the processing of streams of data samples, so their source codes can be imagined as surrounded by an implicit loop having the *time* as iterator. Consequently, each signal in the algorithm has an *implicit* extra dimension corresponding to the *time* axis. This is why the code often contains *delayed* signals, that is, signals produced in previous data processings, which are consumed during the current execution. An illustrative example is shown in Figure 3.2. The delayed signals are the ones followed by the delay operator "@," the next argument signifying the number of *time* iterations in the past when those signals were produced. The delayed signals must be kept "alive" during several *time* iterations; that is, they must be stored during one or several data-sample processings.

Most of the previous non-scalar techniques on the evaluation of storage requirements (see Section 3.2) considered a fixed execution ordering prior to the

---

[4]Like other past works [23–29], as well. In spite of considering parametrized polytopes, Zhao and Malik still consider the parameter values to be known [25], since their estimation results are *numeric*, not *parametric*. On the other hand, researchers from the compiler community did compute the memory size for parametric examples, but these were restricted in size and complexity (e.g., only one—usually, perfect—nest of loops, with one single parameter), which is not always sufficient for multimedia applications.

```
T[0] = 0 ;                    // A[10][529] : input
for ( j=16 ; j<=512 ; j++)
{ S[0][j-16][0] = 0 ;
    for ( k=0 ; k<=8 ; k++)
      for ( i=j-16 ; i<=j+16 ; i++)
        S[0][j-16][33*k+i-j+17] = A[4][j] - A[k][i]@16
                               + S[0][j-16][33*k+i-j+16] ;
      T[j-15] = S[0][j-16][297] + T[j-16] ;
}
for( j=16 ; j<=512 ; j++)
{ S[1][j-16][0] = 0 ;
    for( k=1 ; k<=9 ; k++)
      for( i=j-16 ; i<=j+16 ; i++)
        S[1][j-16][33*k+i-j-16] = A[5][j] - A[k][i]@32
                               + S[1][j-16][33*k+i-j-17] ;
      T[j+482] = S[1][j-16][297] + T[j+481] ;
}
out = T[994];                 // out : output
```

Figure 3.2 Illustrative example of affine specification with two delayed array references [43]. (© 2007 IEEE.)

memory estimation [23–26]. The exceptions are [27], which allows any ordering not prohibited by data dependencies, and [29], where the designer can specify partial ordering constraints. In this chapter, the algorithmic specifications are considered to be *procedural*; therefore the execution ordering is induced by the loop structure and it is thus fixed, like in most of the previous works.[5] (When the specifications are non-procedural, the approach has also an *estimation operating mode*, computing accurate upper bounds.) Fundamentally different from the previous works [23–26] doing only a *memory size estimation*, the algorithm presented in this chapter is able to perform an *exact computation of the minimum data storage*. This approach uses both algebraic techniques specific to the data-flow analysis used in modern compilers [39], and elements of the theory of n-dimensional polyhedra.

Our polyhedral framework can compute the minimum data storage requirement of the entire (procedural) algorithmic specification, (therefore, the optimal memory sharing between *all* the array elements in the code (*inter-array memory sharing*). Optionally, it can also compute the minimum storage for each individual multidimensional array, therefore, the optimal memory sharing between the elements of the same array (*intra-array memory sharing*). For instance, the signal A from the illustrative example in Figure 3.3 needs a minimum data storage of 1752 memory locations since there are at most 1752 A-elements simultaneously alive—as computed by our tool. Similarly,

---

[5]This assumption is based on the fact that in present industrial design, the design entry usually includes a full fixation of the execution ordering. Even if this is not the case, the designer can still explore different algorithmic specifications that are functionally equivalent.

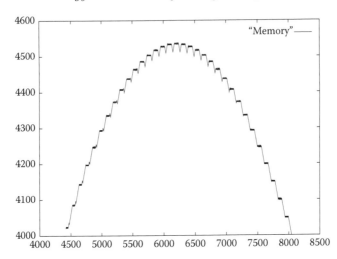

Figure 3.3   Code example and its memory trace illustrating the optimal inter-array memory sharing. (Reprinted from [56] with permission from IOS Press.)

the minimum data storage (or the optimal memory sharing) of signal B is 3104 storage locations. However, since the elements of the arrays A and B can share the same locations if their lifetimes are disjoint, the minimum storage requirement is, actually, 4536 locations, which is less than the sum of the minimum storage requirements of A and B $(1752 + 3104 = 4856)$. In addition to the computation of the minimum data storage, the variation of the memory occupancy during the execution of the algorithmic specification can be displayed as well. Such memory traces are shown in Section 3.6. For instance, Figure 3.3 displays a detail of the memory trace for the illustrative code in the figure.

A limitation of the current implementation is that the input programs are in the *single-assignment* form; that is, each array element is written at most once (but it can be read an arbitrary number of times). In Section 3.6, it will be explained how the algorithm could be modified in order to process specifications that are not in the single-assignment form.

3.3.1   Definitions and concepts

Definitions: A *polyhedron* is a set of points $P \subset \Re^n$ satisfying a finite set of linear inequalities: $P = \{\, \mathbf{x} \in \Re^n \,|\, \mathbf{A} \cdot \mathbf{x} \geq \mathbf{b} \,\}$, where $\mathbf{A} \in \Re^{m \times n}$ and $\mathbf{b} \in \Re^m$. If P is a bounded set, then P is called a *polytope*. If $\mathbf{x} \in \mathbf{Z}^n$, then P is called a $\mathbf{Z}$-polyhedron/polytope. The set $\{\, \mathbf{y} \in \Re^m \,|\, \mathbf{y} = \mathbf{A} \cdot \mathbf{x}, \mathbf{x} \in \mathbf{Z}^n \,\}$ is called the *lattice* generated by the columns of matrix $\mathbf{A}$.

Each *array reference* $M[x_1(i_1, \ldots, i_n)] \cdots [x_m(i_1, \ldots, i_n)]$ of an m-dimensional signal M, in the scope of a nest of n loops having the iterators

$i_1, \ldots, i_n$, is characterized by an *iterator space* and an *index* (or *array*) space. The iterator space signifies the set of all iterator vectors $\mathbf{i} = (i_1, \ldots, i_n) \in \mathbf{Z}^n$ in the scope of the array reference. The index space is the set of all index vectors $\mathbf{x} = (x_1, \ldots, x_m) \in \mathbf{Z}^m$ of the array reference. When the indices of an array reference are linear mappings with integer coefficients of the loop iterators, the index space consists of one or several *linearly bounded lattices* (LBLs) [10]—the image of an affine vector function over the iterator polytope $\mathbf{A} \cdot \mathbf{i} \geq \mathbf{b}$:

$$\{\mathbf{x} = \mathbf{T} \cdot \mathbf{i} + \mathbf{u} \in \mathbf{Z}^m \mid \mathbf{A} \cdot \mathbf{i} \geq \mathbf{b}, \mathbf{i} \in \mathbf{Z}^n\} \tag{3.1}$$

where $\mathbf{x} \in \mathbf{Z}^m$ is the index vector of the m-dimensional signal and $\mathbf{i} \in \mathbf{Z}^n$ is an n-dimensional iterator vector. In our context, the elements of the matrices $\mathbf{T}$, $\mathbf{A}$ and of the vectors $\mathbf{u}$, $\mathbf{b}$ are considered integers.

Example 3.1

```
for (i=0; i<=2; i++)
    for (j=0; j<=3; j++)
        for (k=0; k<=4; k++)
            if ( 6*i+4*j+3*k <=12 )  ···  A[i+2*j+3][j+2*k]  ···
```
$A[i + 2 * j + 3][j + 2 * k]$ has the iterator space

$$P = \left\{ \begin{bmatrix} i \\ j \\ k \end{bmatrix} \in \mathbf{Z}^3 \,\middle|\, \begin{bmatrix} 1 & 0 & 0 \\ 0 & 1 & 0 \\ 0 & 0 & 1 \\ -6 & -4 & -3 \end{bmatrix} \begin{bmatrix} i \\ j \\ k \end{bmatrix} \geq \begin{bmatrix} 0 \\ 0 \\ 0 \\ -12 \end{bmatrix} \right\}.$$

(The inequalities $i \leq 2$, $j \leq 3$, and $k \leq 4$ are redundant and, therefore, eliminated.)

The indices x, y of the A-elements of the array reference can be expressed as the LBL:

$$LBL = \left\{ \begin{bmatrix} x \\ y \end{bmatrix} = \mathbf{T} \cdot \mathbf{i} + \mathbf{u} = \begin{bmatrix} 1 & 2 & 0 \\ 0 & 1 & 2 \end{bmatrix} \begin{bmatrix} i \\ j \\ k \end{bmatrix} + \begin{bmatrix} 3 \\ 0 \end{bmatrix} \,\middle|\, \begin{bmatrix} i \\ j \\ k \end{bmatrix} \in P \right\}.$$

The points of the index space lie inside the $\mathbf{Z}$-polytope $\{x \geq 3, y \geq 0, 3x - 4y \leq 15, 5x + 6y \leq 63, x, y \in \mathbf{Z}\}$, whose boundary is the image of the boundary of the iterator space P (see Figure 3.4).

However, it can be shown—as it will be explained in Section 3.3.2—that only those points (x,y) satisfying also the inequalities $-6x + 8y \geq 19k - 30$, $x - 2y \geq -4k + 3$, and $y \geq 2k \geq 0$, for some positive integer k, belong to the index space; these are the black points in the right quadrilateral from Figure 3.4.

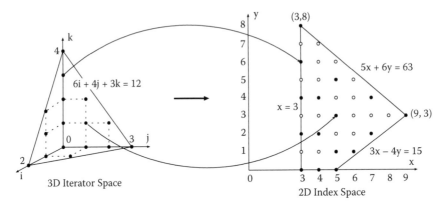

Figure 3.4 The mapping of the iterator space into the index space of the array reference $A[i + 2 * j + 3][j + 2 * k]$. (Reprinted from [56] with permission from IOS Press.)

In this illustrative example, each point in the iterator space is mapped to a distinct point of the index space, but this is not always the case.

3.3.2   The index space of an array reference

Let $\{\, \mathbf{x} = \mathbf{T} \cdot \mathbf{i} + \mathbf{u} \,|\, \mathbf{A} \cdot \mathbf{i} \geq \mathbf{b} \,\}$ be the LBL of a given array reference.[6] This section will show how to model the index space of an array reference, that is, what are the relations satisfied by the coordinates $\mathbf{x}$ of the points in this set. After the theoretical part, illustrative examples will be provided.

For any matrix $\mathbf{T} \in \mathbf{Z}^{m \times n}$ having *rank* $\mathbf{T} = r$, and assuming the first r rows of $\mathbf{T}$ are linearly independent,[7] there exists a unimodular matrix $\mathbf{S} \in \mathbf{Z}^{n \times n}$ such that $\mathbf{T} \cdot \mathbf{S} = \begin{bmatrix} \mathbf{H}_{11} & \mathbf{0} \\ \mathbf{H}_{21} & \mathbf{0} \end{bmatrix}$, where $\mathbf{H}_{11} \in \mathbf{Z}^{r \times r}$ is a lower triangular matrix with positive diagonal elements, and $\mathbf{H}_{21} \in \mathbf{Z}^{(m-r) \times r}$ [40]. The block matrix is called the reduced Hermite form of matrix $\mathbf{T}$.

Let $\mathbf{S}^{-1} \mathbf{i} \overset{def}{=} \mathbf{j} \equiv \begin{bmatrix} \mathbf{j}_1 \\ \mathbf{j}_2 \end{bmatrix}$, where $\mathbf{j}_1$, $\mathbf{j}_2$ are r-, respectively $(n-r)$-, dimensional vectors. Then

$$\mathbf{x} = \mathbf{T}\mathbf{i} + \mathbf{u} = \mathbf{T}\mathbf{S}\mathbf{j} + \mathbf{u} = \begin{bmatrix} \mathbf{H}_{11} \\ \mathbf{H}_{21} \end{bmatrix} \mathbf{j}_1 + \mathbf{u} \qquad (3.2)$$

---

[6]In general, the index space may be a collection of LBLs. For instance, a conditional instruction $if(i \neq j)$ determines two lattices for each array reference within the scope of the condition—one lattice corresponding to $i \geq j + 1$, another one corresponding to $i \leq j - 1$. Without decrease in generality, we assume in this section that the index space of an array reference is represented by a single LBL.

[7]This assumption does not decrease the generality: it is done only to simplify the formulas, affected otherwise by a row permutation matrix.

Denoting $\mathbf{x} = \begin{bmatrix} \mathbf{x}_1 \\ \mathbf{x}_2 \end{bmatrix}$, and $\mathbf{u} = \begin{bmatrix} \mathbf{u}_1 \\ \mathbf{u}_2 \end{bmatrix}$ (where $\mathbf{x}_1, \mathbf{u}_1$ are r-dimensional vectors), it follows that $\mathbf{x}_1 = \mathbf{H}_{11}\mathbf{j}_1 + \mathbf{u}_1$. As $\mathbf{H}_{11}$ is nonsingular (being lower triangular of rank r), $\mathbf{j}_1$ can be obtained explicitly:

$$\mathbf{j}_1 = \mathbf{H}_{11}^{-1}(\mathbf{x}_1 - \mathbf{u}_1) \tag{3.3}$$

The iterator vector $\mathbf{i}$ results with a simple substitution:

$$\mathbf{i} = \mathbf{S}\begin{bmatrix} \mathbf{j}_1 \\ \mathbf{j}_2 \end{bmatrix} = \begin{bmatrix} \mathbf{S}_1 & \mathbf{S}_2 \end{bmatrix} \begin{bmatrix} \mathbf{H}_{11}^{-1}(\mathbf{x}_1 - \mathbf{u}_1) \\ \mathbf{j}_2 \end{bmatrix}$$
$$= \mathbf{S}_1\mathbf{H}_{11}^{-1}(\mathbf{x}_1 - \mathbf{u}_1) + \mathbf{S}_2\mathbf{j}_2$$

where $\mathbf{S}_1$ and $\mathbf{S}_2$ are the submatrices of $\mathbf{S}$ containing the first r, respectively the last $n - r$, columns of $\mathbf{S}$. As the iterator vector must represent a point inside the iterator space $\mathbf{A} \cdot \mathbf{i} \geq \mathbf{b}$, it follows that:

$$\mathbf{A}\mathbf{S}_1\mathbf{H}_{11}^{-1}\mathbf{x}_1 + \mathbf{A}\mathbf{S}_2\mathbf{j}_2 \geq \mathbf{b} + \mathbf{A}\mathbf{S}_1\mathbf{H}_{11}^{-1}\mathbf{u}_1 \tag{3.4}$$

If $r < n$, the $n - r$ variables of $\mathbf{j}_2$ can be eliminated with the Fourier–Motzkin technique [41].

As the rows of matrix $\mathbf{H}_{11}$ are r linearly independent r-dimensional vectors, each row of $\mathbf{H}_{21}$ is a linear combination of the rows of $\mathbf{H}_{11}$. Then from Equation 3.2, it results that there exists a matrix $\mathbf{C} \in \Re^{(m-r)\times r}$ such that[8]

$$\mathbf{x}_2 - \mathbf{u}_2 = \mathbf{C} \cdot (\mathbf{x}_1 - \mathbf{u}_1) \tag{3.5}$$

Taking into account that the elements of $\mathbf{j}_1$ must be integers, it follows (by multiplying and dividing the right member of Equation 3.3 with $det\ \mathbf{H}_{11}$) that the points $\mathbf{x}$ inside the index space must supplementarily satisfy the divisibility constraints

$$det\ \mathbf{H}_{11} \mid \mathbf{h}_i^T(\mathbf{x}_1 - \mathbf{u}_1) \qquad \forall i = 1, \ldots, r \tag{3.6}$$

where $\mathbf{h}_i^T$ are the rows of the matrix with integer coefficients $det\ \mathbf{H}_{11} \cdot \mathbf{H}_{11}^{-1}$, and $a|b$ means "a divides b." According to (Equation 3.6), when $r = n$, the points $\mathbf{x}$ are uniformly spaced along the r linear independent coordinates, the size of gaps in these dimensions being equal to the diagonal elements of $\mathbf{H}_{11}$: if h_{ii} are the diagonal elements of matrix $\mathbf{H}_{11}$, it can be verified that the divisibility constraints (Equation 3.6) are not affected when $\mathbf{x}_1$ is subject to translations of vectors $\mathbf{v}_i = [0 \cdots h_{ii} \cdots 0]$, $\forall i = 1, \ldots, r$. Indeed, $\mathbf{h}_i^T(\mathbf{x}_1 - \mathbf{u}_1 + \mathbf{v}_i) = \mathbf{h}_i^T(\mathbf{x}_1 - \mathbf{u}_1) + \mathbf{h}_i^T\mathbf{v}_i = \mathbf{h}_i^T(\mathbf{x}_1 - \mathbf{u}_1) + det\ \mathbf{H}_{11}$.

The system of inequalities (Equation 3.4), the equations (Equation 3.5), and the divisibility conditions (Equation 3.6) characterize the index space of the given array reference. Several examples will illustrate the generality of this model.

---

[8]The coefficients of matrix $\mathbf{C}$ are determined by backward substitutions from the equations: $\mathbf{H}_{21}.\mathrm{row}(i) = \sum_{j=1}^{r} c_{ij} \cdot \mathbf{H}_{11}.\mathrm{row}(j)$ for any $i = 1, \ldots, m - r$.

Example 3.2

```
for (i=0; i<=2; i++)
    for (j=0; j<=3; j++)      ···   A[3*i][5*i+2*j] ···
```

Since $\mathbf{T} = \mathbf{H}_{11} = \begin{bmatrix} 3 & 0 \\ 5 & 2 \end{bmatrix}$, $\mathbf{H}_{11}^{-1} = \frac{1}{6}\begin{bmatrix} 2 & 0 \\ -5 & 3 \end{bmatrix}$, $\mathbf{u} = \mathbf{u}_1 = \begin{bmatrix} 0 \\ 0 \end{bmatrix}$, $\mathbf{S} = \mathbf{S}_1 = \begin{bmatrix} 1 & 0 \\ 0 & 1 \end{bmatrix}$ ($\mathbf{S}_2$, $\mathbf{j}_2$ do not exist since $n - r = 2 - 2 = 0$; $\mathbf{H}_{21}$, $\mathbf{x}_2$, $\mathbf{u}_2$ do not exist since $m - r = 2 - 2 = 0$), the inequalities (Equation 3.4) with $\mathbf{x}_1 = \begin{bmatrix} x \\ y \end{bmatrix}$ are: $6 \geq x \geq 0$, $18 \geq -5x + 3y \geq 0$, representing the first quadrilateral in Figure 3.5. Not all the integral points inside the quadrilateral have as coordinates the index values of the array reference: only those points satisfying the divisibility conditions (Equation 3.6): $6 \mid 2x$ (or $3 \mid x$) and $6 \mid -5x + 3y$ belong to the index space. These points of integer coordinates, colored *black* in the figure, are uniformly spaced along the two axes Ox and Oy, the size of the gaps in these dimensions being 3 and 2, the diagonal elements of $\mathbf{H}_{11}$. The other integral points in the quadrilateral are "holes" in the index space and they are represented by small circles.

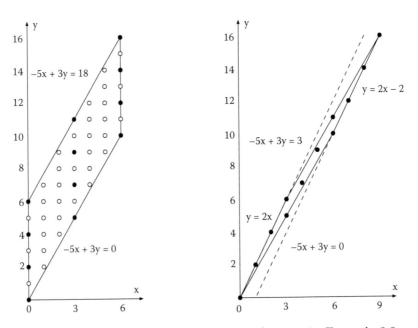

Figure 3.5 The index space of the array references in Example 3.2 and, respectively, in Example 3.3 [57]. (© 2008 IEICE).

Example 3.3

```
for (i=0; i<=2; i++)
    for (j=0; j<=3; j++)      ...   A[3*i+j][5*i+2*j] ...
```

Since $\mathbf{T} = \begin{bmatrix} 3 & 1 \\ 5 & 2 \end{bmatrix}$, $\mathbf{H}_{11} = \mathbf{H}_{11}^{-1} = \begin{bmatrix} 1 & 0 \\ 2 & -1 \end{bmatrix}$, $\mathbf{S} = \mathbf{S}_1 = \begin{bmatrix} 0 & 1 \\ 1 & -3 \end{bmatrix}$, the in-equalities (Equation 3.4) are: $2 \geq 2x - y \geq 0, 3 \geq -5x + 3y \geq 0$. Since $det\ \mathbf{H}_{11} = 1$, there are no divisibility conditions (Equation 3.6). Hence, all the integral points inside the second quadrilateral in Figure 3.5 belong to the index space of the array reference.

Example 3.4

```
for (i=0; i<=2; i++)
    for (j=0; j<=3; j++)      ...   A[3i+j+2][6i+2j+1]   ...
```

Since $\mathbf{T} = \begin{bmatrix} 3 & 1 \\ 6 & 2 \end{bmatrix}$, $\mathbf{u} = \begin{bmatrix} 2 \\ 1 \end{bmatrix}$, $\mathbf{S} = \begin{bmatrix} \mathbf{S}_1 & \mathbf{S}_2 \end{bmatrix} = \begin{bmatrix} 0 & 1 \\ 1 & -3 \end{bmatrix}$, it follows that $\mathbf{TS} = \begin{bmatrix} 1 & 0 \\ 2 & 0 \end{bmatrix}$, $\mathbf{H}_{11} = [1]$, and $\mathbf{H}_{21} = [2]$. Since $m - r = 2 - 1 = 1$, the points (x, y) of the index space satisfy one equation of type (Equation 3.5): $y - 1 = 2(x - 2)$ (see Figure 3.6). The system of inequalities (Equation 3.4) is: $2 \geq t \geq 0, 5 \geq x - 3t \geq 2$, where the vector $\mathbf{j}_2$ has one element (since $n - r = 1$) denoted t. The system of inequalities can be reduced here to $11 \geq x \geq 2$. There are no divisibility conditions (Equation 3.6) since $det\ \mathbf{H}_{11} = 1$. Note that the iterator space contains 12 points, whereas the index space contains only 10 points: two pairs of iterator vectors—$(i = 0, j = 3)$, $(i = 1, j = 0)$ and $(i = 1, j = 3)$, $(i = 2, j = 0)$—are mapped to same points in the index space. Indeed, taking $x = 5$ or $x = 8$ in the inequalities (Equation 3.4) shown previously, there are two possible values for t; whereas taking any other value for x between 2 and 11, only one value for t is possible.

Example 3.5

```
for (i=0; i<=2; i++)
    for (j=0; j<=1; j++)      ...   A[3i+j+2][6i+2j+1]   ...
```

This example differs from Example 3.4 only in the upper bound of the inner loop. Hence, the points of the index space are on the same line $y - 1 = 2(x - 2)$. The system of inequalities (3.4) is: $2 \geq t \geq 0, 3 \geq x - 3t \geq 2$, where the vector $\mathbf{j}_2$ has one element (since $n - r = 1$) denoted t. The system of inequalities can be reduced here to $9 \geq x \geq 2$. Taking $x = 4$ or $x = 7$ in the inequalities (Equation 3.4) shown above, there are no possible values for t (see the 'holes' corresponding to these values in Figure 3.6); whereas taking any other value for x between 2 and 9, only one value for t is possible.

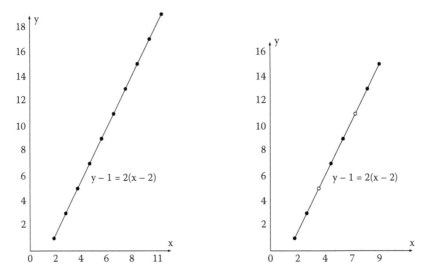

Figure 3.6 The index space of the array references in Example 3.4 and, respectively, in Example 3.5.

3.4　Operations with Linearly Bounded Lattices

This section will present two operations relevant in the context of data storage computation: the *intersection* and the *difference* of two LBLs. While the intersection of two LBLs was addressed in different contexts by other works as well (e.g., [10]), the difference operation was studied more recently [42].

Afterward, this section will present the decomposition of array references into *disjoint* linearly bounded lattices. The motivation of this operation relies on the following intuitive idea: the memory size computation problem would be significantly simplified if the array references in the code were disjoint. But since, in general, they are not, we can transform each array reference in the code into a union of LBLs, disjoint from each other.

Finally, the computation of the number of points (index vectors) contained in an LBL will be presented.

3.4.1　The intersection of two linearly bounded lattices

Let $Lbl_1 = \{\mathbf{x} = \mathbf{T}_1 \cdot \mathbf{i}_1 + \mathbf{u}_1 \mid \mathbf{A}_1 \cdot \mathbf{i}_1 \geq \mathbf{b}_1\}$, $Lbl_2 = \{\mathbf{x} = \mathbf{T}_2 \cdot \mathbf{i}_2 + \mathbf{u}_2 \mid \mathbf{A}_2 \cdot \mathbf{i}_2 \geq \mathbf{b}_2\}$ be two LBLs derived from the same indexed signal, where $\mathbf{T}_1$ and $\mathbf{T}_2$ have obviously the same number of rows—the signal dimension. Intersecting the two LBLs means, first of all, solving a linear Diophantine system[9]

---

[9]Finding the integer solutions of the system. Solving a linear Diophantine system was proven to be of polynomial complexity, all the known methods being based on bringing the system matrix to the Hermite normal form [40].

$\mathbf{T_1} \cdot \mathbf{i_1} - \mathbf{T_2} \cdot \mathbf{i_2} = \mathbf{u_2} - \mathbf{u_1}$ having the elements of $\mathbf{i_1}$ and $\mathbf{i_2}$ as unknowns. If the system has no solution, the intersection is empty. Otherwise, let

$$\begin{bmatrix} \mathbf{i_1} \\ \mathbf{i_2} \end{bmatrix} = \begin{bmatrix} \mathbf{V_1} \\ \mathbf{V_2} \end{bmatrix} \mathbf{t} + \begin{bmatrix} \mathbf{v_1} \\ \mathbf{v_2} \end{bmatrix} \tag{3.7}$$

be the solution of the Diophantine system. If the set of coalesced constraints of the two LBLs

$$\mathbf{A_1 V_1} \cdot \mathbf{t} \geq \mathbf{b_1} - \mathbf{A_1 v_1}$$
$$\mathbf{A_2 V_2} \cdot \mathbf{t} \geq \mathbf{b_2} - \mathbf{A_2 v_2} \tag{3.8}$$

has integer solutions, then the intersection is a new LBL:

$$Lbl_1 \cap Lbl_2 = \{\mathbf{x} = \mathbf{T_1 V_1} \cdot \mathbf{t} + (\mathbf{T_1 v_1} + \mathbf{u_1}) \mid \text{ s.t. (3.8)}\} \tag{3.9}$$

3.4.2 The difference of two linearly bounded lattices

The difference $Lbl_1 - Lbl_2$ (and similarly, $Lbl_2 - Lbl_1$) can be decomposed by generating a set of LBLs, not necessarily disjoint, covering the difference. These LBLs will be further intersected recursively as well, till all the components are disjoint. The LBLs covering the difference are generated as follows.

Suppose the minuend and subtrahend LBLs have the general form (Equation 3.1), and the iterator vectors of the subtrahend have the general form $\mathbf{i} = \mathbf{V} \cdot \mathbf{t} + \mathbf{v}$ as derived from the solution of a Diophantine linear system (Equation 3.7), as explained above. Here, $\mathbf{V}$ is a matrix with integer elements having n rows (n is the dimension of the iterator space) and $p \geq 0$ columns, $\mathbf{t} \in \mathbf{Z}^p$, and $\mathbf{v}$ is a vector with n integer elements. (When $p = 0$, the Diophantine system has the unique solution $\mathbf{i} = \mathbf{v}$.)

Case 1: Assume for the time being that $p > 0$, and matrix $\mathbf{V}$ is transformed into a Hermite reduced form[10] [43]. Then it is easy to detect those rows of $\mathbf{V}$ that are linear combinations of the rows above them. For instance, let row k be such a row. Then we have

$$
\begin{array}{llll}
i_1 = & V_{11}t_1 & & +v_1 \\
i_2 = & V_{21}t_1 & +V_{22}t_2 & +v_2 \\
& & \cdots & \\
i_{k-1} = & V_{k-1,1}t_1 & +\cdots \quad +V_{k-1,k-1}t_{k-1} & +v_{k-1} \\
i_k = & V_{k,1}t_1 & +\cdots \quad +V_{k,k-1}t_{k-1} & +v_k
\end{array}
$$

One can find k integers α_i not all zero such that $\alpha_1(i_1 - v_1) + \cdots + \alpha_k(i_k - v_k) = 0$ (the k-th row of $\mathbf{V}$ is a linear combination of the rows $1, \ldots, k-1$). Since an

---

[10]A Hermite reduced form of $\mathbf{V}$ is a matrix having all the nonzero elements on or below the main diagonal. It can be obtained by post-multiplying $\mathbf{V}$ with a sequence of unimodular matrices and pre-multiplying it with a row permutation matrix.

LBL in the difference cover must be disjoint from the subtrahend LBL where the equality above is satisfied, the LBLs

$$\{\; \mathbf{x} = \mathbf{T} \cdot \mathbf{i} + \mathbf{u} \in \mathbf{Z}^m \,|\, \mathbf{A} \cdot \mathbf{i} \geq \mathbf{b},$$
$$\alpha_1 i_1 + \cdots + \alpha_k i_k \geq \alpha_1 v_1 + \cdots + \alpha_k v_k + 1\}\; \text{and}$$

$$\{\; \mathbf{x} = \mathbf{T} \cdot \mathbf{i} + \mathbf{u} \in \mathbf{Z}^m \,|\, \mathbf{A} \cdot \mathbf{i} \geq \mathbf{b},$$
$$\alpha_1 i_1 + \cdots + \alpha_k i_k \leq \alpha_1 v_1 + \cdots + \alpha_k v_k - 1\}$$

are included in the minuend LBL, but disjoint from the subtrahend LBL. Hence, they are included in the difference.

Note that this case covers also the situations when $p = 0$ (i.e., $i_k = v_k$ for all k), or when matrix $\mathbf{V}$ has null rows (i.e., $i_k = v_k$ for some k): the added inequalities are $i_k \geq v_k + 1$ and $i_k \leq v_k - 1$.

Case 2: Other LBLs covering the difference are derived negating one by one the constraints (Equation 3.8) that define the iterator polytope of the intersection. For instance, for the difference $Lbl_1 - (Lbl_1 \cap Lbl_2)$ the constraints $\mathbf{A_2 V_2} \cdot \mathbf{t} \geq \mathbf{b_2} - \mathbf{A_2 v_2}$ derived from Lbl_2 are negated, and subsequently translated into inequalities between the iterators in Lbl_1. In this way, the resulting LBL (which is, basically, Lbl_1 with an additional constraint between the iterators) will be included in Lbl_1, but will be disjoint from the intersection $Lbl_1 \cap Lbl_2$. This operation is described below.

Let $\beta_1 t_1 + \cdots + \beta_p t_p \geq \beta_0$ be the inequality considered for negation, where β_k are integer coefficients. Since matrix $\mathbf{V}$ is now a reduced Hermite form, we have

$$
\begin{aligned}
i_1 &= V_{11}t_1 & & & &+ v_1 \\
i_2 &= V_{21}t_1 & +V_{22}t_2 & & &+ v_2 \\
& & \cdots & & & \\
i_p &= V_{p,1}t_1 & + \cdots & +V_{p,p}t_p & & + v_p
\end{aligned}
$$

where $i_1, \ldots, i_p$ are p of the n iterators (after a possible reordering). Then, $\sum_k \beta_k t_k$ is a linear combination of $(i_k - v_k)$, that is, $\sum_k \beta_k t_k = \sum_k \alpha_k (i_k - v_k)$. The inequality expressed in terms of the iterators becomes $\sum_k \alpha_k i_k \geq \beta_0 + \sum_k \alpha_k v_k$. The LBL considered for the difference cover will contain this inequality negated, that is:

$$\{\; \mathbf{x} = \mathbf{T} \cdot \mathbf{i} + \mathbf{u} \in \mathbf{Z}^m \,|\, \mathbf{A} \cdot \mathbf{i} \geq \mathbf{b},$$
$$\alpha_1 i_1 + \cdots + \alpha_p i_p \leq \beta_0 + \alpha_1 v_1 + \cdots + \alpha_p v_p - 1\}.$$

Case 1 and *Case 2* generate LBLs that are *outside* the intersection boundary, but still *inside* the minuend LBL. The next case will build LBLs *inside* the boundary of the intersection when this intersection is not *dense* (i.e., it does not contain all the lattice points inside or on its boundary).

Case 3: Let $\mathbf{x} = \mathbf{T_1 V_1} \cdot \mathbf{t} + \cdots$ be the vector function of the intersection (3.9). Assume the matrix $\mathbf{T_1 V_1}$ has linearly independent rows (otherwise, the linear dependent ones will be eliminated), and bring this matrix to the Hermite normal form [40] $\mathbf{H}$. Assuming that *rank* $\mathbf{H}$ is equal to its number of columns, if all the diagonal coefficients $h_{ii} = 1$, then the intersection LBL is dense [44], and we are done with this case. But if $h_{ii} > 1$, to cover the holes inside the boundary of the intersection, we build LBLs having the same mapping as the intersection except a shift of $1, 2, \dots, h_{ii} - 1$ of the i-th index.

Example 3.6

```
for (i=0; i<=5; i++)
    for (j=0; j<=5; j++)
        if ( i ≥ j )   ··· A[2*i+3][j] ···
for (k=0; k<=13; k++)
    for (l=1; l<=7; l++)   ··· A[k][l] ···
```

The lattices of the two array references are:

$$Lbl_1 = \left\{ \begin{bmatrix} x \\ y \end{bmatrix} = \begin{bmatrix} 2 & 0 \\ 0 & 1 \end{bmatrix} \begin{bmatrix} i \\ j \end{bmatrix} + \begin{bmatrix} 3 \\ 0 \end{bmatrix} \ \middle| \ 5 \geq i \geq j \geq 0 \right\}$$

$$Lbl_2 = \left\{ \begin{bmatrix} x \\ y \end{bmatrix} = \begin{bmatrix} 1 & 0 \\ 0 & 1 \end{bmatrix} \begin{bmatrix} k \\ l \end{bmatrix} + \begin{bmatrix} 0 \\ 0 \end{bmatrix} \ \middle| \ 13 \geq k \geq 0, 7 \geq l \geq 1 \right\}$$

where the iterator spaces were written in non-matrix format for economy of space. The elements of the array A covered by the two LBLs are shown with black dots in Figure 3.7a. The two array references intersect and their intersection is:

$$Lbl_1 \cap Lbl_2 = \left\{ \begin{bmatrix} x \\ y \end{bmatrix} = \begin{bmatrix} 2 & 0 \\ 0 & 1 \end{bmatrix} \begin{bmatrix} t_1 \\ t_2 \end{bmatrix} + \begin{bmatrix} 3 \\ 0 \end{bmatrix} \ \middle| \ 5 \geq t_1 \geq t_2 \geq 1 \right\}.$$

The difference $Lbl_1 - Lbl_2$ results from *Case 2* to be composed of one lattice—denoted L_1:

$$L_1 = \left\{ \begin{bmatrix} x \\ y \end{bmatrix} = \begin{bmatrix} 2 \\ 0 \end{bmatrix} \begin{bmatrix} t_1 \end{bmatrix} + \begin{bmatrix} 3 \\ 0 \end{bmatrix} \ \middle| \ 5 \geq t_1 \geq 0 \right\}.$$

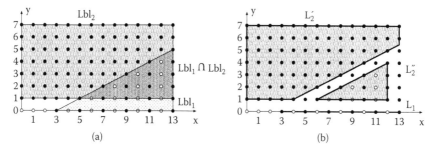

(a)   (b)

Figure 3.7 (a) Example: Lbl_1, Lbl_2, and their (darker) intersection. (b) The LBL L_1 is equal to the difference $Lbl_1 - Lbl_2$; the LBLs L_2' and L_2'', the latter covering the holes of the intersection, make together the difference $Lbl_2 - Lbl_1$ [43]. (© 2007 IEEE).

The difference $Lbl_2 - Lbl_1$ is composed of two disjoint lattices—L'_2 resulted from *Case 2* and L''_2 resulted from *Case 3*:

$$L'_2 = \left\{ \begin{bmatrix} x \\ y \end{bmatrix} = \begin{bmatrix} 1 & 0 \\ 0 & 1 \end{bmatrix} \begin{bmatrix} t_1 \\ t_2 \end{bmatrix} \middle| \ 13 \ge t_1 \ge 0, 7 \ge t_2 \ge 1, 2t_2 + 2 \ge t_1 \right\}$$

$$L''_2 = \left\{ \begin{bmatrix} x \\ y \end{bmatrix} = \begin{bmatrix} 2 & 0 \\ 0 & 1 \end{bmatrix} \begin{bmatrix} t_1 \\ t_2 \end{bmatrix} + \begin{bmatrix} 4 \\ 0 \end{bmatrix} \middle| \ 9 \ge 2t_1 + 1 \ge 2t_2 \ge 2 \right\}.$$

The decomposition of the two differences is shown in Figure 3.7b.

Example 3.7

```
for (k=0; k<=6; k++)
    for (l=0; l<=18; l++)    ··· A[k][l] ···
for (i=0; i<=2; i++)
    for (j=0; j<=3; j++)    ··· A[3*i][5*i+2*j] ···
```

The lattices (in non-matrix format) of the two array references are:

$$Lbl_1 = \{x = k, \ y = l \mid 6 \ge k \ge 0, 18 \ge l \ge 0\}.$$
$$Lbl_2 = \{x = 3i, \ y = 5i + 2j \mid 2 \ge i \ge 0, 3 \ge j \ge 0\}$$
$$= \{6 \ge x \ge 0, 18 \ge -5x + 3y \ge 0, \ 3 \mid x, 6 \mid -5x + 3y \ \}.$$

The index space of Lbl_2 (the second array reference) was computed at Example 3.2 and it is shown in Figure 3.8a. Taking one of the four inequalities of Lbl_2 and adding its negated inequality to the minuend Lbl_1 will create an LBL that (if not empty) is disjoint from Lbl_2 and is included in Lbl_1. For instance, negating the inequality $18 \ge -5x + 3y$ from Lbl_2, we obtain $19 \le -5x + 3y$, or $19 \le -5k + 3l$ with the iterators of Lbl_1. Adding it to the iterator space of Lbl_1, we obtain

$$L_1 = \{x = k, \ y = l \mid 6 \ge k \ge 0, 18 \ge l, 3l \ge 5k + 19\} \text{ shown in Figure 3.8b.}$$

Negating the inequality $-5x + 3y \ge 0$, we obtain

$$L_2 = \{x = k, \ y = l \mid 6 \ge k \ge 0, l \ge 0, 5k - 1 \ge 3l \ \}.$$

Negating the other inequalities ($6 \ge x \ge 0$) yields empty LBLs.

Similarly, taking one of the equalities (Equation 3.5) of the form $\mathbf{a}^T \mathbf{x} = b$, at most two LBLs included in $Lbl_1 - Lbl_2$ can be created, adding the inequalities $\mathbf{a}^T \mathbf{x} \ge b+1$ and, respectively, $\mathbf{a}^T \mathbf{x} \le b-1$ to the iterator space of Lbl_1. (There is no equality (Equation 3.5) in this case.)

The other LBLs in the difference must violate at least one of the divisibility conditions $3 \mid x$ and $6 \mid (-5x + 3y)$. To obtain them, we simply replace the vector $\mathbf{u}_2$ in Lbl_2, keeping the same periodicity of the index space from Lbl_2

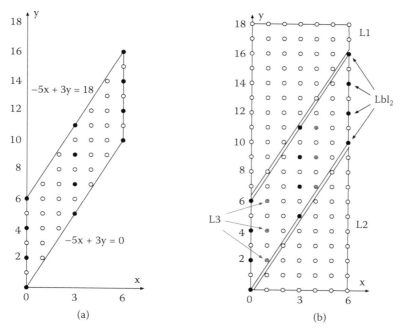

Figure 3.8 (a) The index space of the array reference in Example 3.2. (b) The difference $Lbl_1 - Lbl_2$, where the two lattices correspond to the array references in Example 3.7. Lbl_1 contains all the lattice points in the rectangle and Lbl_2 contains the 12 black points. The difference has (at least) seven LBL components. Two components are L_1 (all the lattice points in the upper quadrilateral) and L_2 (all the lattice points in the lower triangle). Another LBL component L_3 covers the six gray points in the middle area. (Reprinted from [56] with permission from IOS Press.)

(which here is 3 and 2 along the two axes), but doing a translation along the axes. Therefore, taking $\mathbf{u_2} = \begin{bmatrix} k_1 \\ k_2 \end{bmatrix} \neq \begin{bmatrix} 0 \\ 0 \end{bmatrix}$, where $k_1 = 0, 1, 2$ and $k_2 = 0, 1$, five new LBLs are obtained. For instance, choosing $(k_1, k_2) = (1, 0)$, replacing $x = 3i + k_1$ and $y = 5i + 2j + k_2$ in the inequalities $6 \geq x \geq 0, 18 \geq -5x + 3y \geq 0$ of Lbl_2, we get $1 \geq i \geq 0, 3 \geq j \geq 1$. Therefore,

$$L_3 = \{x = 3i + 1, \ y = 5i + 2j \mid 1 \geq i \geq 0, 3 \geq j \geq 1\}$$

is included in $Lbl_1 - Lbl_2$, and it is shown in Figure 3.8b with 6 *gray* points. The other four LBLs are:

$$
\begin{aligned}
L_4 &= \{x = 3i + 2, \ y = 5i + 2j \mid 1 \geq i \geq 0, 4 \geq j \geq 2\} \\
L_5 &= \{x = 3i, \ y = 5i + 2j + 1 \mid 2 \geq i \geq 0, 2 \geq j \geq 0\} \\
L_6 &= \{x = 3i + 1, \ y = 5i + 2j + 1 \mid 1 \geq i \geq 0, 3 \geq j \geq 1\} \\
L_7 &= \{x = 3i + 2, \ y = 5i + 2j + 1 \mid 1 \geq i \geq 0, 4 \geq j \geq 2\}
\end{aligned}
$$

Note that the decomposition is minimal[11] (although not unique): it is not possible to obtain a decomposition of $Lbl_1 - Lbl_2$ with less than 7 LBLs for this example—which is a "difficult" one! In most of the practical cases encountered, the difference can be represented as only one LBL and, if this is the case, the algorithm (informally explained above) will find it. Otherwise, in the general case, the minimality cannot be guaranteed, unless all the combinations of inequalities (Equation 3.4) can be negated, instead of selecting only one at a time. This would increase the computation time though, without practical benefits.

3.4.3   Decomposition of array references into disjoint linearly bounded lattices

This operation partitions the index space of a signal into a collection of disjoint lattices, such that each array reference in the code is a union of lattices in the collection. This decomposition can be performed analytically, by recursively intersecting the array references of every indexed (multidimensional) signal in the code.

Let A be an indexed signal in the algorithmic specification. An inclusion graph is gradually built; this is a directed acyclic graph whose nodes are lattices of A and whose arcs denote inclusion relations between the respective sets. (e.g., an arc from node X to node Y shows that the lattice X is included in Y.) A high-level pseudo-code of the decomposition of A's array references into disjoint lattices is given below:

Algorithm decomposing the array references of a multidimensional signal into a collection of disjoint LBLs.

> let $\mathcal{L}_A$ be the initial collection of lattices of signal A ;
> // these are the lattices of the array references of A
> initialize the inclusion graph creating one node for each lattice ;
> // the graph is initially only a set of nodes
> **repeat** {
> > **for** (*each pair of lattices* (L_1, L_2) *in the collection* $\mathcal{L}_A$)
> > > **if** (($(L_1 \not\subset L_2)$ && $(L_2 \not\subset L_1)$)) {
> > > > // if there is a path in the graph from node L_1 to L_2 then $L_1 \subset L_2$
> > > > compute $L = L_1 \cap L_2$;
> > > > **if** ($L \neq \emptyset$)

(1)
> > > > > **if** ($L == L_1$) {
> > > > > > set $L_1 \subset L_2$;
> > > > > > // add an arc in the graph from node L_1 to L_2

---

[11]For computational reasons, in order to work with only integer arrays, our definition of LBL (Equation 3.1) is slightly different from [45].

compute $L_2 - L_1$;
each lattice in the difference will be added to $\mathcal{L}_A$
(unless it belongs already to it),
    a new node and an inclusion arc from the node
    to L_2 will be added to the graph;
}

(1)   **else**   **if** ($L == L_2$) {
set $L_2 \subset L_1$;
// add an arc in the graph from node L_2 to L_1
compute $L_1 - L_2$;
each lattice in the difference will be added to $\mathcal{L}_A$
(unless it belongs already to it),
    a new node and an inclusion arc from the node
    to L_1 will be added to the graph;
}
else {

(2)   **if** (*there exists already* $L' \in \mathcal{L}_A$ *equivalent*[12] *to* L)
{   **if** ($L' \not\subset L_1$) set $L' \subset L_1$;   **if** ($L' \not\subset L_2$)
set $L' \subset L_2$;   }

(3)   **else** { add the new lattice L to the collection $\mathcal{L}_A$
and add a node L in the graph ;
    set $L \subset L_1$;   set $L \subset L_2$; }
    // add arcs from L to L_1 and L_2
compute $L_1 - L_2$;
each lattice in the difference will be added to $\mathcal{L}_A$
(unless it belongs already to it),
    a new node and an inclusion arc from the node
    to L_1 will be added to the graph;
compute $L_2 - L_1$;
each lattice in the difference will be added to $\mathcal{L}_A$
(unless it belongs already to it),
    a new node and an inclusion arc from the node
    to L_2 will be added to the graph;
}
}
} **until** (no new lattice can be added to the collection $\mathcal{L}_A$) ;

At the beginning, the inclusion graph of the indexed signal contains only the lattices of the corresponding array references as nodes. Every pair of lattices—between which no inclusion relation is known so far (i.e., between which there is currently no path in the graph)—is intersected. If the intersection produces

---

[12] Two LBLs of the same indexed signal are equivalent if they represent the same set of indices. For instance, $\{x = i + j \mid 0 \le i \le 2, 0 \le j \le 2\}$ and $\{x = i \mid 0 \le i \le 4\}$ are equivalent.

a nonempty lattice, there are basically three possibilities: (1) the resulting lattice is one of the intersection operands: in this case, inclusion arcs between the corresponding nodes must be added to the graph; (2) an *equivalent* LBL exists already in the collection: in this case, arcs must be added (only if such arcs are missing) between the nodes corresponding to the equivalent lattice to the operands; (3) the resulting lattice is a new element of the collection $\mathcal{L}_A$: a new node is appended to the graph, along with arcs toward the two operands of the intersection.

The inclusion graph is used on one hand to speed up the decomposition (e.g., if the intersection $L_1 \cap L_2$ results to be empty, there is no sense of trying to intersect L_1 with the lattices included in L_2 since those intersections will be empty as well) and, on the other hand, to determine the structure of each array reference in terms of disjoint lattices: at the end, all the nodes in the graph without incident arcs represent disjoint lattices.

Figure 3.9a shows the result of this decomposition for the two-dimensional signal A in the illustrative example from Figure 3.2. The graph displays the inclusion relations (*arcs*) between the LBLs of A (*nodes*). The four "bold" nodes are the four array references of signal A in the code. The nodes are also labeled with the size of the corresponding LBL—that is, the number of lattice points (i.e., points having integer coordinates) in those sets. The *inclusion graph* is gradually constructed by partitioning analytically the initial (four) array references using LBL *intersections* and *differences*. While the intersection of two non-disjoint LBLs is an LBL as well [10], the difference is

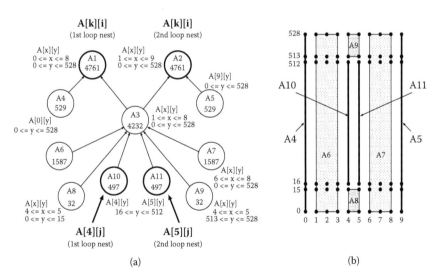

(a)   (b)

Figure 3.9 (a) Decomposition of the index space of signal A from the code example in Figure 3.2 into disjoint LBLs; the arcs in the inclusion graph show the inclusion relations between lattices. (b) The partitioning of A's index space according to the decomposition (a) [43]. (© 2007 IEEE).

not necessarily an LBL—and this latter operation makes the decomposition difficult. In this example, $A1 \cap A2 = A3$ and $A1 - A3$, $A2 - A3$ are also LBLs (denoted $A4$, $A5$ in Figure 3.9a). However, the difference $A3 - A10$ is not an LBL due to the non-convexity of this set. At the end of the decomposition, the nodes without any incident arc represent disjoint LBLs. The partitioning of A's index space is displayed in Figure 3.9b. Every array reference in the code is now either a disjoint LBL itself (like $A10$ and $A11$), or a union of disjoint LBLs (e.g., $A1 = A4 \cup A3 = A4 \cup \bigcup_{i=6}^{11} A_i$).

Example 3.8

```
for (j=0; j<=1; j++)      // 1st loop nest
   for (i=j+1; i<=12; i++)
      if ( 2*i+3*j ≤ 24 )  A[i][j]= ···
for (i=0; i<=2; i++)      // 2nd loop nest
   for (j=i; j<=(24-2*i)/3; j++)  A[i][j]= ···
for (j=0; j<=1; j++)      // 3rd loop nest
   for (i=0; i<=(24-3*j)/2; i++)   ···= A[i][j];
for (i=3; i<=9; i++)      // 4th loop nest
   for (j=2; j<=(24-2*i)/3; j++)  A[i][j]= ···
for (i=1; i<=4; i++)      // 5th loop nest
   for (j=2; j<=(24-2*i)/3; j++)   ···= A[3*i][j] + A[3*i+2][j];
```

Figure 3.10a shows the result of this decomposition for the six array references of signal A from Example 3.8. The resulting lattices have the following expressions (in non-matrix format, in order to save space):

$$L_1 = \{x = i, \ y = j \mid i - j \geq 1, j \geq 0, -j \geq -1, -2 * i - 3 * j \geq -24\}$$
$$L_2 = \{x = i, \ y = j \mid -i + j \geq 0, i \geq 0, -j \geq -1\}$$
$$L_3 = \{x = i, \ y = j \mid i \geq 0, -i \geq -2, j \geq 2, -2 * i - 3 * j \geq -24\}$$
$$L_4 = \{x = 3 * i, \ y = j \mid i \geq 1, -i \geq -3, j \geq 2, -2 * i - j \geq -8\}$$
$$L_5 = \{x = 3 * i + 1, \ y = j \mid i \geq 1, -i \geq -2, j \geq 2, -6 * i - 3 * j \geq -22\}$$
$$L_6 = \{x = 3 * i + 2, \ y = j \mid i \geq 1, -i \geq -2, j \geq 2, -6 * i - 3 * j \geq -20\}$$

While the first array reference in Example 3.8 is the lattice L_1, the second array reference is $L_2 \cup L_3$ and the third is $L_1 \cup L_2$. (Note that the second and third array references have in common the array elements covered by the lattice L_2.) The array reference in the forth loop nest is the union of the lattices L_4, L_5, and L_6. Assuming that the elements covered by L_1 and L_2 are accessed as operands for the last time (i.e., *consumed*) in the third loop nest, and the elements covered by L_4 and L_6 are consumed in the fifth loop nest, we can find the elements alive at the borderline between these blocks of code. For instance, the array elements covered by the lattices L_3, L_4, L_5, and L_6 are alive after the forth loop nest (see Figure 3.10e); the elements covered by L_3 and L_5 are still alive after the fifth loop nest (see Figure 3.10f).

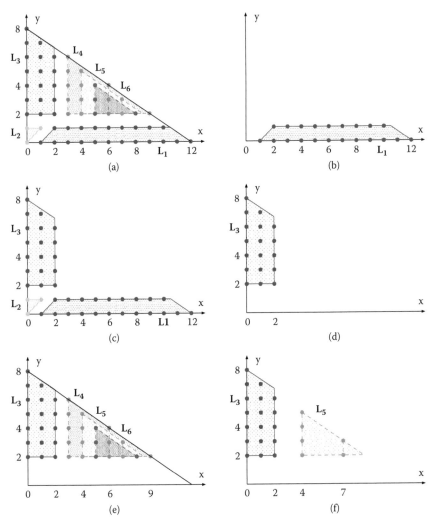

Figure 3.10 (a) The decomposition of the array space of signal A (Example 3.8) in 6 disjoint lattices $L_1, \ldots, L_6$. (b–f) The live lattices of signal A after the first, respectively, second, third, forth, and fifth loop nest (Example 3.8).

3.4.4   Computation of the lattice size

The size of an LBL is the number of m-dimensional points in the lattice. If the LBL represents an array reference, the size of an LBL is a tight (reachable) lower bound of the data memory needed to store the lattice, assuming the array elements are simultaneously alive.

There are two distinct situations to be considered:

Case 1: The affine vector function $\mathbf{i} \longmapsto \mathbf{T} \cdot \mathbf{i} + \mathbf{u}$ of the lattice is a one-to-one mapping.

A sufficient condition is that the rank of matrix $\mathbf{T}$ in Equation 3.1 be equal to its number of columns. Then, the computation of the number of distinct signal indices (i.e., the amount of memory necessary to store the array elements covered by the lattice) is hence reduced to counting the number of iterator vectors in the iterator polytope $\mathbf{A} \cdot \mathbf{i} \geq \mathbf{b}$ in Equation 3.1. In such a situation, a computation technique based on Barvinok's decomposition of a simplicial cone into unimodular cones [46,47] is used. Note that counting the points having integer coordinates in $\mathbf{Z}$-polyhedra can be done much simpler[13]—adapting the Fourier-Motzkin technique [41]. We decided to employ the current approach because of its scalability, which will become apparent by the end of this section.

An example illustrating both the concepts and the technique is given below. Although the theoretical background is not explained in detail (see [47,50] for more theoretical insight), this example is intended to illustrate the main steps of the computation flow. Moreover, it will clearly show the *scalability* of the approach, which makes it appropriate for algorithmic specifications typical to multidimensional signal processing applications.

Example 3.9

```
for (i=2; i<=7; i++)
    for (j=1; j<=-2*i+15; j++)
        if ( j<=i+1 )   · · ·  = A[2*i-j][3*i+2*j];
```

Since the rank of matrix $\mathbf{T}$ is 2, the vector function $\mathbf{i} \longmapsto \mathbf{T} \cdot \mathbf{i} + \mathbf{u} = \begin{bmatrix} 2 & -1 \\ 3 & 2 \end{bmatrix} \begin{bmatrix} i \\ j \end{bmatrix}$ is a one-to-one mapping of the iterator space into the index space. It follows that the computation of the number of array elements covered by $A[2 * i - j][3 * i + 2 * j]$ is equivalent to counting the number of points of integer coordinates in the iterator polytope $\mathbf{A} \cdot \mathbf{i} \geq \mathbf{b}$ shown in Figure 3.11. This latter operation is done as explained below.

Step 1 Find the vertices of the iterator polytope $\mathbf{A} \cdot \mathbf{i} \geq \mathbf{b}$ (see Figure 3.11) and their supporting cones.

Let $r_1, \ldots, r_d \in \mathbf{Z}^d$ be linearly independent integer vectors. The (rational polyhedral) *cone* generated by the *rays* $r_1, \ldots, r_d$ is the set $C(r_1, \ldots, r_d) = \{\sum_1^d \alpha_i r_i, \ \alpha_i \geq 0\}$. For instance, the set of points inside the angle xV_4y (see Figure 3.11) is a 2D cone generated by the rays $r_1 = [-1 \ 0]^T$ and $r_2 = [-1 \ 2]^T$ with the point V_4 taken as origin. There is a *supporting* cone corresponding to each vertex of a polyhedron. For instance, the supporting cone of the vertex V_4 (Figure 3.11), denoted $C(V_4)$, is the one generated by the rays r_1 and r_2. A cone is called *unimodular* if the matrix of the rays $[r_1 \cdots r_d]$ is unimodular. Given the inequalities $\{2 \leq i \leq 7, \ 1 \leq j \leq i + 1, \ j \leq -2i + 15\}$ defining the iterator space, the vertices and the rays are computed using the *reverse search*

---

[13]For parametric $\mathbf{Z}$-polyhedra, there are methods based on Ehrhart polynomials [48] e.g., [49].

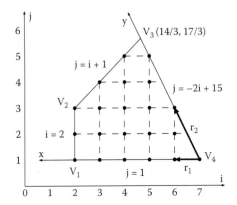

Figure 3.11 2D polytope (quadrilateral) representing the iterator space in Example 3.9 [58]. (© 2009 IEEE).

algorithm [51]. The supporting cones corresponding to the vertices $V_1, \ldots, V_4$ of the iterator polytope, as well as their generating rays shown below as column vectors, are:

$$C(V_1) = \left\{ \begin{pmatrix} 1 \\ 0 \end{pmatrix}, \begin{pmatrix} 0 \\ 1 \end{pmatrix} \right\}, \quad C(V_2) = \left\{ \begin{pmatrix} 1 \\ 1 \end{pmatrix}, \begin{pmatrix} 0 \\ -1 \end{pmatrix} \right\} \tag{3.10}$$

$$C(V_3) = \left\{ \begin{pmatrix} -1 \\ -1 \end{pmatrix}, \begin{pmatrix} 1 \\ -2 \end{pmatrix} \right\}, \quad C(V_4) = \left\{ \begin{pmatrix} -1 \\ 0 \end{pmatrix}, \begin{pmatrix} -1 \\ 2 \end{pmatrix} \right\}$$

Step 2 Decompose the supporting cones into unimodular cones [47].

The last two cones in our example—$C(V_3)$ and $C(V_4)$—are not unimodular. Their decomposition is given below:[14]

$$C(V_3) = \oplus \left\{ \begin{pmatrix} 0 \\ -1 \end{pmatrix}, \begin{pmatrix} -1 \\ -1 \end{pmatrix} \right\} \ominus \left\{ \begin{pmatrix} -1 \\ 2 \end{pmatrix}, \begin{pmatrix} 0 \\ -1 \end{pmatrix} \right\} \tag{3.11}$$

$$C(V_4) = \oplus \left\{ \begin{pmatrix} 0 \\ 1 \end{pmatrix}, \begin{pmatrix} -1 \\ 0 \end{pmatrix} \right\} \oplus \left\{ \begin{pmatrix} -1 \\ 2 \end{pmatrix}, \begin{pmatrix} 0 \\ -1 \end{pmatrix} \right\} \tag{3.12}$$

Step 3 Find out the generating function of the iterator polytope $\mathbf{A} \cdot \mathbf{i} \geq \mathbf{b}$.

Given a **Z**-polytope P in the d-dimensional space, a term $z^{\mathbf{P}} = x_1^{p_1} x_2^{p_2} \ldots x_d^{p_d}$ can be introduced for each point $\mathbf{p} = (p_1, p_2, \ldots, p_d)$ in P. The function $F(P) = \sum_p z^{\mathbf{P}}$ (or, simply, F) is defined as the generating function of the polytope P [47]. For example, the generating function of the

---

[14]The computation is complex: the cones are polarized, then decomposed, then polarized back. This double dualization process eliminates the lower dimension component cones [50].

triangle with the vertices $(0, 0)$, $(1, 2)$, $(2, 0)$ is $F = x^0y^0 + x^1y^0 + x^1y^1 + x^1y^2 + x^2y^0 = 1 + x + xy + xy^2 + x^2$, each monomial term corresponding to one of the lattice points inside or on the border of the triangle: for example, x^1y^2 corresponds to the point $(1,2)$, and so on. By evaluating F at $z = 1$, we get the number of points in P [47]. For instance, if $x = 1, y = 1$ then $F = 5$, which is the number of points of integer coordinates in the triangle.

Writing F as a sum of monomial terms would be impractical for large polytopes. Fortunately, F can be compactly written as an algebraic sum of rational functions, each term corresponding to one of the unimodular cones in the decomposition of the supporting cones $C(V_i)$ $(i = 1, \ldots, 4)$ of the polytope vertices (*Steps 1* and *2*). Every unimodular cone $C(V)$ whose vertex V has integer coordinates has associated a generating function [47] of the form

$$F(V) = \frac{z^V}{\Pi_i(1 - z^{r_i})} \tag{3.13}$$

where $z^V = x^a y^b$ (a, b being the $d = 2$ coordinates of V), and the product is over all the generating rays r_i. For instance, if the vertex of the cone is $V = V_4(7, 1)$, then $z^V = x^7y$; if the ray $r = [-1 \;\; 2]^T$, then $z^r = x^{-1}y^2$. The generating function of any cone is obtained by making the summation of the functions of all the unimodular component cones. From Equation 3.10 and 3.12, the generating functions of the cones $C(V_1)$, $C(V_2)$, and $C(V_4)$ are, respectively:

$$F(V_1) = \frac{x^2y}{(1 - x)(1 - y)}, \quad F(V_2) = \frac{x^2y^3}{(1 - xy)(1 - y^{-1})}$$

$$F(V_4) = \frac{x^7y}{(1 - y)(1 - x^{-1})} + \frac{x^7y}{(1 - x^{-1}y^2)(1 - y^{-1})}$$

When not all the coordinates of vertex V are integer (as in the case of vertex V_3), but still rational numbers (which is the case in practically all the multimedia algorithms), the generating function has the same form (Equation 3.13), but V is the lattice point in the fundamental parallelepiped of the unimodular cone [47]. In this case, V can be determined as follows [50].

Let $\mathbf{R} = [r_1 \cdots r_d]$ be the matrix whose columns are the rays of the cone. Solve the linear system of equations $\mathbf{R} \cdot \mu = \mathbf{x}_0$, where $\mathbf{x}_0$ is the coordinate vector of the vertex ($\mathbf{x}_0 = [14/3 \;\; 17/3]^T$ for V_3). Let $\mu_1, \ldots, \mu_d$ be the solution of the system. Then the coordinates of vertex V in the generating function are the elements of the vector $\mathbf{R} \cdot \overline{\mu}$, where $\overline{\mu} = [\lceil\mu_1\rceil \cdots \lceil\mu_d\rceil]^T$. This technique applied to the two unimodular cones (Equation 3.11) components of $C(V_3)$ yields $V = V_3'(4, 5)$ and $V = V_3''(4, 7)$, respectively. The generating function of the cone $C(V_3)$ is

$$F(V_3) = \frac{x^4y^5}{(1 - y^{-1})(1 - x^{-1}y^{-1})} - \frac{x^4y^7}{(1 - x^{-1}y^2)(1 - y^{-1})}$$

The sum of the rational functions $F(V_1), \dots, F(V_4)$ is the generating function F of the whole quadrilateral in Figure 3.11.

Step 4 Compute the number of points having integer coordinates from the generating function $F = \sum_i F(V_i)$ of the whole polytope.

In order to obtain a one-variable generating function F, we make the substitutions $x \to t^{\lambda_1}$ and $y \to t^{\lambda_2}$, where λ_1 and λ_2 are integers chosen such that no factor $(1 - x^a y^b)$ in the denominators of the terms of F becomes zero. In this example, we choose $\lambda_1 = \lambda_2 = 1$. With the substitution $x = t, y = t$, the generating function $\sum_1^4 F(V_i)$ of the iterator polytope in Figure 3.11 becomes:

$$F = \frac{t^3}{(1-t)^2} + \frac{t^5}{(1-t^2)(1-t^{-1})} + \frac{t^9}{(1-t^{-1})(1-t^{-2})}$$
$$- \frac{t^{11}}{(1-t)(1-t^{-1})} + \frac{t^8}{(1-t)(1-t^{-1})} + \frac{t^8}{(1-t)(1-t^{-1})}$$

After eliminating the negative exponents in the denominators and after factorizing t^3, we substitute $t = s + 1$, obtaining rational terms of the form $\frac{P(s)}{s^d Q(s)}$, where $P(s)$ and $Q(s)$ are polynomials and $d = 2$ is the dimension of the iterator space:

$$F = \frac{1}{s^2} - \frac{(s+1)^3}{s^2(s+2)} + \frac{(s+1)^9}{s^2(s+2)} + \frac{(s+1)^9}{s^2} - \frac{(s+1)^6}{s^2} - \frac{(s+1)^6}{s^2}$$

If $P(s) = a_0 + a_1 s + a_2 s^2 + \dots$ and $Q(s) = b_0 + b_1 s + b_2 s^2 + \dots$, the coefficients of the quotient $P(s)/Q(s) = c_0 + c_1 s + c_2 s^2 + \dots$ can be obtained recursively as follows:

$$c_0 = \frac{a_0}{b_0} \quad \text{and} \quad c_k = \frac{1}{b_0}(a_k - b_1 c_{k-1} - b_2 c_{k-2} - \dots - b_k c_0) \qquad \text{for } k \geq 1$$

It can be proven [50] that the algebraic sum of the coefficients c_2 (since here the space dimension is 2) after the polynomial divisions in all the terms of F is the number of points in the **Z**-polytope. In this example, the six coefficients c_2 (one for each term of F) are $\{0, -\frac{7}{8}, \frac{127}{8}, 36, -15, -15\}$. Their sum is 21, which is indeed the number of lattice points inside or on the border of the quadrilateral in Figure 3.11, and it is also the number of memory locations to store the array reference $A[2 * i - j][3 * i + 2 * j]$ since the vector function $\mathbf{i} \longmapsto \mathbf{T} \cdot \mathbf{i} + \mathbf{u}$ is a one-to-one mapping from the iterator space to the index space, as already explained.

Assume now that the line $V_4 y$ is shifted to the right, keeping its slope unchanged, until the coordinates of V_4 become $(107,1)$, instead of $(7,1)$. The equation of that line becomes $j = -2i + 215$ (instead of $j = -2i + 15$), and the new vertex V_3 will have the coordinates $(214/3, 217/3)$. The computation effort necessary to count the number of points in the new, much larger, **Z**-polytope is not affected by the significant increase in its size. Indeed, the four supporting cones are generated by the same rays, the decompositions are the same, the generating functions are almost the same. The only difference appears at the numerators of $F(V_3)$ and $F(V_4)$ due to the modifications of the coordinates of these vertices. For instance, the numerator of $F(V_4)$ becomes $x^{107}y$ since the new coordinates of V_4 are $(107,1)$, while the numerators of the two terms of $F(V_3)$ become $x^{71}y^{72}$ and $x^{71}y^{73}$, respectively. The number of points in the new **Z**-polytope is 3921.

Besides its excellent *scalability*, the technique sketched above, illustrated for a 2D polytope (a quadrilateral), *works for an arbitrary number of dimensions*. Therefore, it is well-suited to address the large size of array references typical to telecom and multimedia applications.

Case 2: The affine vector function $\mathbf{i} \longmapsto \mathbf{T} \cdot \mathbf{i} + \mathbf{u}$ of the lattice is not a one-to-one mapping.

When the rank r of matrix $\mathbf{T}$ is smaller than n (the number of columns of $\mathbf{T}$), then $\mathbf{T}$ is brought to the Hermite normal form [40] postmultiplying it by a unimodular matrix $\mathbf{S}$.[15] The size of the lattice is, afterward, obtained counting the number of points having integer coordinates from the projection of the **Z**-polytope $\mathbf{A} \cdot \mathbf{S} \cdot \mathbf{i} \geq \mathbf{b}$ on $\mathcal{R}^r$ along the first r coordinates [35,52].

Example 3.10

```
for (i=0; i<=4; i++)
    for (j=0; j<=2*i; j++)
        if ( j<=-i+6 ) ··· = A[3*i+j];
```

The rank of matrix $\mathbf{T}$ is $r = 1$, hence less than the number of columns $n = 2$ of $\mathbf{T}$; in this case, the vector function $\mathbf{i} \longmapsto \mathbf{T} \cdot \mathbf{i} + \mathbf{u}$ may not be a one-to-one mapping. Indeed, the iterator vectors $[i\ j]^T = [2\ 3]^T$ and $[3\ 0]^T$ are mapped to the same index $3i + j = 9$. The unimodular transformation $\mathbf{S} = \begin{bmatrix} 0 & 1 \\ 1 & -3 \end{bmatrix}$ bringing $\mathbf{T} = \begin{bmatrix} 3 & 1 \end{bmatrix}$ at the Hermite normal form $\mathbf{T} \cdot \mathbf{S} = \begin{bmatrix} 1 & 0 \end{bmatrix}$ modifies the initial iterator space into $\{\mathbf{A} \cdot \mathbf{S} \cdot [k\ l]^T \geq \mathbf{b}\} = \{3l \leq k \leq 5l,\ k \leq 2l + 6,\ l \leq 4\}$ (where $[k\ l]^T = \mathbf{S}^{-1}[i\ j]^T$ is the new iterator vector after the transformation $\mathbf{S}$) whose exact 1D projection (since $r = 1$) is $\{0 \leq k \leq 14\}$, obtained eliminating l in the inequalities above [41]. The points in the "dark shadow" [52] $4 \leq k \leq 14$ correspond to lattice points (k,l) in the modified iterator space. Checking the values of k outside the "dark shadow", $k = 1$ and 2 result to be invalid projections: replacing these values in the modified iterator space, no integer solution for l can be found. Conversely, $k = 0$ and 3

---

[15] While the Hermite normal form is canonical, the factorizing matrix $\mathbf{S}$ is not necessarily unique when $\mathbf{T}$ is not square and non-singular. Most of the algorithms computing the Hermite normal form propose also techniques to build the factorizing (unimodular) matrix. In our context, any such an approach will do, provided it has polynomial complexity for reason of efficiency. (Such algorithms are designed to prevent the so-called "coefficient swell" [40].)

are valid projections, since $(k,l) = (0,0)$ and $(3,1)$ belong to the modified iterator space. Therefore, storing $A[3*i+j]$ requires 13 locations: the index can take all the values between 0 and 14, except 1 and 2.

3.5   Computation of the Minimum Data Storage

Section 3.5.1 presents in an intuitive way the general flow of the algorithm computing the minimum data storage for affine specifications [42]. Section 3.5.2 explains how this approach can handle delayed signals. Section 3.5.3 details another algebraic technique used by the computation algorithm.

3.5.1   The flow of the algorithm

Algorithm computing the minimum data storage of a given affine specification.

Step 1 *Extract the array references from the given algorithmic specification and decompose the array references of every indexed signal into disjoint linearly bounded lattices.*

This step applies the technique described in Section 3.4.3 to decompose the index space of each signal in the code into disjoint lattices. The algorithm flow will be illustrated using the example in Figure 3.12, identical to the code in Figure 3.2, but without the delay operators addressed in the next section. The inclusion graph and the decomposition of the array space of signal A are

```
Block 1:     T[0] = 0 ;          // A[10][529] : input
            - - - - - - - - - - - - - - - - - - - - - - - - - - -       Memory: 5291
            for ( j=16 ; j<=512 ; j++)
       (1)  { S[0][j-16][0] = 0 ;
                for ( k=0 ; k<=8 ; k++)
Block 2:            for ( i=j-16 ; i<=j+16 ; i++)
       (2)             S[0][j-16][33*k+i-j+17] = A[4][j] - A[k][i] + S[0][j-16][33*k+i-j+16] ;
       (3)          T[j-15] = S[0][j-16][297] + T[j-16] ;
            }
            - - - - - - - - - - - - - - - - - - - - - - - - -       Memory: 4762
            for( j=16 ; j<=512 ; j++)
            { S[1][j-16][0] = 0 ;
                for( k=1 ; k<=9 ; k++)
Block 3:            for( i=j-16 ; i<=j+16 ; i++)
                       S[1][j-16][33*k+i-j-16] = A[5][j] - A[k][i] + S[1][j-16][33*k+i-17] ;
                    T[j+482] = S[1][j-16][297] + T[j+481] ;
            }
            - - - - - - - - - - - - - - - - - - - - - - - - - - -       Memory: 1
Block 4:     out = T[994];        // out : output
```

Figure 3.12 Illustrative example similar to the code in Figure 3.2, but without delayed array references. The total number of scalars is 302,498; the storage requirement is 5292 [43]. (©2007 IEEE).

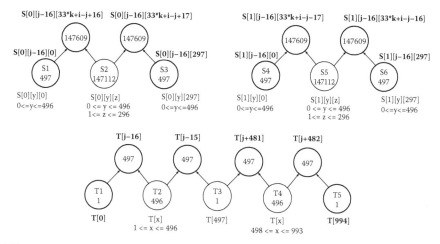

Figure 3.13 Inclusion graphs of the signals S and T from the code in Figure 3.12 [43]. (© 2007 IEEE).

the same as in Figure 3.9a–b. The inclusion graphs of the signals S and T are shown in Figure 3.13.

Step 2 *Determine the memory size at the boundaries between the blocks of code.*

The algorithmic specification is, basically, a sequence of nested loops. (Single instructions outside nested loops are actually nests of depth zero.) We refer to these loop nests as *blocks* of code. After the decomposition of the array references in the specification code, it is determined in which block each of the disjoint LBLs is created (i.e., *produced*), and in which block it is used as an operand for the last time (i.e., *consumed*). Based on this information, the memory size between the blocks can be determined *exactly*, by adding the sizes of every live lattice.

The memory sizes at the beginning/end of the specification code are the storage requirements of the inputs/outputs, therefore $size(A) = 5290$ and, respectively, $size(out) = 1$ for this example. *Block 1* produces the LBL $T1$ (just one scalar); hence the memory size after *Block 1* is: $size(A) + size(T1) = 5291$. *Block 2* consumes only the LBL $A4$ from signal A; it both produces and consumes the LBLs $S1$, $S2$, and $S3$ of signal S; it also produces the LBLs $T2$ and $T3$ consuming $T1$ and $T2$ (see Figure 3.13). The memory size after *Block 2* is: $size(A) - size(A4) + size(T3) = 4762$. (The LBLs both produced and consumed cancel each other.) Similarly, *Block 3* consumes all the signal A still alive; both produces and consumes the LBLs $S4$, $S5$, and $S6$ of signal S. Since only $T5$ is still alive, the memory size after *Block 3* is: $size(T5) = 1$.

Step 3 *Compute the storage requirement inside each block.*

Step 3.1 *Determine the* characteristic *memory variation for each assignment instruction in the current block.*

Take, for instance, the first loop nest from the illustrative example in Figure 3.12. The assignment Equation 3.1 produces at each iteration a new element $S[0][j-16][0]$ of the array S. We say that the *characteristic* memory variation of this assignment is +1 since each time the instruction is executed the memory size will increase by one location. Similarly, the assignment (3) has a characteristic memory variation of -1 (i.e., $+1-1-1$) since at each iteration one scalar signal $T[j-15]$ is produced and two other scalars—$S[0][j-16][297]$ and $T[j-16]$—are consumed (used for the last time).

In general, the characteristic memory variation of an assignment is easy to compute: a *produced* array reference having a bijective vector function of its LBL(s) has a contribution of +1; an entirely *consumed* array reference having a one-to-one mapping has a contribution of -1; an array reference having no component LBL consumed in the block has a zero contribution, independent of its mapping. The rest of the array references in the assignment are ignored for the time being, being dealt with at *Step 3.3*.

For instance, at assignment (Equation 3.2), the two array references S bring a contribution of +1 and -1, respectively; the array reference $A[4][j]$ brings a zero contribution since it contains only the LBL $A10$ (see Figure 3.9a), which is not consumed in this block ($A10$ is also covered by the array reference $A[k][i]$ from the second loop nest, so it will be consumed in *Block 3*). Only part of the array reference $A[k][i]$ is consumed in this block, that is, the LBL $A4$ in Figure 3.9a. Therefore, the characteristic memory variation of assignment (Equation 3.2) is zero, meaning that the *typical* variation is of zero locations; however, for some iterations the memory variation may be -1 due to the consumption of the elements in the LBL $A4$ covered by the array reference $A[k][i]$.

Note that testing whether the vector function $\mathbf{i} \longmapsto \mathbf{T} \cdot \mathbf{i} + \mathbf{u}$ of an array reference is a one-to-one (i.e., bijective) mapping or not is very simple at this phase. It is sufficient to compare the size of its LBL(s) (the vector function is bijective if the two are equal), whereas it is not bijective if the former is smaller.

Step 3.2 *Check whether the maximum storage requirement could occur in the current block or not.*

The maximum possible memory increase in the block is the number of executions of the assignment instructions with positive characteristic memory variations. If this maximum possible increase added to the amount of memory at the beginning of the block is not larger than the maximum storage at the block boundaries (known from *Step 2*), then the maximum storage requirement cannot occur in this block, which hence can be safely skipped from further analysis. In particular, the blocks in which no signals are consumed (used for the last time) can be skipped since the memory will only increase to the amount at the end of the block (known from *Step 2*).

The memory size at the beginning of the first loop nest is 5291 (and this is the largest value among the memory sizes at the block boundaries). The maximum possible memory increase is due only to assignment (Equation 1), having a characteristic memory variation of $+1$, executed 497 times. So, theoretically, the memory size could reach (but not exceed) the value $5291 + 497$ (although it will not). The maximum storage requirement could occur in this loop nest (*Block 2*), so its analysis should continue.

On the other hand, the memory size at the beginning of the second loop nest is 4762. The memory size within that block could reach at most $4762 + 497 = 5259$ locations, a value already smaller than 5291. It follows that the maximum storage requirement cannot occur in the second loop nest, so this block can be skipped from further analysis (unless the tool is in the *trace* mode, as shown in Section 3.6). In such a situation, *Step 3* is resumed from the beginning for the next block of code.

This code pruning enhances the running times, concentrating the analysis on those portions of code where the memory increase is likely to happen.

Remark: The memory computation tool can be also used to *estimate* the storage requirements for *non-procedural* specifications by finding an upper bound of the memory size large enough for any possible execution ordering *inside* the blocks of code (the block organization being considered fixed though). Still assuming that any scalar must be stored only during its lifetime, it was shown above that the execution of *Block 2* (and, actually, of the whole code) cannot necessitate more than $5291 + 497 = 5788$ locations even if the computation order is changed. This upper bound is only 9.4% larger than the *exact* minimum storage (5292) for the procedural code and it is a reasonably good estimation. When the tool runs in the *estimation mode*, processing non-procedural specifications, it resumes *Step 3* from the beginning for the next block of code.

Step 3.3 *Determine and sort the iterator vectors of the consumed elements covered by array references that (a) are only partly consumed, or (b) their mappings (between iterators and indexes) are not bijective.*

Case 1: The vector function $\mathbf{i} \longmapsto \mathbf{T} \cdot \mathbf{i} + \mathbf{u}$ of the partly consumed array reference is a one-to-one mapping.

Each consumed LBL is intersected with the LBL of the array reference (in which it is included). By doing this, the general expression of the iterator vectors addressing only the elements of the consumed LBL is obtained. At each such iteration vector, one array element will be consumed (since the mapping of the array reference is bijective).

Case 2: The vector function $\mathbf{i} \longmapsto \mathbf{T} \cdot \mathbf{i} + \mathbf{u}$ of the array reference is not bijective.

This case is more difficult since distinct iterator vectors can access a single array element, whereas we are interested in only that unique iterator vector accessing the array element *for the last time*. This is what we call the *maximum (lexicographic) iterator vector*.

Definition: Let $\mathbf{i} = [i_1, \ldots, i_n]^T$ and $\mathbf{j} = [j_1, \ldots, j_n]^T$ be two iterator vectors in the scope of n nested loops, which may be assumed "normalized" (i.e., all the iterators are increasing with the step size of 1). The iterator vector $\mathbf{j}$ is larger lexicographically than $\mathbf{i}$ (written $\mathbf{j} \succ \mathbf{i}$) if $(j_1 > i_1)$, or $(j_1 = i_1$ and $j_2 > i_2)$, ..., or $(j_1 = i_1, \ldots, j_{n-1} = i_{n-1}$, and $j_n > i_n)$. The *minimum/maximum* iterator vector from a set of such vectors is the smallest/largest vector in the set relative to the lexicographic order. For instance, in the loop nest

```
for (i=0; i<=3; i++)
    for (j=0; j<=3; j++)
        for (k=0; k<=3; k++)    ... A[i+j+k] ...
```

the *maximum* iterator vector such that the element, say, $A[5]$ be accessed is $[i\ j\ k]_{max}^T = [3\ 2\ 0]^T$, while the *minimum* iterator vector is $[0\ 2\ 3]^T$. Section 3.5.3 will present the algorithm computing the maximum and minimum iterator vectors.

$A4$ is the only LBL consumed in the first loop nest satisfying the conditions of *Step 3.3*. It is part of the array reference $A[k][i]$, the elements of $A4$ having the indexes in the set $\{x = 0, 0 \le y \le 528\}$. The maximum iterator vectors of these elements indicate the iterations when they are consumed. For instance, the element $A[0][0]$ is consumed in the iteration $[j\ k\ i]^T = [16\ 0\ 0]^T$; $A[0][1]$ is consumed in $[17\ 0\ 1]^T$, while $A[0][528]$ is consumed in the iteration $[512\ 0\ 528]^T$. In general, the maximum iterator vectors of $A4$'s elements are: $[j\ k\ i]_{max}^T = [t + 16\ 0\ t]^T$ when $0 \le t \le 496$, and $[512\ 0\ t]^T$ when $497 \le t \le 528$.

Remark: If the consumed LBL is included in several array references within the block, the maximum iterator vectors are taken over by *all* these array references. For instance, if in the loop

```
for (i=0; i<=10; i++) {    ...    = X[i];    ...    = X[10-i];    }
```

all the elements X are consumed, the maximum iterator vector of $X[0]$ is $[i]_{max} = [0]$ relative to the array reference $X[i]$, whereas it is $[i]_{max} = [10]$ relative to the array reference $X[10 - i]$. We take, obviously, $[i]_{max} = [10]$ since $X[0]$ is accessed the last time in that iteration. Also, $X[5]$ is accessed twice when $i = 5$; we take $[i]_{max} = [5]$ relative the array reference $X[10 - i]$, since $X[5]$ is accessed for the last time in the second assignment of the loop.

Step 3.4 *Compute the storage requirement in the current block.*

In this moment, there is enough information to compute the memory size after any assignment. For instance, the memory size after the assignment (Equation 3.1) for the iterator $j = 100$ is:

$init\_memory + \left(\sum_{i=1}^{3} characteristic\_memory\_variation(assignment_i) * num\_executions(assignment_i)\right) - num\_consumed\_elements = 5291+$

$1 \cdot (100 - 15) + 0 \cdot (99 - 15) \cdot 9 \cdot 33 + (-1) \cdot (99 - 15) - (99 - 15) = 5208.$

Here, the initial memory at the beginning of the block is 5291; the characteristic memory variations of the three assignments are $+1$, 0, and -1; the other consumed elements are given by maximum iterator vectors (determined in *Step 3.3*) lexicographically smaller than $j = 100$, and they are: $[j \ \ k \ \ i]^T = [16 \ \ 0 \ \ 0]^T, \ldots, [99 \ \ 0 \ \ 83]^T$. Their number is the last term in the sum above.

The algorithm computes the memory size only *after* the assignments increasing the memory (with positive characteristic memory variation) and *before* the assignments decreasing the memory in order to identify the local maxima of the memory variation. The maximum storage requirement (5292) is reached in the first iteration ($j = 16$) after assignment (Equation 3.1). It is the maximum value for the entire code. Therefore, the minimum data memory necessary to ensure the code execution is 5292 locations.

The main ideas of the algorithm presented and illustrated in this section are summarized below. As already mentioned, the algorithm is actually computing the maximum number of scalars (array elements) *simultaneously alive* since this number corresponds to the minimum data storage necessary for the code execution.

Firstly (at *Steps 1* and *2*), the algorithm computes the number of simultaneously alive scalars at the borderline between the blocks of code (nests of loops). The lattices of array elements created in blocks *before* a certain borderline and, also, included in operands from blocks *after* the borderline contain the live array elements. (The rest of the array elements are either already "dead", or still "unborn.") The total size of these lattices is the exact storage requirement at the respective borderline.

Afterward (at *Step 3*), the computation focuses on each block of code, one by one. Some assignment instructions inside the block have a constant memory variation for each execution, since each time they create and consume the same amount of scalars. For the other assignments, the "time" of death (in terms of executed datapath instructions) of the array elements covered by the "dying" lattices is precisely determined, using *maximum (lexicographic) iterator vectors*. Based on these data, all the local maxima of the memory variation are exactly determined. In particular, the global maximum at the level of the entire code is accurately computed as well.

The algorithm can also determine the data memory variation for each individual signal: it is sufficient to take into account only the lattices corresponding to that signal (derived from its array references).

3.5.2 Handling array references with delays

The current implementation supports constant-value delays, the typical situation in most of the multimedia applications. The presence of the delay operators does not affect the partitioning of the signals' index spaces (*Step 1* of the algorithm). The LBLs will get attached an additional piece of information— their maximum delays in the specification code. These data can be easily determined with a top-down traversal of the inclusion graphs, a lower-level node getting the maximum delay from the nodes of the sets containing it. For instance, in the inclusion graph in Figure 3.9a the maximum delay is 16 for the nodes $A1$ and $A4$, and it is 32 for all the other nodes.

The memory size at the beginning of the code in Figure 3.2 is the total storage requirement of the current input and of the previous 32 inputs; but the older 16 inputs (delays 17–32) without their LBLs $A4$ since its maximum delay is 16:

$$size(A)+size(A@1)+\cdots+size(A@32) - (size(A4@17)+\cdots+size(A4@32))$$
$$= 33 \cdot size(A) - 16 \cdot size(A4)$$
$$= 33 \cdot 5290 - 16 \cdot 529 = 166,106.$$

Here, we denoted $A@k$ and $A4@k$ the input signal A and, respectively, the lattice $A4$ from k sample processings in the past. (Obviously, their sizes are the same as in the current code execution.) The memory size at the end of the code is 160,817 since $A4@16$ and $(A-A4)@32$ (that is, 5290 scalars in total) were consumed during the execution. The maximum storage requirement is 166,108; it is reached in the iteration $(j = 16)$ of the first loop nest, after the first assignment.

In order to handle delays, the *Steps 2* and *3* of the algorithm are only slightly modified: they only have to take into account that an LBL is consumed when it belongs to an array reference whose delay is equal to the LBL's maximum delay.

3.5.3   Computation of the maximum iterator vector of an array element

Given an array reference $M[x_1(i_1,\ldots,i_n)] \cdots [x_m(i_1,\ldots,i_n)]$ in the scope of the iterator polytope $\mathbf{A} \cdot [i_1 \ldots i_n]^T \geq \mathbf{b}$ derived from loop boundaries and conditional instructions, find the maximum iterator vector (the loops being normalized) accessing a given element $M[x_1^0] \cdots [x_m^0]$.

Step 1 Solve the Diophantine system [40] of equations $x_j(i_1,\ldots,i_n) = x_j^0$, $j = 1,\ldots,m$. If the system has no solution, there is no iterator vector $\mathbf{i} = [i_1 \ldots i_n]^T$ accessing the element; if it has a unique solution $\mathbf{i} = \mathbf{u} \in \mathbf{Z}^n$, then there is a unique iterator vector accessing the element, provided that $\mathbf{A} \cdot \mathbf{u} \geq \mathbf{b}$. If the solution of the system is not unique, it has the general form $\mathbf{i} = \mathbf{V} \cdot \mathbf{t} + \mathbf{u}$; assume that the $\mathbf{Z}$-polytope $\mathbf{A}(\mathbf{V} \cdot \mathbf{t} + \mathbf{u}) \geq \mathbf{b}$ is not empty (otherwise, there is no solution either).

Step 2 Bring matrix $\mathbf{V}$ to a reduced Hermite form [43] by post-multiplying it with a unimodular matrix $\mathbf{S}$ (less a possible row permutation): $\mathbf{V}' = \mathbf{V} \cdot \mathbf{S} = \begin{bmatrix} \mathbf{V_1} \\ \mathbf{V_2} \end{bmatrix}$, where matrix $\mathbf{V_1}$ is lower-triangular, with positive diagonal elements.

Step 3 Project the new $\mathbf{Z}$-polytope $\mathbf{A}(\mathbf{V}' \cdot \mathbf{t}' + \mathbf{u}) \geq \mathbf{b}$ on the first coordinate axis and find the maximum coordinate $t_1' \in \mathbf{Z}$ in the "exact shadow" [53]. Replace its value in the $\mathbf{Z}$-polytope and repeat this operation, finding

t'_2, t'_3, ... Then, $\mathbf{i}_{max} = \mathbf{V}' \cdot \mathbf{t}' + \mathbf{u}$, where $\mathbf{t}'$ is the vector of the t'_k values.

Example 3.11

```
for (i=0; i<=5; i++)
    for (j=0; j<=5; j++)
        for (k=0; k<=5; k++)    ···   A[i-3*j+2*k] ···
```

Assume we need to compute the *maximum* iterator vector $[i \ j \ k]^T_{max}$ such that the element $A[5]$ is accessed.

The solution of the Diophantine equation $i - 3j + 2k = 5$ has the general form $\mathbf{i} = \mathbf{V} \cdot \mathbf{t} + \mathbf{u}$ [40]:

$$\begin{bmatrix} i \\ j \\ k \end{bmatrix} = \begin{bmatrix} 3 & -2 \\ 1 & 0 \\ 0 & 1 \end{bmatrix} \begin{bmatrix} t_1 \\ t_2 \end{bmatrix} + \begin{bmatrix} 5 \\ 0 \\ 0 \end{bmatrix}$$

Post-multiplying $\mathbf{V}$ with the unimodular matrix $\mathbf{S} = \begin{bmatrix} 1 & 2 \\ 1 & 3 \end{bmatrix}$,

$$\begin{bmatrix} i \\ j \\ k \end{bmatrix} = \mathbf{V}' \cdot \mathbf{t}' + \mathbf{u} = \begin{bmatrix} 1 & 0 \\ 1 & 2 \\ 1 & 3 \end{bmatrix} \begin{bmatrix} t'_1 \\ t'_2 \end{bmatrix} + \begin{bmatrix} 5 \\ 0 \\ 0 \end{bmatrix}$$

where the matrix $\mathbf{V}' = \mathbf{V} \cdot \mathbf{S}$ is a reduced Hermite form [43]. It is quite obvious that choosing t'_1 as large as possible, the iterator i will be maximized; afterward, choosing t'_2 as large as possible, the iterator j will be maximized as well. Replacing the iterators from the matrix equation above in the iterator polytope derived from the loop boundaries: $\{0 \leq i, j, k \leq 5\}$, a $\mathbf{Z}$-polytope in t'_1 and t'_2 is obtained: $\{0 \geq t'_1 \geq -5, 5 \geq t'_1 + 2t'_2 \geq 0, 5 \geq t'_1 + 3t'_2 \geq 0\}$. Projecting this $\mathbf{Z}$-polytope on the first axis, the maximum lattice point of the projection (*exact shadow* [53]) is $t'_1 = 0$. For this value, the maximum t'_2 is 1. Replacing in the above matrix equation, the maximum iterator vector is obtained:

$$[i \ j \ k]^T_{max} = \mathbf{V}' \begin{bmatrix} 0 \\ 1 \end{bmatrix} + \mathbf{u} = [5 \ 2 \ 3]^T.$$

Remark: Taking the minimum values of t'_1 and t'_2, the *minimum* iterator vector is obtained:

$$[i \ j \ k]^T_{min} = \mathbf{V}' \begin{bmatrix} -5 \\ 3 \end{bmatrix} + \mathbf{u} = [0 \ 1 \ 4]^T.$$

3.6 Experimental Results

A memory management framework containing also a data storage computation tool[16] has been implemented in C++, incorporating the ideas and algorithms described in this chapter. For the syntax of the algorithmic specifications, we adopted a subset of the C language, "enriched" with the delay operator "@" (see, e.g., the illustrative example in Figure 3.2). This is not a restrictive feature of the theoretical model since any modification in the specification language would affect only the front end of the framework. In addition to the computation of the minimum data storage for the entire code or for specific signals, the variation of the memory occupancy during the execution of the algorithmic specification can be displayed as well. In this mode, the tool generates a data file used as an input by GnuPlot—which actually draws the memory trace.

The memory trace generated in Figure 3.14 shows the variation of the storage during the execution of a 2D Gaussian blur filter algorithm from a medical image processing application that extracts contours from tomograph images in order to detect brain tumors. The abscissae are the numbers of datapath instructions in the code; the ordinates are the memory locations in use. While the first graph represents the entire trace, the second graph is a detailed trace in the interval [23800, 24900], which corresponds to the end of the "horizontal Gaussian blur" and the start of the "vertical Gaussian blur." The third graph is a detailed trace in the interval [25100, 25550] covering the fourth inner-loop iteration in the part of the code performing the "vertical Gaussian blur."

The memory trace generated in Figure 3.15 shows the variation of the storage during the execution of a singular value decomposition (SVD) updating algorithm [54]—algebraic kernel used in spatial division multiplex access (SDMA) modulation in mobile communication receivers, in beamforming, and Kalman filtering.

The memory traces generated in Figure 3.16 display the data storage variation for Durbin's algorithm, an algebraic kernel used in many signal processing applications for solving Toeplitz systems of equations.

Table 3.1 summarizes the results of our experiments carried out on a PC with a 1.85 GHz Athlon XP processor and 512 MB memory. The benchmarks used are: (1) a motion detection algorithm used in the transmission of real-time video signals on data networks [1]; (2) a real-time regularity detection algorithm used in robot vision; (3) Durbin's algorithm for solving Toeplitz systems with N unknowns; (4) the kernel of a motion estimation algorithm for moving objects (MPEG-4); (5) an SVD updating algorithm [54]; (6) the

---

[16]Named $K2$ after the famous peak whose climbing adversity intends to suggest the difficulty of the implementation.

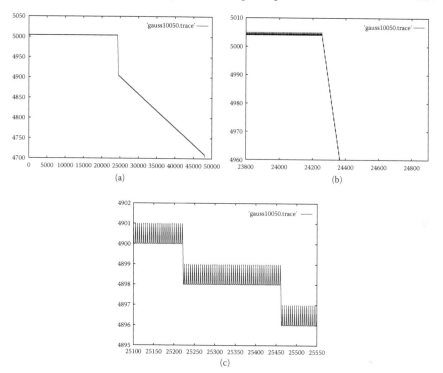

Figure 3.14 Memory trace for a 2D Gaussian blur filter ($N = 100$, $M = 50$) algorithm from a medical image processing application. (a) The entire trace in the interval $[0, 48025]$. The global maximum is at the point ($x = 5$, $y = 5005$). (b) Part of memory trace focusing on the end of the "horizontal Gaussian blur" and the start of the "vertical Gaussian blur." (c) Detailed trace in the part of the code performing the "vertical Gaussian blur [43]." (© 2007 IEEE).

kernel of a voice coding application—essential component of a mobile radio terminal.

Columns "Num. Array References" and "Num. Scalars" display the numbers of array references and, respectively, scalar signals (array elements) in the specifications. Column "Memory Size" displays the exact storage requirements (numbers of memory locations) and column "CPU" displays the corresponding running times.

This tool can process large specifications in terms of number of loop nests, lines of code, and number of array references. For instance, the voice coding application contains 232 array references organized in 40 loop nests. In one of our experiments, the tool processed a difficult example of about 900 lines of code, with 113 loop nests 3-level deep, and a total of 906 array references (many having complex indices), yielding a total of 3159 LBLs, in less than 7 minutes. The example in Figure 3.2, which is quite complex though not very large, was processed in less than 2 seconds.

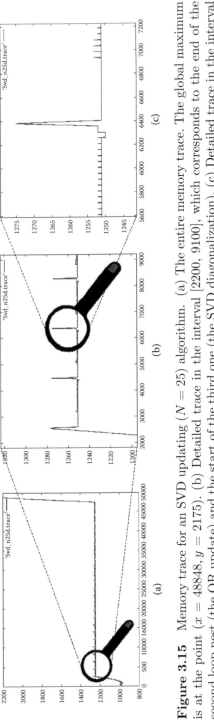

Figure 3.15 Memory trace for an SVD updating ($N = 25$) algorithm. (a) The entire memory trace. The global maximum is at the point ($x = 48848, y = 2175$). (b) Detailed trace in the interval [2200, 9100], which corresponds to the end of the second loop nest (the QR update) and the start of the third one (the SVD diagonalization). (c) Detailed trace in the interval [5595, 7215] covering the end of the second and start of the third iterations in the SVD diagonalization block [43]. (© 2007 IEEE).

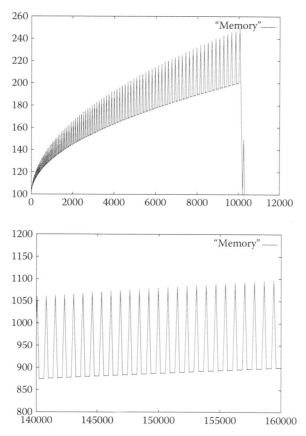

Figure 3.16 Memory traces for the execution of Durbin's algorithm ($N = 500$). The top graph is the entire trace, the bottom graph is a detail. (Reprinted from [56] with permission from IOS Press.)

The running time depends not only on the number of scalars, but also on many other factors like the code organization (e.g., number and depth of the loop nests, complexity of the conditionals), the number of array references and the complexity of their indices, and also on unquantifiable aspects during the computation (like the amount of LBLs whose affine functions are not one-to-one mappings, or the density or sparsity of the arrays' index spaces, or whether the LBLs overlap in a complex way generating many disjoint LBLs). For instance, the computation times for the motion detection benchmark are typically higher than those for other examples in spite of a lower number of LBLs. The reason is the benchmark contains array references with very complex indices that generate LBLs having higher-dimension iterator spaces and intersecting each other in complex ways.

As already mentioned, the current implementation works for the time being only for single-assignment specifications. Checking whether the code is in the

Table 3.1   Experimental results [43] (© 2007 IEEE)

Application (parameters)	Num. array references	Num. scalars	Memory size	CPU (sec)
Motion detection				
($M = N = 32$, $m = n = 4$)		72,543	2,740	2
($M = N = 64$, $m = n = 4$)		318,367	9,524	6
($M = N = 120$, $m = n = 8$)	11	3,749,063	33,284	16
Regularity detection	19	4,752	2,304	1
(MaxGrid = 8, L = 64)				
Durbin alg.	27	252,499	1,249	15
($N = 500$)				
MPEG-4 Motion estimation	68	265,633	2,465	18
SVD updating	87	3,045,447	34,950	26
($N = 100$)				
Vocoder kernel	232	33,706	11,890	2

single-assignment form is quite easy in our framework: each LBL must be written at most once, and each produced array reference must have a bijective mapping. However, it would be possible to process also programs that are not in the single-assignment form: using minimum iterator vectors in *Step 3.3*, one can find out the iterations when the array elements are written for the first time, exactly the opposite of using the maximum iterator vectors to find out when the array elements are read for the last time. Here, we assumed that an array element is alive from the first time it is written to the last time it is read, even if it is "dead" for some parts of this period and is written again.

It must be emphasized that previous memory estimation techniques yield storage results that were sometimes quite inaccurate. Zhao and Malik obtained an estimation of 1372 memory locations for the motion detection kernel, when the set of parameters is $M = N = 32$, $m = n = 4$ [25]. This was a rather poor estimation since the correct result for the same set of parameters is 2740 storage locations (see Table 3.1). One could argue that it may be better to obtain fast estimations, even not very accurate, rather than exact determinations with a significantly higher computation effort. This argument is fair enough, but it does not apply to the present situation. The aforementioned estimation result was obtained in 21 seconds on a Sun Enterprise 4000 machine with 4 (336 MHz UltraSparc) processors and 4 GB memory [25], whereas our computation time was only 2 seconds on the Athlon XP PC and 7 seconds on a Sun Ultra 20. Therefore, not only did our approach find the exact result, but it did it faster than the estimation technique from [25].

3.7 Conclusions

This chapter has presented a non-scalar approach for computing the memory size for data-intensive telecom and multimedia applications, where the storage of large multidimensional signals causes a significant cost in terms of both area and power consumption. This method uses modern elements in the theory of polyhedra and algebraic techniques specific to the data-flow analysis used nowadays in compilers. Different from past works that perform memory size *estimations*, this approach does *exact evaluations* for high-level procedural specifications.

The computation of the minimum data storage is useful in evaluating the impact of different code (and, in particular, loop) transformations on the data storage. In addition, different variants of code of a same application can be compared one against another in storage point of view, without the need for performing a proper memory allocation—a significantly more expensive solution—for each variant.

References

1. F. Catthoor, S. Wuytack, E. De Greef, F. Balasa, L. Nachtergaele, and A. Vandecappelle, *Custom Memory Management Methodology: Exploration of Memory Organization for Embedded Multimedia System Design*, Boston: Kluwer Academic Publishers, 1998.

2. A. Macii, L. Benni, and M. Poncino, *Memory Design Techniques for Low Energy Embedded Systems*, Boston: Kluwer Academic Publishers, 2002.

3. F. Catthoor, K. Danckaert, C. Kulkarni, E. Brockmeyer, P.G. Kjeldsberg, T.-V. Achteren, and T. Omnes, *Data Access and Storage Management for Embedded Programmable Processors*, Springer, 2010.

4. M.E. Wolf and M.S. Lam, "A data locality optimization algorithm," *Proc. ACM SIGPLAN'91 Conf.*, pp. 30–44, Toronto, Canada, June 1991.

5. A. Darte, "On the complexity of loop fusion," *Parallel Computing*, vol. 26, no. 9, pp. 1175–1193, 2000.

6. Q. Hu, A. Vandecappelle, M. Palkovic, P.G. Kjeldsberg, E. Brockmeyer, and F. Catthoor, "Hierarchical memory size estimation for loop fusion and loop shifting in data-dominated applications," in *Proc. Asia & S. Pacific Design Automation Conference*, pp. 606–611, Yokohama, Japan, Jan. 2006.

7. F. Balasa, P.G. Kjeldsberg, M. Palkovic, A. Vandecappelle, and F. Catthoor, "Loop transformation methodologies for array-oriented memory management," *Proc. 17th IEEE Int. Conf. Application-Specific Systems, Architectures, and Processors*, pp. 205–212, Steamboat Springs CO, Sep. 2006.

8. G. Chen, M. Kandemir, N. Vijaykrishnan, M.J. Irwin, and W. Wolf, "Energy savings through compression in embedded Java environments," *Proc. CODES*, 2002.

9. I.I. Luican, H. Zhu, and F. Balasa, "Signal-to-memory mapping analysis for multimedia signal processing," in *Proc. Asia & South-Pacific Design Automation Conf.*, Yokohama, Japan, 2007, pp. 486–491.

10. L. Thiele, "Compiler techniques for massive parallel architectures," in *State-of-the-Art in Computer Science*, P. Dewilde, Ed., Kluwer Academic Publishers, 1992.

11. A.V. Aho, R. Sethi, and J.D. Ullman, *Compilers—Principles, Techniques, and Tools*, Reading, MA: Addison-Wesley, 1986.

12. C.J. Tseng and D. Siewiorek, "Automated synthesis of data paths in digital systems," *IEEE Trans. on Comp.-Aided Design of ICs and Syst.*, vol. CAD-5, no. 3, pp. 379–395, July 1986.

13. G. Goossens, J. Rabaey, J. Vandewalle, and H. De Man, "An efficient microcode compiler for custom DSP processors," in *Proc. IEEE Int. Conf. Comp.-Aided Design*, Santa Clara CA, Nov. 1987, pp. 24–27.

14. A. Hashimoto and J. Stevens, "Wire routing by optimizing channel assignment within large apertures," in *Proc. 8th Design Automation Workshop*, 1971, pp. 155–169.

15. F.J. Kurdahi and A.C. Parker, "REAL: A program for register allocation," in *Proc. 24th ACM/IEEE Design Automation Conf.*, 1987, pp. 210–215.

16. G. Goossens, *Optimization Techniques for Automated Synthesis of Application-Specific Signal-Processing Architectures*, Ph.D. thesis, K.U. Leuven, Belgium, 1989.

17. P.G. Paulin and J.P. Knight, "Force-directed scheduling for the behavioral synthesis of ASIC's," *IEEE Trans. on Comp.-Aided Design of ICs and Syst.*, vol. 8, no. 6, pp. 661–679, June 1989.

18. C.H. Gebotys and M.I. Elmasry, *Optimal VLSI Architectural Synthesis*, Boston: Kluwer Academic Publishers, 1992.

19. L. Stok and J. Jess, "Foreground memory management in data path synthesis," *Int. J. Circuit Theory and Appl.*, vol. 20, pp. 235–255, 1992.

20. K.K. Parhi, "Calculation of minimum number of registers in arbitrary life time chart," *IEEE Trans. Circ. & Syst. - II: Analog and Digital Signal Processing*, vol. 41, no. 6, pp. 434–436, 1994.

21. S.Y. Ohm, F.J. Kurdahi, and N. Dutt, "Comprehensive lower bound estimation from behavioral descriptions," in *Proc. IEEE/ACM Int. Conf. on Computer-Aided Design*, 1994, pp. 182–187.

22. D. Gajski, F. Vahid, S. Narayan, and J. Gong, *Specification and Design of Embedded Systems*, Englewood Cliffs, NJ: Prentice Hall, 1994.

23. I. Verbauwhede, C. Scheers, and J.M. Rabaey, "Memory estimation for high level synthesis," in *Proc. 31st ACM/IEEE Design Automation Conf.*, 1994, pp. 143–148.

24. P. Grun, F. Balasa, and N. Dutt, "Memory size estimation for multimedia applications," in *Proc. 6th Int. Workshop on Hardware/Software Co-Design*, 1998, pp. 145–149.

25. Y. Zhao and S. Malik, "Exact memory size estimation for array computations," *IEEE Trans. VLSI Syst.*, vol. 8, no. 5, pp. 517–521, Oct. 2000.

26. J. Ramanujam, J. Hong, M. Kandemir, and A. Narayan, "Reducing memory requirements of nested loops for embedded systems," in *Proc. 38th ACM/IEEE Design Automation Conf.*, 2001, pp. 359–364.

27. F. Balasa, F. Catthoor, and H. De Man, "Background memory area estimation for multi-dimensional signal processing systems," *IEEE Trans. VLSI Syst.*, vol. 3, no. 2, pp. 157–172, June 1995.

28. F. Balasa, F. Catthoor, and H. De Man, "Practical solutions for counting scalars and dependences in ATOMIUM—a memory management system for multi-dimensional signal processing," *IEEE Trans. on Comp.-Aided Design of Integrated Circuits and Systems*, vol. 16, no. 2, pp. 133–145, Feb. 1997.

29. P.G. Kjeldsberg, F. Catthoor, and E.J. Aas, "Data dependency size estimation for use in memory optimization," *IEEE Trans. Comp.-Aided Design of ICs and Syst.*, vol. 22, no. 7, pp. 908–921, July 2003.

30. F. Catthoor, E. Brockmeyer, K. Danckaert, C. Kulkarni, L. Nachtergaele, and A. Vandecappelle, "Custom memory organization and data transfer: architecture issues and exploration methods," in *VLSI section of Electrical and Electronics Engineering Handbook*, M. Bayoumi (ed.), Academic Press, 2000.

31. P.R. Panda, F. Catthoor, N. Dutt, K. Dankaert, E. Brockmeyer, C. Kulkarni, and P.G. Kjeldsberg, "Data and memory optimization techniques for embedded systems," *ACM Trans. Design Automation of Electronic Syst.*, vol. 6, no. 2, pp. 149–206, April 2001.

32. Ph. Clauss and V. Loechner, "Parametric analysis of polyhedral iteration spaces," *J. VLSI Signal Processing*, vol. 19, no. 2, pp. 179–194, 1998.

33. P. D'Alberto, A. Veidembaum, A. Nicolau, and R. Gupta, "Static analysis of parametrized loop nests for energy efficient use of data caches," in *Proc. Workshop on Compilers and Operating Systems for Low Power*, 2001.

34. V. Loechner, B. Meister, and P. Clauss, "Precise data locality optimization of nested loops," *J. Supercomputing*, vol. 21, no. 1, pp. 37–76, 2002.

35. S. Verdoolaege, K. Beyls, M. Bruynooghe, and F. Catthoor, "Experiences with enumeration of integer projections of parametric polytopes," in *Compiler Construction: 14th Int. Conf.*, vol. 3443, R. Bodik (ed.), Springer, 2005, pp. 91–105.

36. W. Pugh, "Counting solutions to Presburger formulas: How and why," in *SIGPLAN Conf. Programming Language Design and Implementation*, 1994, pp. 121–134.

37. P. Feautrier, "Parametric integer programming," *Operations Research*, vol. 22, no. 3, pp. 243–268, 1988.

38. F. Balasa, F. Franssen, F. Catthoor, and H. De Man, "Transformation of nested loops with modulo indexing to affine recurrences," *J. Parallel Processing Letters*, vol. 4, no. 3, pp. 271–280, 1994.

39. S.S. Muchnick, *Advanced Compiler Design and Implementation*, San Francisco: Morgan Kaufmann, 1997.

40. A. Schrijver, *Theory of Linear and Integer Programming*, New York: John Wiley, 1986.

41. G.B. Dantzig, B.C. Eaves, "Fourier-Motzkin elimination and its dual," *J. Combinatorial Theory (A)*, vol. 14, pp. 288–297, 1973.

42. F. Balasa, H. Zhu, and I.I. Luican, "Computation of storage requirements for multi-dimensional signal processing applications," *IEEE Trans. on VLSI Systems*, vol. 15, no. 4, pp. 447–460, April 2007.

43. M. Minoux, *Mathematical Programming—Theory and Algorithms*, New York: John Wiley, 1986.

44. W. Li and K. Pingali, "A singular loop transformation framework based on non-singular matrices," in *Proc. 5th Annual Workshop on Languages and Compilers for Parallelism*, Aug. 1992.

45. D. Wonnacott, *Constraint-Based Array Dependence Analysis*, Ph.D. thesis, 1995.

46. A.I. Barvinok, "A polynomial time algorithm for counting integral points in polyhedra when the dimension is fixed," *Mathematics of Operations Research*, vol. 19, no. 4, pp. 769–779, Nov. 1994.

47. A.I. Barvinok and J. Pommersheim, "An algorithmic theory of lattice points in polyhedra," in *New Perspectives in Algebraic Combinatorics*, Cambridge Univ. Press, 1999, pp. 91–147.

48. E. Ehrhart, "Polynômes arithmétiques et méthode des Polyèdres en combinatoire, in *International Series of Numerical Mathematics*, vol. 35, Basel/Stuttgart: Birkhäuser-Verlag, 1977.

49. Ph. Clauss, "Counting solutions to linear and nonlinear constraints through Ehrhart polynomials: Applications to analyze and transform scientific programs," *Proc. ACM Int. Conf. on Supercomputing*, pp. 278–285, 1996.

50. J.A. De Loera, R. Hemmecke, J. Tauzer, and R. Yoshida, "Effective lattice point counting in rational convex polytopes," *The Journal of Symbolic Computation*, vol. 38, no. 4, pp. 1273–1302, 2004.

51. D. Avis, "lrs: A revised implementation of the reverse search vertex enumeration algorithm," in *Polytopes—Combinatorics and Computation*, G. Kalai and G. Ziegler, Eds. Birkhäuser-Verlag, 2000, pp.177–198.

52. W. Pugh, "A practical algorithm for exact array dependence analysis," *Comm. of the ACM*, vol. 35, no. 8, pp. 102–114, Aug. 1992.

53. W. Pugh and D. Wonnacott, "Experiences with constraint-based array dependence analysis," in *Principles and Practice of Constraint Programming*, 1994, pp. 312–325.

54. M. Moonen, P. V. Dooren, and J. Vandewalle, "An SVD updating algorithm for subspace tracking," *SIAM J. Matrix Anal. Appl.*, vol. 13, no. 4, pp. 1015–1038, 1992.

55. I. I. Luican, H. Zhu, and F. Balasa, "Computation of the minimum data storage and applications in memory management for multimedia signal processing," *Integrated Computer-Aided Engineering*, IOS Press, vol. 15, no. 2, pp. 181–196, 2008.

56. H. Zhu, I. I. Luican, F. Balasa, and D. K. Pradhan, "Formal model for the reduction of the dynamic energy consumption in multi-layer memory subsystems," *IEICE Trans. on Fundamentals of Electronics*, Communications and Computer Sciences, vol. E91-A, no. 12, pp. 3559–3567, December 2008.

57. F. Balasa, H. Zhu, and I. I. Luican, "Signal assignment to hierarchical memory organizations for embedded multidimensional signal processing systems." *IEEE Trans. on VLSI Systems*, vol. 17, no. 9, pages 1304–1317, September 2009.

Chapter 4

Polyhedral Techniques for Parametric Memory Requirement Estimation

Philippe Clauss
Team CAMUS, INRIA
University of Strasbourg, France

Diego Garbervetsky
Universidad de Buenos Aires, Argentina

Vincent Loechner
Team CAMUS, INRIA
University of Strasbourg, France

Sven Verdoolaege
INRIA Saclay, France

Contents

4.1   Estimating Memory Requirements of a Loop Nest

Memory requirement estimation is an important issue in the development of embedded systems, since memory directly influences performance, cost, and power consumption. It is therefore crucial to have tools that automatically compute accurate estimates of the memory requirements of programs to better control the development process and avoid some catastrophic execution exceptions.

Compute-intensive applications often spend most of their execution time in nested loops. The polyhedral model provides a powerful abstraction to reason about analyses and transformations on such loop nests by viewing an instance, or iteration, of each statement as an integer point in a polyhedron. From such a representation and a precise characterization of inter- and intra-statement dependences, it is possible to analyze loop nests statically in a completely mathematical setting relying on machinery from linear algebra, integer linear programming, and polynomial algebra. The analyses made on integer points in polyhedra correspond to effective quantitative characteristics of the original code such as the variable liveness or the amount of consumed memory.

The polyhedral model is applicable to loop nests in which the data access functions and loop bounds are affine combinations of the enclosing loop indices and parameters. While a precise characterization of data dependences is feasible for programs with static control structures and affine references and loop bounds, codes with non-affine array access functions or code with dynamic control can also be handled, but either with conservative assumptions on some dependences, or with additional knowledge coming from developers, or from advanced profiling/modeling techniques.

In this chapter, we present several techniques using the polyhedral model and providing automatic estimations of memory requirements in nested-loops programs. In the next section, useful background on the polyhedral model is provided. Section 4.3 relates the computation of the exact number of touched memory locations during the execution of a loop nest with the general issue of computing the number of integer points in the affine transformation of a polyhedron. Section 4.4 considers the computation of the maximum memory amount required during the execution of loop nests, while using temporary variables or dynamically allocated memory. The proposed technique is related to the general issue of maximizing a parametric multivariate polynomial defined over a polyhedron. Finally, conclusions are given in Section 4.5.

4.2 The Polyhedral Model of Loop Nests

4.2.1 Loop nests and polyhedra

In this chapter, we will consider loop nests with increments of one, and the statements not modifying the loop indices, nor exiting prematurely the loop. The loop bounds of each loop are affine functions of the outer loop indices and of some parameters (variables with unknown values, but that can be affinely constrained to each other).

Example 4.1 Consider for example the following loop nest of depth two:

```
for (i = 0; i < n; i++)
{
    /* S1(i) */
    for (j = i; j < i+n; j++)
    {
        /* S2(i,j) */
    }
    /* S3(i) */
}
```

where S1, S2, and S3 are three statements, containing no instructions `break`, `continue`, and `exit`, using but not modifying i and j.

The i-loop bounds $min_i = 0$ and $max_i = n$ are affine functions of a parameter n, and the j-loop bounds $min_j = i$ and $max_j = i + n$ are affine functions of the parameter and the outer loop index i.

Definition 4.1 (iteration domain) *An iteration domain is associated to each statement: it is the set of iterations of the loop nest enclosing the statement.*

The following definitions introduce some geometric concepts. We start with the concept of a parametric polyhedron, already briefly touched upon in Chapter 2.

Definition 4.2 (parametric polyhedron) *A parametric polyhedron is the intersection of a finite number of affine half-spaces, defined as a set of affine inequalities on the indices and parameters:*

$$\mathcal{P}(\mathbf{p}) = \{\mathbf{z} \mid A\mathbf{z} + B\mathbf{p} + \mathbf{c} \geq 0\}$$

where z is the vector of indices, p is a vector of parameters, A and B are matrices, c is a vector.

The iteration domain of each statement of an affine loop nest as defined so far in this section is the set of integer points contained in a parametric polyhedron.

Example 4.1 (continued) The iteration domain of statement S1 is:

$$\mathcal{D}_{S1}(n) = \{i \in \mathbb{Z} \mid i \geq 0 \wedge i < n\}$$

The iteration domain of statement S2 is:

$$\mathcal{D}_{S2}(n) = \{(i,j) \in \mathbb{Z}^2 \mid 0 \leq i < n \wedge i \leq j < i + n\}$$

Definition 4.3 (polyhedral integer set) A polyhedral integer set *is a subset of $\mathbb{Z}^d$ defined as a set of integer vectors constrained by affine inequalities on the indices, on the parameters, and on integer valued existentially quantified variables.*

The class of nonparametric polyhedral integer sets is the same as that of the linearly bounded lattices used in Chapter 2.

Property 4.1 A *polyhedral integer set* geometrically is a finite union of parametric polytopes intersected with integer lattices [1,2].

The set of references accessed through an affine array in a loop nest is a polyhedral integer set.

Example 4.2 Consider the following loop nest:

```
for (i = 0; i < n; i++)
{
    for (j = i; j < i+n; j++)
    {
        A[2*j-i] = i+j ; /* S2(i,j) */
    }
}
```

The elements of array A being accessed by this loop nest in statement S2 is the following polyhedral integer set, where the affine access reference is denoted by x, and the loop iterators i and j are existentially quantified variables:

$$\{x \in \mathbb{Z} \mid \exists (i,j) \in \mathbb{Z}^2 \text{ such that } 0 \le i < n \wedge i \le j < i + n \wedge x = 2j - i\}$$

The iteration domain of a statement in a loop nest with *quasi-affine bounds* (using floor, ceil, modulo, or fractional part of division) is a polyhedral integer set. Indeed, these functions can be transformed using an extra existentially quantified variable.

Example 4.3 The equality

$$x = \lfloor y/z \rfloor$$

with $x, y, z \in \mathbb{Z}$ is equivalent to:

$$\exists t \in \mathbb{Z} \text{ such that } 0 \le t < z \wedge zx = y + t$$

The following definition introduces the concept of relation in our context. Usually, a binary relation on a set S is defined as a collection of ordered pairs of elements of S, or equivalently as a subset of the cartesian product $S \times S$. We will define it in a very similar way, but making it depend on a vector of parameters.

Definition 4.4 (polyhedral relation) *A polyhedral relation is a function:*

$$R : \mathbb{Z}^m \quad \rightarrow \quad \mathbb{Z}^{d_1} \times \mathbb{Z}^{d_2}$$
$$p \quad \mapsto \quad \{z_1 \rightarrow z_2 \mid (z_1, z_2) \in \mathcal{D}(p)\}$$

where p is the parameters vector and $\mathcal{D}$ is a polyhedral integer set.

Example 4.4 Consider the loop nest of Example 4.2. The access relation of array A in statement S2 is:

$$R(n) = \{(i,j) \rightarrow x \mid 0 \le i < n \wedge i \le j < i + n \wedge x = 2j - i\}$$

4.2.2 Data-flow analysis

The data-flow analysis consists of computing which resource access is dependent on which other. A resource is usually a variable or an array in the source code, compiled as a memory location or a register in assembly.

The data-flow graph is a subgraph of the control flow graph of the program (the graph of all paths that might be traversed through a program during its execution), since the origin of the dependence executes before the endpoint of the dependence.

Definition 4.5 (dependence) *Let S1 and S2 be two statement instances of a program. There is a dependence from S1 to S2 if:*

- *S1 is executed before S2;*
- *S1 and S2 access the same resource, and at least one of the statements modifies the resource;*
- *there is no statement executed between S1 and S2 that modifies the resource.*

Notice that there is no dependence if both statements read the same resource: reading the same data twice can be done in any particular order. However, this kind of *input "dependence"* can be used by the compiler to make some optimizations (in particular cache optimizations).

If S1 modifies data being read by S2, it is said to be a *flow (or true) dependence*. If S1 reads data being modified afterwards by S2, it is an *antidependence*. If both S1 and S2 modify the same data, it is an *output dependence*.

The following definition formalizes this notion of dependence between array accesses in loop nests as a quasi-affine polyhedral relation.

Definition 4.6 (dependence relation) *Let S1 and S2 be two statements accessing a given array, executed at iteration z_1 and z_2 of their enclosing loop nests respectively, without any other intermediate access from any other statement. The dependence relation between those two array accesses is a polyhedral relation $z_1 \rightarrow z_2$, where z_1 is the last iteration of statement S1 accessing to the array, included in a polyhedral integer set defined by:*

- *z_1 and z_2 are iterations of the loop nests enclosing S1 and S2 respectively;*
- *z_1 is lexicographically smaller than z_2 if S1 and S2 are in the same loop nest, else the loop nest enclosing S1 is executed before the loop nest enclosing S2: this ensures $S1(z_1)$ executes before $S2(z_2)$, as stated in the first item of Definition 4.5;*
- *$a_1(z_1) = a_2(z_2)$, with a_1 and a_2 the arrays access functions in S1 and S2: the two statements access the same resource (second item of Definition 4.5).*

Notice that the third item of Definition 4.5 is ensured by computing the last iteration z_1 of this polyhedral integer set.

An example of such a computation is given in Section 4.4.1, Example 4.8.

4.2.3   Software tools

4.2.3.1   PipLib

PipLib[1] is a parametric integer linear programming solver library. It finds the lexicographic minimum (or maximum) in the set of integer points belonging to

---

[1] First release of PIP 1988 - Development: http://piplib.org

a convex polyhedron. It is based on parametric Gomory cuts and on the parametric dual simplex method. It can be used to solve the dependence system.

4.2.3.2 PolyLib

PolyLib[2] is a library of polyhedral functions, that can manipulate unions of rational polyhedra of any dimension as described in Chapter 2. It was the first to provide an implementation of the computation of parametric vertices of a parametric polyhedron, and the computation of an Ehrhart polynomial (expressing the number of integer points contained in a parametric polytope, as presented in Section 4.3), based on an interpolation method.

4.2.3.3 Omega

The Omega+ Library[3] provides all basic operations on polyhedral integer sets and relations. It includes the *Omega Test*, a decision test for the existence of integer solutions to affine constraints, that can be used for dependence analysis. The key transformation implemented in this project is the elimination of integer existentially quantified variables. It also contains a code generation library, to generate loop nests from polyhedral integer sets.

4.2.3.4 PPL: The parma polyhedra library

The Parma Polyhedra Library[4] can manipulate partially open rational polyhedra. It also provides a mixed integer linear programming problem solver using an exact-arithmetic version of the simplex algorithm, and a parametric integer programming solver. It is user friendly, portable, and has a very clean design.

4.2.3.5 The integer set library (isl)

The Integer Set Library[5] manipulates polyhedral integer sets and relations by exclusively using a constraints based representation. It provides computation of the integer projection, the integer affine hull, the lexicographic minimum using parametric integer programming, and Bernstein expansion.

4.2.3.6 CLooG

CLooG generates code for scanning the elements of a collection of polyhedral integer sets based on an extension of the algorithm for a single parametric polytope presented in Chapter 2.

---

[2]First release 1993 - http://icps.u-strasbg.fr/PolyLib

[3]First release of the Omega Project 1992 - http://www.cs.utah.edu/chunchen/omega/

[4]First release 2001 - http://www.cs.unipr.it/ppl/

[5]First release 2009 - http://freshmeat.net/projects/isl/

4.2.3.7 Barvinok

The barvinok library[6] implements the computation of the number of elements in parametric integer sets, based on Barvinok's counting algorithm. The same algorithm, restricted to nonparametric polytopes, was first implemented in LattE and was later also reimplemented in K2 (see Chapter 3). The Barvinok distribution also includes an interactive tool called iscc that exposes some of the functionality of isl, Barvinok, and CLooG, and that will be used in some of the examples in this chapter.

4.2.3.8 PoCC and PLuTo

The PoCC package (Polyhedral Compiler Collection)[7] is a source-to-source iterative compiler, which contains Clan (the Chunky loop analyzer) to extract a polyhedral intermediate representation from the source code; the Chunky analyzer for dependences in loops (Candl) to compute polyhedral dependences; LetSee (Legal transformation Space explorator) for finding different affine multidimensional schedules; the automatic parallelizer and locality optimizer PLuTo[8], for finding a valid tiling and parallel loop nest in the polyhedral model; CLooG (Chunky Loop Generator), to generate loop nests code; PipLib; PolyLib; and FM (Fourier-Motzkin library).

4.3 Counting the Elements in a Polyhedral Set

In many memory requirement estimation problems, the memory requirements fluctuate during the execution of the program. Solving such a problem then involves two steps. In the first step, the memory requirements of the program at any given point during the execution are computed. In the second step, the ultimate memory requirements of the complete program are obtained as a bound on this local memory requirement over the entire execution of the program. The first step is usually a counting problem and these counting problems are the subject of this section. The prototypical example is that of computing the maximal number of live memory elements in a program, where the first step consists of computing the number of live memory elements at any given point during the program execution.

Before we discuss this prototypical example, we first consider the counting problem in a slightly more general setting. Next, we explain how to compute reuse distances, and we finish this section with a discussion on the

---

[6]First release 2003 - http://freshmeat.net/projects/barvinok/

[7]First release 2009 - http://pocc.sourceforge.net/

[8]First release 2008 - http://pluto-compiler.sourceforge.net/

generalized problem of weighted counting. Throughout this section, we will focus on how to *model* counting problems. For details on how to *solve* these counting problems, we refer to the literature: [3] for an overview; [4–6] for non-parametric problems; [7–12] for parametric problems; and [13] for weighted counting problems.

4.3.1 Definitions and examples

As an introduction to counting, we will consider a few simple loop nests and ask ourselves a very simple question: how many times is the body of the loop nest executed?

The first loop nest is shown in Figure 4.1. Let us first assume that n has a fixed value, say 5. The iteration domain of the loop nest can be described as

$$\{\,(i,j) \mid 1 \le i \le 5 \wedge 1 \le j \le i\,\}$$

and is shown in Figure 4.2. The number of iterations of the loop nest is equal to the number of elements in this iteration domain, i.e.,

$$\#\{\,(i,j) \mid 1 \le i \le 5 \wedge 1 \le j \le i\,\} = 15,$$

which the reader can easily verify by hand. Usually, however, we are interested in solving such counting problems for arbitrary values of the parameters, i.e., in this case, for arbitrary values of n. The answer is then not just a number,

```
1  for (i = 1; i <= n; ++i)
2      for (j = 1; j <= i; ++j)
3          /* S */
```

Figure 4.1 A simple loop nest.

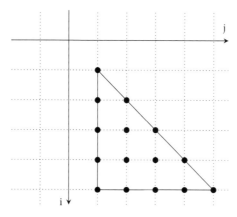

Figure 4.2 The iteration domain of the loop nest in Figure 4.1.

but a formula in terms of the parameters. For the loop nest in Figure 4.1, we obtain

$$\#\{(i,j) \mid 1 \le i \le n \wedge 1 \le j \le i\} = \frac{n(n+1)}{2}. \tag{4.1}$$

The previous result is indeed a special case of this formula, as can be verified by plugging in the value $n = 5$.

The formula in Equation 4.1 describing the number of iterations of the loop nest in Figure 4.1 is a polynomial in the parameter(s). In general, however, the result of such a counting problem is not representable using a single polynomial. Assume, for example, that the upper bound on the i-loop in Figure 4.1 is not simply n, but min(n, m), where m is a second parameter. The number of iterations can now be described as

$$\#\{(i,j) \mid 1 \le i \le n, m \wedge 1 \le j \le i\} = \begin{cases} \dfrac{n(n+1)}{2} & \text{if } 1 \le n \le m \\ \dfrac{m(m+1)}{2} & \text{if } 1 \le m \le n. \end{cases} \tag{4.2}$$

Notice that two polynomials are needed to describe the number of iterations, one that is valid when n is smaller than m and one that is valid in the other case. For this simple example, this result can easily be explained by the fact that the smaller of the two parameters determines the number of iterations in the outer loop and therefore also the total number of iterations. In principle, there is also a third case, where either n or m is smaller than 1, where the loop is not executed and the number of iterations is zero. We will usually omit such cases and assume that the result of the counting problem is zero for all cases that are not explicitly mentioned. The different cases will be called *cells*.

Let us try the example above in iscc. The set $\{(i,j) \mid 1 \le i \le m, n \wedge 1 \le j \le i\}$ is represented as

```
[m,n] -> { [i,j] : 1 <= i <= m,n and 1 <= j <= i }
```

with the list of parameters separated from the main set description by a ->. Computing

```
card [m,n] -> { [i,j] : 1 <= i <= m,n and 1 <= j <= i };
```

results in

```
[m, n] -> { (1/2 * n + 1/2 * n^2) : n <= m and n >= 1;
            (1/2 * m + 1/2 * m^2) : m >= 1 and n >= 1 + m }
```

Notice that the second domain has a constraint $m \le n - 1$ instead of $m \le n$. The reason is that we usually prefer to have disjoint cells. This preference will be partly explained in Section 4.3.4 where we discuss incremental counting. In Equation 4.2, the two cells overlap at $n = m$, and on this overlap the two polynomials are equivalent. Here, the overlap has been removed from one of the two cells.

```
1 for (i = 1; i <= n; ++i)
2     for (j = 1; j <= n - 2 * i; ++j)
3         /* S */
```

Figure 4.3 A slightly less simple loop nest.

Besides polynomials and cells, we need one final ingredient to describe the number of elements in a polyhedral set in general. To show the need for this final ingredient, consider the loop nest in Figure 4.3. The iteration domain can be described as

$$\{ (i,j) \mid 1 \leq i \leq n \wedge 1 \leq j \leq n - 2i \}.$$

This iteration domain is shown in Figure 4.4 for varying values of n. Note that the upper bound on the i-loop is slightly misleading, as the j-loop is only executed for values of i that result in an upper bound that is at least as large as the lower bound, i.e., for $1 \leq n - 2i$ or $i \leq (n - 1)/2$. Also note that the vertices of the iteration domains in Figure 4.4 are not always integral. In fact, they are only integral for even values of n. This nonintegrality results in "jumps" in the number of iterations when going from an odd value of n to an even value of n. The effect of these jumps is that two polynomials are required to describe the number of iterations in the loop nest, one for even n and one for odd n, namely,

$$\begin{cases} -\frac{n}{2} + \frac{n^2}{4} & \text{if } \exists \alpha : n \geq 4 \wedge n = 2\alpha \\ \frac{1}{4} - \frac{n}{2} + \frac{n^2}{4} & \text{if } \exists \alpha : n \geq 3 \wedge n = 2\alpha + 1. \end{cases} \tag{4.3}$$

This pair of polynomials is shown in Figure 4.5. They can be captured by a *single* polynomial expression if we allow this expression to contain floors of

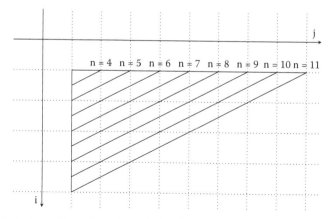

Figure 4.4 Iteration domains of the loop nest in Figure 4.3 for varying values of n.

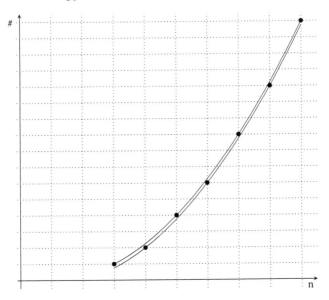

Figure 4.5 The number of elements in the iteration domains in Figure 4.4.

affine expressions. In particular, the function in Equation 4.3 can be described as

$$-\frac{n}{4} + \frac{n^2}{4} - \frac{1}{2}\left\lfloor\frac{n}{2}\right\rfloor \quad \text{if } n \geq 3. \tag{4.4}$$

The two polynomials in Equation 4.3 can easily be recovered from this expression if we note that $\left\lfloor\frac{n}{2}\right\rfloor = \frac{n}{2}$ for even n, while $\left\lfloor\frac{n}{2}\right\rfloor = \frac{n}{2} - \frac{1}{2}$ for odd n. We generally prefer the representation in Equation 4.4 because it is usually more compact than a representation using only polynomials, as in Equation 4.3. This more compact representation is also what iscc will give you by default. That is,

```
card [n] -> { [i,j] : 1 <= i <= n and 1 <= j <= n - 2i };
```

results in

```
[n] -> { ((-1/4 * n + 1/4 * n^2) - 1/2 * [(n)/2]) : n >= 3 }
```

We are now ready to describe the counting problem on polyhedral sets and relations. We first define step-polynomials and piecewise step-polynomials.

Definition 4.7 (Step-polynomial) *A step-polynomial $q(\mathbf{x})$ in the integer variables $\mathbf{x}$ is a rational polynomial expression in greatest integer parts of rational affine expressions in the variables, i.e., $q(\mathbf{x}) \in \mathbb{Q}\left[\lfloor\mathbb{Q}[\mathbf{x}]_{\leq 1}\rfloor\right].$*

A small note on the notation: $\mathbb{Q}[\cdot]$ represents polynomial expressions in "·" and $\mathbb{Q}[\cdot]_{\leq d}$ represents polynomial expressions in "·" of degree at most d. We will usually consider step-polynomials in either both parameters and variables or in only parameters.

Definition 4.8 (Piecewise step-polynomial) *A piecewise step-polynomial $q(\mathbf{x})$, with $\mathbf{x} \in \mathbb{Z}^d$ consists of a finite set of pairwise disjoint polyhedral sets $K_i \subseteq \mathbb{Z}^d$, called cells, each with an associated step-polynomial $q_i(\mathbf{x})$. The value of the piecewise step-polynomial at $\mathbf{x}$ is the value of $q_i(\mathbf{x})$ with K_i the cell containing $\mathbf{x}$ or zero if no cell contains $\mathbf{x}$, i.e.,*

$$q(\mathbf{x}) = \begin{cases} q_i(\mathbf{x}) & \text{if } \mathbf{x} \in K_i \\ 0 & \text{otherwise.} \end{cases}$$

The operation of counting the number of elements in a set then simply takes a parametric polyhedral set as input and produces a piecewise step-polynomial in the parameters describing the number of elements in the set.

Operation 4.1 (Number of elements in a set)
Input: a polyhedral set $S : \mathbb{Z}^n \to \mathbb{Z}^d$
Output: a piecewise step-polynomial $q : \mathbb{Z}^n \to \mathbb{Q} : (\mathbf{s}) \mapsto q(\mathbf{s}) = \#S(\mathbf{s})$

Note that the type of q is that of a function that returns a rational number, but the actual function values will always be integers.

There are different ways of defining a counting operation on a relation. One possibility would be to count the total number of pairs of domain and image elements. However, for our applications, it is more convenient to define an operation that counts the number of elements in the image of an arbitrary domain element.

Operation 4.2 (Number of image elements in a relation)
Input: a polyhedral relation $R : \mathbb{Z}^n \to \mathbb{Z}^{d_1} \times \mathbb{Z}^{d_2}$
Output: a piecewise step-polynomial $q : \mathbb{Z}^n \times \mathbb{Z}^{d_1} \to \mathbb{Q} : (\mathbf{s}, \mathbf{t}) \mapsto q(\mathbf{s}, \mathbf{t}) = \#R(\mathbf{s}, \mathbf{t})$

Note that this operation is equivalent to treating the domain dimensions as extra parameters and then counting the number of elements in the resulting set.

4.3.2 The number of live memory elements

In this chapter, we extend the computation of data storage requirements of Chapter 3 to programs that may involve parameters and that need not necessarily be in single-assignment form. As explained before, this computation involves two steps. We first compute the number of memory elements live at any given point during the execution and then compute an upper bound on this number over the entire execution of the program, which is then also an upper bound on the minimal memory requirements of the program. The second step involves an approximation and will be explained in Section 4.4. The first step can be performed exactly and is explained in this section.

Recall from Section 4.2.2 that dataflow analysis determines for each read access, the write access that wrote the value being read. The corresponding memory element is then clearly live at execution points between the write and the read. However, for *counting* the number of live elements, the standard dataflow relations are not very practical, because a given memory element may be involved in more than one of these relations at the same time if the result of a write is read multiple times. Counting can be made substantially easier by constructing injective relations that map the statement instance that produces an element to the statement instance that kills it, i.e., where it is used for the last time. Below, we describe two ways of constructing such relations. For simplicity, we assume here that each statement instance produces at most one element, so that we can use the statement instance as an identifier for the element being produced.

- One way is to essentially apply dataflow analysis twice. In the first application, we determine for each *write*, what is *next* write to the same memory element. In the second application, we determine for each write, what is the *last* subsequent read to the same memory element, *before* the next write, or, for those writes without a next write, simply the last subsequent read. The resulting relations clearly capture the births and deaths of memory elements. An alternative method of computing the same relation is to first perform standard dataflow analysis, relating each read to the last preceding write, and to then compute the last read associated to each write in the resulting relation.

- The second way is to determine for each read, what is the previous write *or read*. In this case, a memory element is considered dead after the first read, but is immediately revived if there is a second read. This resuscitation process then continues until the final read.

Example 4.5 Consider the sequence of writes and reads to a single memory element shown in Figure 4.6. The top row illustrates the first way of computing liveness relations. First, the next write is computed. This results in the relation

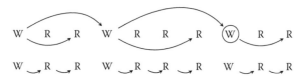

Figure 4.6 Writes and reads of a single memory element and the corresponding liveness relation.

shown above the first row and the encircled last write, which by definition has no next write. Based on this information the last read before the next write is computed and the result is shown below the top row. The second row illustrates the second way of computing liveness relations.

In general, the first way is more expensive, because dataflow analysis is applied twice, but the resulting relations have fewer elements (i.e., fewer arrows in Figure 4.6). Note, though, that this does not necessarily mean that computing the number of elements will be cheaper, as this computation is performed entirely symbolically.

Once a suitable injective relation R has been computed, there are again at least two ways of computing the number of elements live at iteration $\mathbf{i}$.

- The first way is to count all elements in the relation such that the source precedes $\mathbf{i}$ and the sink follows $\mathbf{i}$, i.e.,

$$L(\mathbf{i}) = \#\{(\mathbf{s}, \mathbf{t}) \in R \mid \mathbf{s} \prec \mathbf{i} \preccurlyeq \mathbf{t}\}.$$

- The second way is to count both the sources and the sinks that precede $\mathbf{i}$. The difference is equal to those preceding sources of which the corresponding sink does not precede $\mathbf{i}$, i.e.,

$$L(\mathbf{i}) = \#\{\mathbf{s} \in \operatorname{dom} R \mid \mathbf{s} \prec \mathbf{i}\} - \#\{(\mathbf{s}, \mathbf{t}) \in R \mid \mathbf{t} \prec \mathbf{i}\}.$$

Note that we cannot use $\mathbf{t} \in \operatorname{ran} R$ in the second counting problem because a given statement instance may kill more than one element.

It is difficult to predict which of these two methods will be more efficient. Since the lexicographical order needs to be linearized, the number of basic relations may grow by a factor of d^2 in the first method (with d the loop nest depth), while it only grows by a factor of up to $2d$ in the second method. On the other hand, basic relations that lie entirely before $\mathbf{i}$ are counted twice in the second method, while they are not considered at all in the first method.

4.3.3 Reuse distances

Modern computer architecture typically has a hierarchy of caches. Each cache stores recently used data. If this data is later reused, an access to a slower cache or the main memory can be avoided. Whenever a new value is stored in the cache, some other value needs to be evicted. There are many policies for deciding which value to evict. One such policy is to evict the least recently used (LRU) value. If the number of distinct data elements accessed between two consecutive accesses to the same data element is smaller than the cache size, then this this data element will still be in the cache, assuming the cache is LRU and fully associative. This number of distinct elements is called the reuse distance and its computation is useful for analyzing and optimizing cache behavior [14].

Pairs of consecutive accesses to the same memory element can be obtained from dataflow analysis (Section 4.2.2), where now no distinction is made between read and write accesses and all accesses are treated as both potential sources and potential sinks. Depending on the desired accuracy of the analysis, the access relations may need to be composed with allocation relations that map array indices to memory elements or even cache lines. The dataflow analysis yields a set of relations $R_{r \to s}$ of consecutive accesses, one for each pair of references r and s. Note that the domains of these relations for fixed r but varying s are disjoint because any given access has exactly one next access, which may be an instance of any, but only one, reference s. The *forward* reuse distance, i.e., the distance to the *next* access, can then be computed as

$$F_r(\mathbf{i}) = \sum_s \left(\# \bigcup_t \{ \mathbf{i} \to A_t(\mathbf{k}) \mid \exists \mathbf{j} \in S_s, \mathbf{k} \in S_t : \mathbf{i} \to \mathbf{j} \in R_{r \to s} \wedge \mathbf{i} \preccurlyeq \mathbf{k} \preccurlyeq \mathbf{j} \} \right),$$
$$(4.5)$$

where both s and t range over all references and A_t is the access relation of reference t. The *backward* reuse distance can be defined in a similar way.

Let us now consider how one of the relations in Equation 4.5 might be constructed using primitive operations. We first construct a relation that maps the domain elements of $R_{r \to s}$ to later iterations of t, i.e.,

$$R_1 = (\text{dom } R_{r \to s} \to S_t) \cap L_{r,t},$$

with $L_{r,t}$ the lexicographic order on r and t. This computation yields the relation

$$R_1 = \{ \mathbf{i} \to \mathbf{k} \mid \mathbf{i} \in \text{dom } R_{r \to s} \wedge \mathbf{k} \in S_s \wedge \mathbf{i} \preccurlyeq \mathbf{k} \}.$$

We can similarly construct a relation that maps the range elements to earlier iterations and then pull it back to the domain of $R_{r \to s}$. That is, we compute

$$R_2 = \left((\text{ran } R_{r \to s} \to S_t) \cap L_{t,s}^{-1} \right) \circ R_{r \to s}.$$

which yields the relation

$$R_2 = \{ \mathbf{i} \to \mathbf{k} \mid \exists \mathbf{j} \in S_s : \mathbf{i} \to \mathbf{j} \in R_{r \to s} \wedge \mathbf{k} \in S_s \wedge \mathbf{k} \preccurlyeq \mathbf{j} \}.$$

Finally, the relation in Equation 4.5 can be constructed as

$$A_t \circ (R_1 \cap R_2).$$

4.3.4 Weighted counting

So far, we have assumed that all points in the sets we want to enumerate are equal and then the enumeration simply counts the number of elements. In some cases, however, different points may have different *weights* and then we want to compute the sum of the weights of all elements in the set. For example, consider once more the loop nest in Figure 4.1. Let us now first compute the number of iterations of the j-loop for any iteration of the i-loop:

```
card [n] -> { [i] -> [j] : 1 <= i <= n and 1 <= j <= i };
```

The result is

```
[n] -> { [i] -> i : i <= n and i >= 1 }
```

This function assigns the weight i to each iteration i of the outer loop. The total number of iterations is then given by the sum of all these weights over all iterations of the outer loop:

```
sum [n] -> { [i] -> i : i <= n and i >= 1 };
```

and the result is

```
[n] -> { (1/2 * n + 1/2 * n^2) : n >= 1 }
```

As expected, this result is the same as that computed before in Equation 4.1. In this example, there is then also no point in performing the computation incrementally, because the results of one counting problem are used directly as the input of the next counting problem. There are some applications, however, where the result of one counting problem is first manipulated using some other operations before being used as an input to a second counting problem. For example, the input to the second counting problem may be the sum of several counting problems and/or a reformulation in terms of different variables. We will come across such an example in Section 4.4.4. As we will see below, there are also some applications where the counting problem is naturally weighted.

Since the output of an (unweighted) counting problem is a piecewise step-polynomial, the input of a weighted counting problem should be generic enough to include such piecewise step-polynomials. Similarly, since an unweighted counting problem is a special case of a weighted counting problem with weights 1, so should the output of the weighted counting problem. In fact, both input and output are exactly piecewise step-polynomials. More precisely:

Operation 4.3 (Weighted counting)
Input: a piecewise step-polynomial $q : \mathbb{Z}^n \times \mathbb{Z}^d \to \mathbb{Q}$, with (disjoint) cells K_i and associated step-polynomials q_i
Output: a piecewise step-polynomial

$$r : \mathbb{Z}^n \to \mathbb{Q} : (\mathbf{s}) \mapsto r(\mathbf{s}) = \sum_{\mathbf{t} \in \mathrm{dom}\, q(\mathbf{s})} q(\mathbf{s}, \mathbf{t}) = \sum_i \sum_{\mathbf{t} \in K_i(\mathbf{s})} q_i(\mathbf{s}, \mathbf{t})$$

Note that the final equality only holds if the cells K_i are indeed disjoint.

```
p = a;
for (i = 0; i < N; ++i)
    for (j = i; j < N; ++j) {
        p += j * ((j-i)/4);
        *p = hard_work(i,j);
    }
```

Figure 4.7   Pointer conversion example.

Example 4.6   Consider the program in Figure 4.7. We would like to parallelize this code, but there is a (false) dependence through the pointer p because it is updated in every iteration and so each iteration depends on the previous iteration. The dependence can be removed by computing the sum of all updates in any previous iteration, i.e.,

$$p = a + \sum_{\substack{(i',j') \in S \\ (i',j') \preccurlyeq (i,j)}} j' \left\lfloor \frac{j' - i'}{4} \right\rfloor$$

with $S = \{ (i', j') \in \mathbb{Z}^2 \mid 0 \le i' < N \wedge i' \le j' < N \}$, which is clearly a weighted counting problem.

Example 4.7   Consider the program in Figure 4.8, where first some memory is allocated, then some other code is executed, and finally the memory is freed. Suppose we want to know how much memory is allocated in total. Each iteration of the first loop allocates an amount of memory that depends on the values of the iterators. The total amount of memory can then again be computed as a weighted counting problem,

$$T(N) = \sum_{(i,j) \in S} (ij + i - N + 1),$$

with $S = \{ (i, j) \in \mathbb{Z}^2 \mid 0 \le i < N \wedge i \le j < N \}$.

```
for (i = 0; i < N; ++i)
    for (j = i; j < N; ++j)
        p[i][j] = malloc(i * j + i - N + 1);
/* ... */
for (i = 0; i < N; ++i)
    for (j = i; j < N; ++j)
        free(p[i][j]);
```

Figure 4.8   Memory allocation example.

4.4 Memory Requirement Estimates Based on Maximization Problems

4.4.1 Introduction

Recall that in this chapter, we focus on parametric estimation where some parameters values stay entirely or partially unknown while estimating the maximum required memory amount. For nonparametric cases, the reader is invited to refer to Chapter 3, although the techniques presented in this chapter can also be applied efficiently for nonparametric cases.

As it was presented in the previous section, many memory requirement estimation problems can be handled using the same common strategy: compute the number of elements that satisfy some conditions and then compute an upper bound of the resulting expression. For instance, the data storage requirements of some loop nests can be evaluated by first determining the amount of memory "in use" at a given point during the execution of the program and then by computing the maximum of the resulting expression over all such points. The memory in use at a given loop iteration is expressed as a piecewise step-polynomial in both the loop iterators and the structural parameters. The problem of calculating the memory requirements of a program then reduces to computing the maximum of such a polynomial over all integer points contained in parametric polytopes, resulting in an expression that only depends on the structural parameters. Notice that the general problem of maximizing a polynomial over a parametric polytope also has applications in extending static analysis beyond the polytope model [15].

As a typical example of the kind of memory requirement estimation problems that can be handled, we consider the problem of finding the maximal number of live elements during the course of a program, where an element is "live" at a given point in the program if it has been defined (written) and still needs to be used (read). A bound on the maximal number of live elements is an indication of the amount of memory required for the execution of the program.

Example 4.8 Consider the code fragment of Figure 4.9 where each statement has been labeled Sx, and assume that array t is a temporary array that is only used inside the given loop nests. The number of live elements of array t is obviously all the n elements, since every element is updated in the first loop and read in the last loop. However, we use this simple example to show how such a problem must generally be handled.

Each array element that is defined in the first loop is read and written several times in the second loop nest, and read exactly once in the last loop. As explained in Section 4.3.2, the number of live elements has to be computed for any iteration of the statements of each loop nest. The memory requirement

```
for (i = 0; i < n; ++i)
    S1: t[i] = a[i];
for (i = 0; i < n; ++i)
    for (j = 0; j < n - i; ++j)
        S2: t[j] = f(t[j], t[j+1]);
for (i = 0; i < n; ++i)
    S3: b[i] = t[i];
```

Figure 4.9 Two nested loops with temporary array t.

is then estimated by computing an upper bound of this number over the entire execution of the program.

The three iteration domains defined by each statement are:

$$S_1 = \{i \mid 0 \le i < N\}$$
$$S_2 = \{(i,j) \mid 0 \le i < N, 0 \le j < n - i\}$$
$$S_3 = \{i \mid 0 \le i < N\}$$

By using the first way of computing the number of live elements (Section 4.3.2), we first determine for each read of an array element, what is the previous write of the same element. Such a relation represents a data dependence between the update of an element and its use (a *flow dependence*, Section 4.2.2). For each read, the corresponding last previous write of the same element occurs at a previous iteration, which is the last executed iteration between all the previous iterations where writes of the same element occur.

For instance, consider the reading access t[j] in statement S2. For a given value of j, let us call it j', array element t[j'] is accessed when j = j', i.e., at every iteration (i,j') with $0 \le i < n$. Previous writes of t[j'] occur through the writing access t[j] in S2 and t[i] in S1. In the same way as for the reading access t[j], array element t[j'] is updated at every iteration (i,j') with $0 \le i < n$, while it is updated only once through statement S1 when $i = j'$. Hence for any read of an element t[j'] through reference t[j] in S2 at a given iteration (i',j'), the previous writes are given by:

$$\{i \in S_1 \mid i = j'\} \cup \{(i,j) \in S_2 \mid j = j', (i,j) \prec (i',j')\}$$

It is obvious that the last previous write is performed through statement S1 at iteration $i = j'$ if $i' = 0$ and through statement S2 at iteration $(i = i' - 1, j = j')$ if $i' > 0$. Then we can deduce the relations linking the corresponding writes and reads in the following way:

$$R_1 = \{j \to (0,j) \mid j \in S_1 \wedge (0,j) \in S_2\}$$
$$R_2 = \{(i - 1, j) \to (i,j) \mid (i,j) \in S_2 \wedge i > 0\}$$

Having determined in the same way the previous last writes (the sources) for all the remaining reads (the sinks), the next step is to count the sources and the sinks that precede a given iteration, and make the difference between both. For instance, if we only consider the reading accesses `t[j]` in statement S2 and their sources, this computation would be, for any iteration (i', j') of the second loop nest:

$$\# \{i \in \mathrm{dom}\, R_1\} - \#\{(j, 0, j) \in R_1 \mid (0, j) \prec (i', j')\}\, \text{if } i' = 0$$

$$\#\{(i, j) \in \mathrm{dom}\, R_2 \mid (i, j) \prec (i', j')\} - \#\{(i, j, k, l) \in R_2 \mid (k, l) \prec (i', j')\}\text{if } i' > 0$$

resulting in the number of live elements for any iteration (i, j) in $S_2 = n - i$. Hence the maximal number of live elements is equal to n.

The whole analysis considering all the statements and accesses to array elements can be achieved using `iscc`:

- Definition of the iteration domains:

  ```
  D := [n] -> { S1[i] : 0 <= i < n;
                S2[i,j] : 0 <= i < n and 0 <= j < n - i;
                S3[i] : 0 <= i < n };
  ```

- Definitions of the maps associated to the write accesses and the maps associated to the read accesses:

  ```
  W := { S1[i] -> t[i]; S2[i,j] -> t[j]; S3[i] -> b[i] } * D;
  R := { S1[i] -> a[i]; S2[i,j] -> t[j]; S2[i,j] -> t[j+1];
         S3[i] -> t[i] } * D;
  ```

- Mapping of the statements onto a common iteration space:

  ```
  S := { S1[i] -> [0,i,0]; S2[i,j] -> [1,i,j]; S3[i] -> [2,i,0] };
  ```

- Computing the relations between the writes and all corresponding reads:

  ```
  Dep := (last W before R under S)[0];
  ```

- Computing the last reads:

  ```
  M := (lexmax (Dep . S)) . S^-1;
  ```

- Setting the lexicographic order:

  ```
  LGT := S >> S;
  ```

- Computing the number of live elements for any iteration (i, j):

  ```
  Live := (card (LGT * (D -> (dom M)))) - (card ((LGT . (M^-1)) * D));
  Live;
  ```

- Resulting in:

  ```
  [n] -> { S2[i, j] -> n : j <= -1 + n - i and i >= 2 and j >= 1;
           S2[i, j] -> ((1/2 + n) * i - 1/2 * i^2) :
                       i = 1 and j <= -2 + n and j >= 1;
  ```

```
S2[i, j] -> n : j = 0 and i <= -1 + n and i >= 2;
S2[i, j] -> ((1/2 + n) * i - 1/2 * i^2) :
             i = 1 and j = 0 and n >= 2;
S2[i, j] -> n : i = 0 and j <= -1 + n and j >= 1;
S2[i, j] -> n : i = 0 and j = 0 and n >= 1;
S1[i] -> i : i <= -1 + n and i >= 1;
S3[i] -> (n - i) : i >= 2 and i <= -1 + n and n >= 2;
S3[i] -> (-1 + n) : i = 1 and n >= 2;
S3[i] -> n : i = 0 and n >= 2;
S3[i] -> 1 : n = 1 and i = 0 }
```

- Computing the maximum, *i.e.*, an upper bound of the data storage requirement:

```
ub Live;
```

- Resulting in:

```
([n] -> { max(n) : n >= 4; max(n) : n = 3;
          max(n) : n = 2; max(n) : n = 1 }, True)
```

The computed bound is the exact bound since the keyword **True** is appearing in the answer, and it is obviously equal to n.

In Section 4.4.4, it is shown how dynamic memory requirements in Java programs can also be estimated using similar techniques. In the following section, an overview of the main approaches and their mathematical concepts is given. Section 4.4.3 deals with the general issue of the problem formulation.

4.4.2 Maximization of polynomials

The exact parametric maximum of polynomials over the integer points in a parametric polytope may not in general be easily computable. In the technique used in the above example [16], the problem is relaxed first by computing the maximum over all *rational* points instead of all integer points and second by computing an *upper bound* rather than the maximum. This approach consists in an extension of *Bernstein expansion* [17–19] to parametric polytopes to compute these upper bounds. The resulting upper bounds are usually fairly accurate and it can be *detected* whether the actual maximum has been computed or not.

Some other techniques to handle polynomials have been proposed by Maslov and Pugh [20], Blume and Eigenmann [21], and Van Engelen et al. [22]. However, these techniques are either strongly restrictive or produce less accurate estimations than the ones produced by the technique presented in [16] and based on Bernstein expansion.

4.4.2.1 Bernstein expansion

Bernstein expansion allows for the determination of bounds on the range of a multivariate polynomial considered over a box [15,23,24]. Numerical applications of this theory have been proposed to the resolution of systems of strict polynomial inequalities [25,26]. A symbolic approach to Bernstein expansion used in program analysis has also been proposed in [15]. It has been shown that Bernstein expansion is generally more accurate than classic interval methods [27]. Moreover, in [28], Stahl has shown that for *sufficiently small* boxes, the exact range is obtained.

Bernstein polynomials are particular polynomials that form a basis for the space of polynomials. Hence any polynomial can be expressed in this basis through coefficients, the Bernstein coefficients, that have interesting properties and that can be computed through a direct formula. Due to the Bernstein convex hull property [29], the value of the polynomial is then bounded by the values of the minimum and maximum Bernstein coefficients. The direct formula allows symbolic computation of these Bernstein coefficients giving a supplementary interest to the use of this theory [15,16].

4.4.2.2 Symbolic range propagation

Symbolic range propagation [?,21] is certainly the most commonly used approach due mainly to its relative simplicity. Even if it usually provides less accurate results than with Bernstein expansion, it can still be useful while considering complex problems inducing many iteration domains and parameters. Indeed, the computation time in such cases can be huge with the Bernstein approach, since it needs several complex computations on parametric polytopes.

In this technique, it is assumed that each variable has a (symbolic) lower and upper bound and these ranges are repeatedly substituted in the polynomial. In each iteration, the expression is simplified using a set of rewrite rules. If a variable occurs multiple times in the same expression, then overly conservative bounds can be generated. However, if it can be determined, by recursively applying the algorithm to the first order forward difference of the polynomial, that the expression is monotonically nonincreasing or nondecreasing in a given variable, then the lower and upper bounds of the variable can safely be substituted simultaneously in the whole expression, leading to a tighter bound. The main disadvantage of this technique is that the accuracy can be very low for non-monotonic polynomials.

4.4.3 General problem formulation

When handling a memory requirement estimation problem, the essential task is to translate it conveniently into a counting problem followed by a maximization problem.

We are interested in the case where the elements of counting are integer vectors and the conditions can be described by *linear* constraints. The first step is then to compute the number of elements $f(\mathbf{p}, \mathbf{q})$ of some set $S(\mathbf{x}, \mathbf{p}, \mathbf{q})$, i.e.,

$$f(\mathbf{p}, \mathbf{q}) = \#\{\mathbf{x} \in \mathbb{Z}^n \mid \exists \mathbf{y} \in \mathbb{Z}^{n'} : p(\mathbf{x}, \mathbf{y}, \mathbf{p}, \mathbf{q})\}, \qquad (4.6)$$

where $p(\mathbf{x}, \mathbf{y}, \mathbf{p}, \mathbf{q})$ is a conjunction of m linear constraints on $\mathbf{x}$, $\mathbf{y}$, $\mathbf{p}$, and $\mathbf{q}$,

$$p(\mathbf{x}, \mathbf{y}, \mathbf{p}, \mathbf{q}) \iff A\mathbf{x} + B\mathbf{y} \geq C\mathbf{p} + D\mathbf{q} + \mathbf{f},$$

with $A \in \mathbb{Z}^{m \times n}$, $B \in \mathbb{Z}^{m \times n'}$, $C \in \mathbb{Z}^{m \times r}$, $D \in \mathbb{Z}^{m \times r'}$, and $\mathbf{f} \in \mathbb{Z}^m$. In the second step, we compute an upper bound on $f(\mathbf{p}, \mathbf{q})$. That is, we compute $U(\mathbf{q})$ such that

$$U(\mathbf{q}) \geq M(\mathbf{q}) = \max_{\mathbf{p} \in Q(\mathbf{q})} f(\mathbf{p}, \mathbf{q}) \qquad \text{for all } \mathbf{q}, \tag{4.7}$$

with Q the domain of f and where we use the shorthand $Q(\mathbf{q}) = \{ \mathbf{p} \mid (\mathbf{p}, \mathbf{q}) \in Q \}$.

The first problem (Equation 4.6) is a counting problem (see Section 4.3) while the second problem (Equation 4.7) is a "maximization" problem. Both of these problems are *parametric*, i.e., the result is not simply a number, but rather an expression in a number of parameters. The variables that act as parameters in one problem are, however, not the same as those that act as parameters in the other problem. In general, we can identify fours sets of variables in the two problems:

- The variables that are existentially quantified in the counting problem Equation 4.6; in the example of Section 4.4.1, there are no such variables.

- The elements that need to be counted; in the example, these are the indices of the array or the iteration in which they are defined or used.

- The variables over which the maximum needs to be taken; in the example, these are the iterations of $D_2(n)$.

- The structural parameters; in the example, there is a single structural parameter $\mathbf{n}$.

The latter two sets of variables will be parameters for the counting problem, while only the structural parameters will be parameters for the maximization problem.

If there are no existentially quantified variables $\mathbf{y}$, i.e., if $n' = 0$, then the set in Equation 4.6 is a *parametric polyhedron* (Section 4.2). If the polyhedron is bounded for each value of the parameters (which will usually be the case when we want to count the number of integer points in the polyhedron), then the set is a *parametric polytope* (Section 4.2).

As it was presented in Section 4.3, the number of integer points in such parametric polytopes are piecewise step-polynomials:

$$f(\mathbf{p}, \mathbf{q}) = \begin{cases} f_1(\mathbf{p}, \mathbf{q}) & \text{if } (\mathbf{p}, \mathbf{q}) \in Q_1 \\ \dots \\ f_M(\mathbf{p}, \mathbf{q}) & \text{if } (\mathbf{p}, \mathbf{q}) \in Q_M, \end{cases} \tag{4.8}$$

i.e., a subdivision of the parameter space Q (of the counting problem), with a step-polynomial $f_i(\mathbf{p}, \mathbf{q})$ associated to each cell Q_i of the subdivision. These piecewise step-polynomials that result from counting problems are also called Ehrhart polynomial by some authors, e.g., [30]. Notice that the cells Q_i in the subdivision are themselves polyhedra.

To compute $U(\mathbf{q})$ in Equation 4.7, we first compute

$$U_i(\mathbf{q}) \geq M_i(\mathbf{q}) = \max_{\mathbf{p} \in Q_i(\mathbf{q})} f_i(\mathbf{p}, \mathbf{q}) \qquad \text{for all } \mathbf{q}.$$

Note that Q_i is interpreted here as a parametric polyhedron with only the $\mathbf{q}$ as parameters. As in the counting problem, we may assume that Q_i is a parametric *polytope*. Finally $U(\mathbf{q})$ is constructed such that

$$U(\mathbf{q}) \geq U_i(\mathbf{q}) \qquad \text{for all } i \text{ and for all } \mathbf{q}. \tag{4.9}$$

4.4.4 Estimating dynamic memory requirements

The techniques mentioned so far can be also used to compute parametric bounds of dynamic memory requirements. As an example we will analyze a technique that computes dynamic memory requirements for Java programs [31].

Given a method m with parameters $p_1, \ldots, p_k$, the technique computes a parametric polynomial in $p_1, \ldots, p_k$ over-approximating the amount of dynamic memory *required* to execute m.

Java is an object oriented language with automatic memory management. Memory allocation is controlled by the programmer, who creates objects when executing a **new** statement. Memory deallocation is not controlled by the programmer but by a special agent, a *garbage collector* (GC), which takes care of collecting objects when they are no longer referenced.

In order to compute accurate bounds, it is very important to analyze program allocations and deallocations. Therefore, an analysis should consider both the program and GC behaviors. One alternative used to approximate GC behavior is to statically compute object lifetimes (a sequence of program statements starting at the moment of object creation and finishing at the point when the object can be collected) in order to predict at compile time, for every object, where it can be collected. This can be performed by the aid of escape analysis techniques [32,33] or by performing region synthesis [34,35], which associate the lifetime of sets of objects to the lifetime of computation units (e.g. methods, classes, threads, etc).

Here we will adopt the notion of regions by associating object lifetimes to methods. We will use ret$_\mathbf{m}$ to denote the size of the objects returned by (or escaping) a method m. That is, the objects that should live even when method m finishes its execution. We will use cap$_\mathbf{m}$ to denote the size of the objects allocated by method m that can be safely collected at the end of method m execution. Another way to see cap$_\mathbf{m}$ is thinking they are auxiliary objects created during method m execution that are not longer required by

```
void m0(int m) {                  void m1(int k) {
   for(c=1;c<= m;c++) {              for(i=1;i<=k;i++) {
      m1(c);                            A a = new A();
      B[] m2Arr=m2(2*m-c);              B[] captArr=m2(i);
   }                                 }
}                                 }
B[] m2(int n) {
   B[] arrB = new B[n];
   for(j=1;j<= n;j++) {
      B b=new B();
      C c=new c();
      arrB[j-1]=b;
   }
   return arrB;
}
```

Figure 4.10   Dynamic memory allocation example.

a caller of m. The amount of memory required to run a method m (denoted memRq$_m$) is composed by both cap$_m$ and ret$_m$. This is because in order to run m the system will require enough memory to allocate the objects that will be created by m and collected when it finishes (cap$_m$) and the objects allocated by m that live longer (ret$_m$). Nevertheless, we will see later that objects returned by methods tend to be captured by other methods in the call stack, meaning that eventually returned objects are considered when computing captured objects. Therefore, for convenience, we will consider only captured objects when computing memRq$_m$ and we will add ret$_m$ to the estimation only when m represents the application's *main* method.

Consider the example of Figure 4.10 and assume for simplicity that all objects are of size 1. Method m2 does not call any other methods. All allocations assigned to variable c can be captured by m2 since they are neither returned nor linked to parameters or static variables. The method returns arrB, which makes the array and objects assigned to variable b live longer than method m2. By describing the iteration space of the loop by $1 \leq j \leq n$, we can apply a parametric counting technique to count the number of visits to each new statement and approximate its consumption. We therefore have:

$$\text{cap}_{m2}(n) = \#\{ (j) \mid 1 \leq j \leq n\} = n$$

$$\text{ret}_{m2}(n) = n + \#\{ (j) \mid 1 \leq j \leq n\} = 2n$$

$$\text{memRq}_{m2}(n) = \text{cap}_{m2}(n) = n.$$

Note that we assume here that n is nonnegative.

Method m1 does call another method, namely m2, and it captures all the memory it allocates itself, as well as the memory that escaped (and was returned) from m2. Note that m2 is called several times within a loop which iteration space can be modeled by $(1 \leq i \leq k)$. Escaped objects are accumulative, meaning the space for the objects allocated and returned by m2 at *each* iteration has to be considered when computing m1 requirements.

To compute the amount of memory required to run m1 we need to consider its allocations (assigned to variable a) and the allocations performed by the calls m2. Method m2, in addition to returned objects, requires space for the i objects it captures. Since they are released at the end of its execution, it is only needed to consider the space for the call to m2 that consumes most auxiliary objects.

Therefore, we have:

$$\mathrm{ret}_{m1}(k) = 0$$

$$\mathrm{cap}_{m1}(k) = \#\{\,(i) \mid 1 \leq i \leq k\} + \sum_{1 \leq i \leq k} (\mathrm{ret}_{m2}(i))$$

$$= k + \sum_{1 \leq i \leq k} (2i) = k + k^2 + k = k^2 + 2k$$

$$\mathrm{memRq}_{m1}(k) = \mathrm{cap}_{m1}(k) + \max_{1 \leq i \leq k} \mathrm{memRq}_{m2}(i)$$

$$= k^2 + 2k + k = k^2 + 3k$$

where Bernstein expansion is used to compute $\max_{1 \leq i \leq k} \mathrm{memRq}_{m2}(i)$ and weighted counting is used to compute the sum. We again assume that $k \geq 0$.

Finally, notice that method m0 calls methods m1 and m2 within a loop. We have to consider the objects escaping from both m1 and m2:

$$\mathrm{ret}_{m0}(m) = 0$$

$$\mathrm{cap}_{m0}(m) = \sum_{1 \leq c \leq m} \mathrm{ret}_{m1}(c) + \mathrm{ret}_{m2}(2m - c) = \sum_{1 \leq c \leq m} 0 + 2(2m - c) = 3m^2 - m$$

Similarly, to compute the actual requirements for m0 we need to observe that the space reserved for objects from method m1 and m2 can be shared. Method m0 first calls to method m1 and, when it returns, then it calls m2. Objects required for m1 are no longer needed when it finished its execution, releasing space for objects allocated when executing m2. Thus, it is enough to consider only the maximum between requirements for m1 and m2:

$$\text{memRq}_{\text{mo}}(m)$$

$$= \text{cap}_{\text{mo}}(m) + \max\left(\max_{1 \leq c \leq m} \text{memRq}_{\text{m1}}(c), \max_{1 \leq c \leq m} \text{memRq}_{\text{m2}}(2m - c)\right)$$

$$= \text{cap}_{\text{mo}}(m) + \max\left(\max_{1 \leq c \leq m} c^2 + 3c, \max_{1 \leq c \leq m} 2m - c\right)$$

$$= 3m^2 - m + \max\left(m^2 + 3m, 2m - 1\right)$$

$$= 3m^2 - m + m^2 + 3m$$

$$= 4m^2 + 2m,$$

The general solution is

$$\text{memRq}_{\text{mo}}(m) = \begin{cases} 4m^2 + 2m & \text{if } m \geq 1 \\ 0 & \text{if } m \leq 0. \end{cases}$$

In general, using this notion of captured and returned (escaping) objects, the memory requirements of a method can be computed in terms of memory requirements of the methods it calls. Therefore:

$$\text{memRq}_{\text{m}}(\bar{p}) = \text{cap}_{\text{m}}(p_m) + \max_{m' \text{called by } m}\left(\max_{inv^m_{m'}} \text{memRq}_{\text{m}'}(\bar{a}^m_{m'})\right)$$

where $\bar{p}$ are methods m's formal parameters, $inv^m_{m'}$ is the iteration space defined in the call from method m to method m' and $\bar{a}^m_{m'}$ are the arguments used in that call.

This analysis can be automated using iscc. Notice there are many ways to solve this problem using the calculator (e.g., calling it several times for each symbolic operation, generating a set of operations per method). Here we propose one that makes extensive use of iscc's symbolic capabilities and generates all the results in one script.

- Analysis of method m2. Definition of iteration space for the loop:
  D_L_m2 := {[n] -> [j] : 1<=j<=n};
  The number of objects assigned respectively to c and b are computed counting the number of integer solutions.
  alloc_b := card D_L_m2; alloc_c := card D_L_m2;
  which is {[n] -> n : n>=1};

For the array we use its dimension:
```
alloc_bArr := {[n] -> n : n>=1};
```
Thus, the amount of objects captured, returned, and required are:
```
cap_m2 := alloc_c;
```
which is `{[n] ->  n : n >= 1 }`
```
ret_m2 := alloc_b + alloc_bArr;
```
which is `{[n] ->  2 * n : n >= 1 }`
```
memReq_m2:= cap_m2 +  {[n] -> max(0) };
```
which is `{[n] ->  max(n) : n >= 1}` (we convert `memReq_m2` to the
result of a maximization operation for the sake of compositionality)

- Analysis of method m1. The iteration space for the loop is:
```
D_L_m1 := {[k] -> [i] : 1<=i<=k};
```
Then, the numbers of objects assigned to a are:
```
alloc_a := card D_L_m1;
```
This relation represents the call `m2(i)` in the loop:
```
call_m1m2 := (domain_map D_L_m1)^-1 . { [[k]->[i]] -> [i] };
```
The objects returned from m2 are accumulated:
```
capRet_m2:=  call_m1m2 . ret_m2;
cap_m1 := alloc_a + capRet_m2;
```
which is `{[k] ->  (2 * k + k^2) : k >= 1 }`
```
ret_m1 := {[k] -> 0};
```
And the object required by m2 but collected by itself are maximized
(notice that `memReq` is a fold):
```
max_m1m2 := call_m1m2 . memReq_m2;
memReq_m1 := cap_m1 + max_m1m2;
```
which is `{[k] -> max((3 * k + k^2)) : k >= 1 }`

- Analysis of method m0. The iteration space for the loop is:
```
D_L_m0 := {[m] -> [c] : 1 <= c <= m };
```
These are the relations for the calls `m1(c)` and `m2(2m-c)` in the loop.
```
call_m0m1 := (domain_map D_L_m0)^-1 . { [[m]->[c]] -> [c] };
call_m0m2 := (domain_map D_L_m0)^-1 . { [[m]->[c]] -> [2m-c]};
```
Then:

```
capRet_m1:=  call_m0m1 . ret_m1;
capRet_m2:=  call_m0m2 . ret_m2;
max_m0m1 := call_m0m1 . memReq_m1;
max_m0m2 := call_m0m2 . memReq_m2;
```

Thus, the amount of objects captured, returned, and required are:
```
retM0 := {[m] -> 0};
cap_m0 := capRet_m1 + capRet_m2;
```
which is `{[m] -> (-m + 3 * m^2) : m >= 1}`
```
memReq_m0 := cap_m0 + (max_m0m1 . max_m0m2);
```
which is `{[m] ->  max((2 * m + 4 * m^2)) : m >= 1}` as we manually computed.

4.4.5   Software implementation

The computation of lower and upper bounds on arbitrary multivariate piece-wise step-polynomials, defined over linearly parametrized convex polytopes, has been implemented in the *Integer Set Library (isl)* (Section 4.2.3.5). This includes the computation of the parametric Bernstein coefficients of the polynomials and a procedure reducing the number of potential results. These latter computations were initially implemented as a standalone application named *bernstein* [16].

4.5   Conclusion

Memory requirement evaluation of applications is a major issue in the design of computer systems, and specifically in the case of embedded systems. More-over, evaluation results, when parametrized, can cover all possible execution configurations from only one unique program analysis process. We have shown for several application examples that this problem can often consist in max-imizing a parametric and multivariate polynomial defined over a parametric convex domain. We proposed the use of some advanced mathematical tools to compute accurate bounds for such polynomials, and even exact bounds in some cases. All these tools have been implemented and are freely available.

References

1. David Wonnacott. *Constraint-Based Array Dependence Analysis.* PhD thesis, University of Maryland, August 1995.

2. Rachid Seghir and Vincent Loechner. Memory optimization by counting points in integer transformations of parametric polytopes. In *Proceed-ings of the International Conference on Compilers, Architectures, and Synthesis for Embedded Systems, CASES 2006, Seoul, Korea*, October 2006.

3. Sven Verdoolaege, Kevin M. Woods, Maurice Bruynooghe, and Ronald Cools. Computation and manipulation of enumerators of integer pro-jections of parametric polytopes. Report CW 392, Dept. of Computer Science, K.U.Leuven, Leuven, Belgium, 2005.

4. Bernard Boigelot and Louis Latour. Counting the solutions of Pres-burger equations without enumerating them. *Theoretical Computer Sci-ence*, 313(1):17–29, February 2004.

5. Erin Parker and Siddhartha Chatterjee. An automata-theoretic algorithm for counting solutions to Presburger formulas. In *Compiler Construction 2004*, volume 2985 of *Lecture Notes in Computer Science*, pages 104–119, Berlin, April 2004. Springer-Verlag.

6. Jesús A. De Loera, Raymond Hemmecke, Jeremiah Tauzer, and Ruriko Yoshida. Effective lattice point counting in rational convex polytopes. *The Journal of Symbolic Computation*, 38(4):1273–1302, 2004.

7. William Pugh. Counting solutions to Presburger formulas: How and why. In *SIGPLAN Conference on Programming Language Design and Implementation (PLDI'94)*, pages 121–134, 1994.

8. Philippe Clauss. Counting solutions to linear and nonlinear constraints through Ehrhart polynomials: Applications to analyze and transform scientific programs. In *International Conference on Supercomputing*, pages 278–285, 1996.

9. A. Barvinok and J. Pommersheim. An algorithmic theory of lattice points in polyhedra. *New Perspectives in Algebraic Combinatorics*, 38:91–147, 1999.

10. S. Verdoolaege, R. Seghir, K. Beyls, V. Loechner, and M. Bruynooghe. Counting integer points in parametric polytopes using Barvinok's rational functions. *Algorithmica*, 48(1):37–66, June 2007.

11. Sven Verdoolaege, Kristof Beyls, Maurice Bruynooghe, and Francky Catthoor. Experiences with enumeration of integer projections of parametric polytopes. In Rastislav Bodík, editor, *Proceedings of 14th International Conference on Compiler Construction, Edinburgh, Scotland*, volume 3443 of *Lecture Notes in Computer Science*, pages 91–105, Berlin/Heidelberg, 2005. Springer.

12. Matthias Köppe, Sven Verdoolaege, and Kevin M. Woods. An implementation of the Barvinok–Woods integer projection algorithm. In Matthias Beck and Thomas Stoll, editors, *The 2008 International Conference on Information Theory and Statistical Learning*, July 2008.

13. Sven Verdoolaege and Maurice Bruynooghe. Algorithms for weighted counting over parametric polytopes: A survey and a practical comparison. In Matthias Beck and Thomas Stoll, editors, *The 2008 International Conference on Information Theory and Statistical Learning*, July 2008.

14. Kristof Beyls and Erik D'Hollander. Generating cache hints for improved program efficiency. *Journal of Systems Architecture*, 51(4):223–250, 2005.

15. Ph. Clauss and I. Tchoupaeva. A symbolic approach to Bernstein expansion for program analysis and optimization. In Evelyn Duesterwald,

editor, *13th International Conference on Compiler Construction, CC 2004*, volume 2985 of *LNCS*, pages 120–133. Springer, April 2004.

16. Philippe Clauss, Federico Javier Fernández, Diego Garbervetsky, and Sven Verdoolaege. Symbolic polynomial maximization over convex sets and its application to memory requirement estimation. *IEEE Transactions on Very Large Scale Integration (VLSI) Systems*, 17(8):983–996, 2009.

17. S. Bernstein. *Collected Works*, volume 1. USSR Academy of Sciences, 1952.

18. S. Bernstein. *Collected Works*, volume 2. USSR Academy of Sciences, 1954.

19. M. Zettler and J. Garloff. Robustness analysis of polynomials with polynomial parameter dependency using Bernstein expansion. *IEEE Transactions on Automatic Control*, 43(3):425–431, 1998.

20. Vadim Maslov and William Pugh. Simplifying polynomial constraints over integers to make dependence analysis more precise. In *CONPAR 94 - VAPP VI, Int. Conf. on Parallel and Vector Processing*, September 1994.

21. William Blume and Rudolf Eigenmann. Symbolic range propagation. In *IPPS '95: Proceedings of the 9th International Symposium on Parallel Processing*, pages 357–363, Washington, DC, USA, 1995. IEEE Computer Society.

22. R. A. Van Engelen, K. Gallivan, and B. Walsh. Parametric timing estimation with the Newton-Gregory formulae. *Journal of Concurrency and Computation: Practice and Experience*, 18(10):1434–1464, September 2006.

23. Jakob Berchtold and Adrian Bowyer. Robust arithmetic for multivariate Bernstein-form polynomials. *Computer-Aided Design*, 32:681–689, 2000.

24. R.T. Farouki and V.T. Rajan. On the numerical condition of polynomials in Bernstein form. *Computer Aided Geometric Design*, 4(3):191–216, 1987.

25. J. Garloff. Application of Bernstein expansion to the solution of control problems. In J. Vehi and M. A. Sainz, editors, *Proceedings of MISC'99 - Workshop on Applications of Interval Analysis to Systems and Control*, pages 421–430. University of Girona, Girona (Spain), Springer Netherlands, 1999.

26. J. Garloff and B. Graf. *The Use of Symbolic Methods in Control System Analysis and Design*, Solving Strict Polynomial Inequalities by Bernstein Expansion, pages 339–352. Institution of Electrical Engineers (IEE), London, 1999.

27. R. Martin, H. Shou, I. Voiculescu, A. Bowyer, and G. Wang. Comparison of interval methods for plotting algebraic curves. *Computer Aided Geometric Design*, 19:553–587, 2002.

28. V. Stahl. *Interval Methods for Bounding the Range of Polynomials and Solving Systems of Nonlinear Equations*. PhD thesis, Johannes Kepler University Linz, Austria, 1995.

29. G. Farin. *Curves and Surfaces in Computer Aided Geometric Design*. Academic Press, San Diego, 1993.

30. Ph. Clauss and V. Loechner. Parametric analysis of polyhedral iteration spaces. *Journal of VLSI Signal Processing*, volume 19(2), Kluwer Academic, 1998.

31. Víctor Braberman, Federico Fernández, Diego Garbervetsky, and Sergio Yovine. Parametric prediction of heap memory requirements, *Proceedings of the 7th International Symposium on Memory Management*, ISMM .08, pp. 141–150, ACM, June 2008.

32. A. Salcianu and M. Rinard. Pointer and escape analysis for multi-threaded programs. In *PPoPP '01: Proceedings of the Eighth ACM SIGPLAN Symposium on Principles and Practices of Parallel Programming*, pages 12–23. ACM Press, 2001.

33. B. Blanchet. Escape analysis for object-oriented languages: application to Java. In *OOPSLA '99: Proceedings of the 14th ACM SIGPLAN conference on Object-oriented programming, systems, languages, and applications*, pages 20–34. ACM Press, 1999.

34. Sigmund Cherem and Radu Rugina. Region analysis and transformation for java programs. In *ISMM '04: Proceedings of the 4th International Symposium on Memory Management*, pages 85–96, New York, NY, USA, 2004. ACM Press.

35. Diego Garbervetsky, Chaker Nakhli, Sergio Yovine, and Hichem Zorgati. Program instrumentation and run-time analysis of scoped memory in java. In *RV 2004: International Workshop on Runtime Verification*, volume 113 of *ENTCS*, pages 105–121, Barcelona, Spain, April 2004. ETAPS, Elsevier.

Chapter 5

Storage Allocation for Streaming-Based Register File

Praveen Raghavan
SSET, IMEC vzw, Heverlee, Belgium

Francky Catthoor
SSET, IMEC vzw, Leuven, Belgium

Contents

5.1 Stream Register File: Why and How

5.1.1 Register files and their nonuniformal access patterns

Register files have been known to be a notorious power consuming part of a processor architecture. It was already shown in various works [1–3] that there is a need for a comprehensive treatment of register files such that their power consumption is reduced while still meeting all the realtime requirements of an application. Multi-ported data register files (RF) are one of the most power hungry parts of any processor, especially very long instruction word processors (VLIWs) [4,5]. On average every operation requires three accesses (two reads and one write) to the RF, which make them a very active part of the processor. Current architectures try to achieve a high performance by exploiting parallelism, and therefore perform multiple operations per cycle (e.g., instruction level parallelism or ILP, as used in VLIW processors). This quickly results in a large port requirement for the register file, which is mostly implemented as a single/centralized or distributed large multi-ported register file. A high number of ports has a strong negative impact on the energy efficiency of register files as well as facing strong performance constraints for design. Traditionally, this problem is addressed through various clustering techniques [5] that partition (or bank) the RF. Data can then only be passed from one partition to another through intercluster communication [6,7]. However, as partitions get smaller the cost of intercluster copies quickly grows. In addition, the resulting register files are still multi-ported. For high energy efficiency, it is clearly preferable that the register cells be single ported [8].

Broadly speaking, from the application perspective, variables in an application can be of different types:

- Dynamic data types
- Input/output arrays *with spatial* locality in access

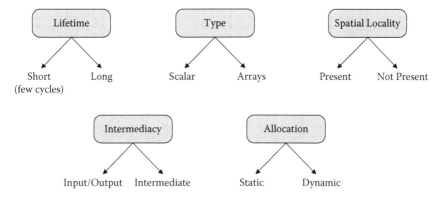

Figure 5.1 Different application variable types.

- Input/output arrays *without spatial* locality
- Intermediate arrays *with spatial locality* (short[1] and long lifetime)
- Intermediate arrays *without spatial locality* (short and long)
- Low lifetime intermediate scalar variables
- High lifetime intermediate scalar variables

These different variables are also shown in Figure 5.1. Dynamic data types are not in the focus of this chapter, and it is assumed that they can be converted into static variables by the time the compiler has to deal with them. This can be achieved by grouping the data into pools under the control of a dynamic memory allocator [9]. However, given the embedded application domain and even in some general purpose domains, there is a set of variables (input/output or intermediate) that exhibit spatial locality. This nature is often not exploited by typical state of the art register file architectures. In contrast there is a small set of variables that exhibit poor spatial locality need to be accessed in random order. This requires a typical register file that can be addressed and written to in irregular order. For an efficient solution each of these sets of variables needs to be treated effectively and their properties need to be exploited. In case the variables do not exhibit spatial locality or their data layout is improper, it is assumed that appropriate data-layout transformations have been done to make the spatial locality exploitable.

Besides looking at the access part of the variables in application, the designer also needs to look into the physical design aspect of the processor architecture. While in most cases during the design phase it may be too early to take into account the physical layout aspect, it is still important that the architecture is defined in a "layout-friendly" way. Given the increasing wire capacitance due to scaling [10–13], it is important that even in the early design

---

[1]Short lifetime implies a few cycles such that it can be handled via the processor's pipeline/forwarding network.

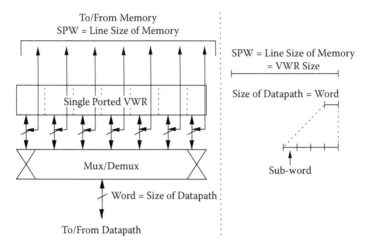

Figure 5.2   Very wide register organization.

phase the cost of wiring is taken into account. However, even if interconnect does not scale worse than logic, the proposed register file would still be efficient but the gains of using such a register file may be lower.

This chapter presents a comprehensive technique for organizing the register file such that all the different types of variables are handled in an efficient way and discusses how to map code efficiently on such foreground memory architectures. More specifically, this chapter presents a novel asymmetric register file organization called *very wide register* or VWR. This VWR, together with its interface to the wide memory, achieves a significantly higher energy efficiency than conventional organizations and forms an efficient and layout-friendly solution for arrays with spatial locality. The proposed register file or foreground memory organization is shown in Figure 5.2. Three aspects are important in the proposed organization: the interface to the memory, single ported cells, and the interface to the datapath. The interface of this foreground memory organization is asymmetric: *wide* toward the memory and *narrower* toward the datapath. The wide interface enables the exploitation the locality of access of applications through wide loads from the memory to the foreground memories (registers). At the same time the datapath is able to access words of a smaller width for the actual computations (further details in Section 5.2). Internally each of these words can consist of sub-words [for a single instruction multiple data (SIMD) datapath]. Furthermore for such foreground memory architectures, an efficient mapping strategy is necessary to obtain the gains of this architectural feature.

5.1.2   Very wide register: a streaming foreground memory architecture

The architectural motivation for the proposed very wide register architecture is derived from various parts of the processor. The following subsections introduce a streaming foreground memory architecture called very wide register. While various other variants exist for streaming foreground memory like Stream [14,15] etc., the following subsections introduce a particular micro-architecture; the compilation and mapping methodology introduced later in this chapter is applicable to other architectures with minor modifications.

5.1.2.1 Data (background) memory organization and interface

Energy consumption in memories can be reduced by improving one or more of three aspects: the memory design (circuit level), the mapping of data onto the memory, and the memory organization (and its interface). This section discusses the background (L1 data) memory organization. A detailed energy breakdown of an SRAM-based scratchpad shows that for a typical size for the level-1 data memory (e.g., 64 KB) about half of the energy is spent in the decoder and the word-line activation [16,17]. The other half is spent in the actual storage cells and in the sense amplifiers. The decode cost is the price that is paid for being able to access words in any given order. The energy consumption in the memory organization can be optimized further by performing as few decodings as possible by reading out more data for every decode. In the embedded systems domain this can be achieved by aggressively exploiting the available spatial locality of data. While this spatial locality is used for DMA transfers for L2 and L1, cache optimization etc., it is not further exploited between the transfer for L1 memory and the register file.

In the proposed architecture (see Figure 5.3), spatial locality of data in the L1 memory is exploited to reduce the decoding overhead. The row address (*Row Addr* in Figure 5.3) selects the desired row in the memory through the pre-decoder. The sense amplifiers and pre-charge lines are only activated for the words that are needed and only these will consume energy and are read out. Figure 5.3 also shows the address organization that has to be provided for such a memory. To be able to handle partial rows (less optimal for energy, but more flexible), the full address contains two additional fields: *Position* decides at which word the read-out will start, while *No. Words* decides the number of words that have to be read out. Hence, at most, a complete row and at least one word of the SRAM can be read out and will be transferred from the scratchpad to the VWR registers. The scratchpad can be internally partitioned or banked and the proposed technique can be applied on top of the banked structure.

This architecture is compatible with almost all existing SRAM generators (e.g., Artisan, Virage), but in actual instantiations, such a wide interface may not yet be available. If the used design library does not contain such a wide memory, it can be composed from multiple narrower memories by connecting them in parallel, but the overhead due to extra decoding would not allow the gains to be maximal. For maximal gains it would be necessary that many words (either from the different banks or the same bank) share the same decode circuitry. However, a case study detailed in [1,18,19] shows that even

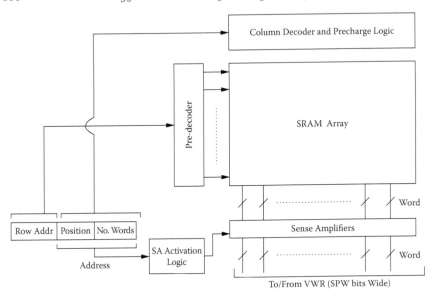

Figure 5.3   VWR and scratchpad organization.

with connecting multiple narrow memories in parallel, the gains are still substantial. Concatenating multiple narrower memories still results in gains at the processor level as wider loads/stores implies fewer address computations at run-time, fewer load/store instructions and fewer decodes required even for the register file.

5.1.2.2   Foreground memory organization

The proposed register file has single ported register cells as shown in Figure 5.2. This register organization is called *very wide register* (VWR). The VWR has asymmetric interfaces: a wide interface toward the memory and a narrow interface to the datapath. Every VWR is as wide as the line size of the scratch pad or background memory, and *complete* or *partial lines* can be read from the scratchpad into these VWRs. The VWRs have only a post-decode circuit, which consists of a multiplexer/De-multiplexer (Mux/Demux). This circuit selects the words that will be read from or written to the VWR. Each VWR has its own Mux and Demux, as shown in Figure 5.2. The controls of the Mux and Demux on which register is to be accessed is derived from the instructions. Because of the single-ported cell design, the read and write access of the registers to the scratchpad memory and access to the datapath cannot happen in parallel. The VWR is part of the datapath pipeline with a single cycle access similar to register files.

Since the VWR is single ported, it is important that data that are needed in the same cycle/operation are placed in different VWRs. Arrays are mapped on the VWR during a separate mapping process (explained further in

Section 5.1.3). Scalar data like scalar constants, iterators, and addresses etc. can be mapped to a separate scalar register file (SRF) in order not to pollute the data in the VWR with intermediate results.

The interface of the VWR to the memory is as wide as a complete VWR, which is of the same width as the memory and the bus. Therefore the load/store unit is also different. It is capable of loading or storing complete (or partial[2]) lines from the scratchpad to the VWRs. Section 5.1.3 shows an example on how the load/store operations are performed between the memory and the VWR. A more clear example of which data are to be loaded/stored and the scheduling and data layout of this data are explained in detail in Section 5.3.

To analytically show the gains, assume that M words which are stored in the memory need to be read and operated on. Assume: N words per line in the memory and also N words in one VWR, where $M \geq N^3$. Conventional register file would require: *M memory pre-decodes + M memory cell activations + M memory post-decodes/column decode + M writes to the register file*, where as in case of the VWR: *M/N memory pre-decodes + M memory cell activations + M memory post-decodes/column decode + M/N wide VWR write*. The ratio between these two can be of the order of 2 to 5, excluding the gains in the instruction memory.

Due to the split interfaces of the register file, other optimizations can be exploited at the physical design level. During *placement and routing*, the cells of the VWR are aligned with pitch of the sense amplifiers of the memory to reduce the amount of interconnect and the related energy. This enables clear direct routing between the wide memory and the VWR without much interconnect overhead. The same optimization cannot be done in the case of a traditional register file, because of the fact that the memory and datapath interfaces of the register file are shared and due to the multi-ported nature of these register files. For a more detailed experimental setup and analysis of placement and routing, the reader is referred to [20].

5.1.2.3 Connectivity between VWR and datapath

The foreground memory consisting of VWRs and SRF can be connected to any datapath organization (consisting of multipliers, adders, accumulators, shifters, comparators, branch-unit etc.) by replacing the register file. Figure 5.4 shows the connectivity between the VWRs, SRF, and the datapath. The datapath may or may not support sub-word parallelism similar to state of the art processor engines like Altivec, MMX, or SSE2.

Once the appropriate data are available in the foreground memory, the decoded instruction steers the read and write operations from and to the foreground memory and the datapath. At a given cycle, one word (consisting

---

[2]Multiple contiguous words in the same row of the SRAM.

[3]Note that this is not a necessary condition, but gains are higher when M is greater than N.

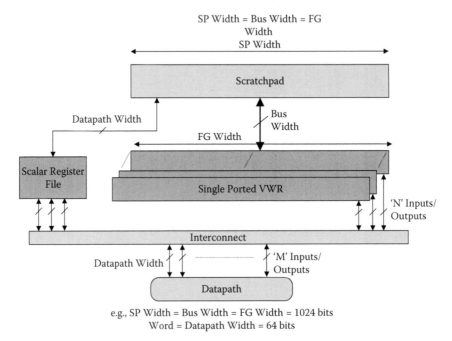

Figure 5.4 VWR and scalar register file connectivity to the datapath and the L1 memory.

of sub-words) will be read from the VWR to the datapath and the result will be written back to a VWR. The foreground VWR organization along with the datapath is shown in Figure 5.4.

In case the processor is designed with a higher ILP (i.e., multiple instructions can be issued in the same cycle), more VWRs are needed. Given that each VWR is single ported, the number of VWRs needed scales linearly with the number of issue slots. Around three VWRs are needed per issue slot. In case not all issues slots write back to the VWR in the same cycle, the number of VWRs per slot can be lower. This would imply that the VWRs are shared over the different issue slots. More complex schemes can also be imagined where parts of the same word are used as the two operands for an operation. It is expected that the number of VWRs needed would be as few as needed since it is more efficient to first exploit the DLP as much as possible and then use ILP to meet the real-time requirements for optimal energy efficiency as well as performance.

5.1.3   VWR operation

Figure 5.5 presents the operation of the VWR on simplified example code, assuming a 32-bit processor datapath and a 256-bit line-size. This means that one VWR at any given point in time can store eight words. For the

Modified Code with VWR:

```
for ( i =0;  i <8;  i++ ) {
    LOAD_row VWR2,  b [ i *8 ];
    LOAD_row VWR1,  c [ i *8 ];
    for ( j =0;  j <8;  i++ ) {
        VWR3[ j ]  = VWR2[ j ]  * VWR1[ j ];
    }
    STORE_row VWR3,  a [ i *8 ];
}
```

Original Code:

```
for ( i =0;  i <64;  i++ ) {
    a [ i ]  = b [ i ]  * c [ i ];
}
```

Figure 5.5   Rewritten C code with very wide registers and load/store operations.

sake of simplicity no sub-word parallelism or vector parallelism is used in this example, of which Figure 5.5 shows the C code (with intrinsics). The asymmetric interface of the VWR results in the following mode of operation: a complete row of the scratchpad is copied to the VWR at once, using a *LOAD_row*. In this example, operands from arrays b and c are allocated to two different rows in the scratchpad and to two different VWRs (VWR 1 and 2). Therefore two rows are loaded. In the next phase these operands are consumed one by one by the inner loop and the results are stored in a third VWR (VWR 3). Only when all computations of the inner loop are finished, the complete VWR 3 is stored back to the scratchpad.

Figure 5.6 shows an illustration of the data allocation in the scratchpad, as well as the data layout in the three VWR registers at the end of the first iteration of the outer loop (i loop). The datapath organization can be generic (single/multi-issue etc.). In the first iteration (i = 0), a row of data b[0]-b[7] is loaded onto *VWR* and a row of data c[0]-c[7] is loaded onto *VWR1*. At each iteration of the the inner loop j, one element of b and one element of c from VWRs 2 and 1, respectively, are consumed to produce one element of a in VWR3. At the end of the inner loop j, produced data a[0]-a[7] in *VWR3* can be stored back onto the L1 background memory. At the beginning of the next iteration (i = 1), the next set of data (b[8]-b[15] and c[8]-c[15]) can be loaded and consumed in the inner j loop and so on till the end. In this example, there is no need for an epilogue as the number of loop iterations is a multiple of the number of locations in the VWR. If this is not the case, a smaller epilogue loop may be needed for the remaining elements.

Because in the embedded signal processing systems domain (including the benchmarks used here), most data are streaming and continuous in the foreground memory, it is reasonable to assume that most of the time complete lines of the scratchpad can be loaded with relevant data. It is still possible to load partial rows if not enough independent data words can be found to fill a complete row (for instance at the end of a loop). Also any sort of buffering can be done in the higher level memories such that when the data come to the L1 data memory and the foreground memory, even non-streaming access can be performed.

Currently the allocation of arrays to the VWR is shown to be done using *intrinsics* (like *LOAD_row*, *STORE_row*, etc.) as the compilation has not been

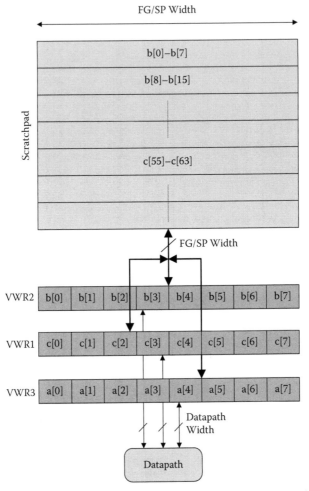

Figure 5.6 Illustration of the data-layout for Figure 5.5 in the different VWR registers and the scratchpad memory.

automated. However, a technique for compiling C code onto VWR is detailed in Section 5.3. For a more detailed overview of the gains possible with such a streaming based register file the reader is referred to [3,18,19].

5.2   Model for Compilation on Stream Register File

Section 5.1 showed that the VWR-based architecture is an efficient alternative to the traditional-register-file-based architecture. The section also showed that the VWR-based architecture is more power efficient, as well as performance efficient, in applications that have spatial locality. However, the complexity of the architecture was pushed to the compiler, which has to ensure that an efficient data layout of the application is exploited both in the L1 SRAM memory and the VWR foreground memory.

Furthermore, current embedded designers prefer the use of languages like C to program. Programming the VWR therefore requires a compiler to compile from the C language onto the VWR architecture to obtain an efficient data layout both in the SRAM L1 data memory and in the VWR register files. Also the compiler has to be flexible enough to cover the wide architecture space offered by the VWR-based architectures. The compiler also needs to be scalable to large applications, which are quite common in today's embedded application space.

This section introduces a formal model that can model that application/ program and the storage (layout aspect) of the program's data set as well as the execution order (scheduling aspect) of the different accesses. It also introduces the necessary and sufficient conditions for correct execution of the application/program. This formal model is used (in Section 5.3) to compile efficiently onto the VWR based architecture.

5.2.1 Basics of the model

This section presents a quick overview of the basics of the geometrical or polyhedral model, which is well known in the compiler domain. Since in literature different terminologies and representations have been used, the basics are presented again to ensure the terminology is consistent.

5.2.1.1 Iteration domain

The term *iteration domain* had been introduced to model the potential parallelism available between statements in a program. The concept of iteration domain has already been introduced in various previous research works [21]. It has also been called *iteration space* or *index space* in other research works.

Iteration domain is the mathematical description of the execution of different statements in the program with respect to the loop iterators in the program. It is a geometrical domain where each point in the domain corresponds to an instance of occurrence of a statement in the program. Each statement has its own iteration domain. A small example of an iteration domain is illustrated below:

Example 5.1

```
    for ( i = 0; i < 5; i++ )
        for ( j = 0; j < 5; j++ )
S1:       A[i][j] = (B[i][j] + C[i]);
```

The dimensions of the iteration domain of statement S1 are denoted by i and j. The constraints on these dimensions can be obtained from the loop

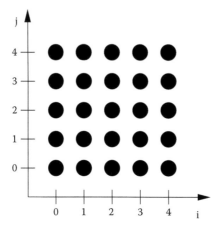

Figure 5.7   Graphical representation of iteration domain of statement S1 in Example 5.1

iterators (also i and j): $0 \leq i \leq 4$ and $0 \leq j \leq 4$. Therefore we have the iteration domain of statement S1 (denoted by $\mathbf{D}_1^{\text{iter}}$) to be as follows:

$$\mathbf{D}_1^{\text{iter}} = \{\ [i,j]\ |\ 0 \leq i \leq 4\ \wedge\ 0 \leq j \leq 4\ \wedge\ [i,j] \in \mathbb{Z}^2\ \}$$

A graphical representation of this domain is given in Figure 5.7.

In the previous example, the iteration domain for statement S1 was square; however, that may not always be the case and the iteration domain can be of different shapes and sizes. Further, the iteration domain can not always be represented by using constraints of the iterators, and may require insertion of extra dimensions also called *wild cards* [22]. While such wild card dimensions are complementary to the model, they are not detailed any further. For more details on extension of the iteration domain with wild card or auxiliary dimensions the reader is referred to [22–24].

5.2.1.2   Variable, definition, and operand domain

In the domain of parallelizing compilers, iteration domain was sufficient to study the parallelism availably; however, in the context of memory optimizations like IMEC's ATOMIUM [25] work, as well as other memory optimization works like [26–28], it is also important to study how the program variables, especially array variables, are accessed and created. This has necessitated the need for other definitions such as *variable domain*, *definition domain*, and *operation domain*.

In most programs, different array variables are read from and written to during the execution of the program. *Variable domain* corresponds to the geometrical description of each of these individual array variables in the program. These arrays can be multidimensional structures that can be addressed in different ways. The array can be of any type of variable (including a **struct** or a SIMDized word or a vector). Each point in this geometrical domain corresponds to a single unique variable in this array.

When each statement of the program is executed, none or one or more array variables may be read from or written to. The *definition* and *operand* domains of a statement represent the complete domain of the variables accessed during *all* executions of that statement. Each point in this domain corresponds to the one variable accessed in the domain (can be read or write) during a particular execution of a statement. The *definition* domain corresponds to the write access (or in other words when a variable is "defined") and the *operand* domain corresponds to a read access (or in other words when a variable forms an operand in a statement).

A relation (mapping) exists between the different statements in a program and the different array variables accessed. The mapping between the iteration domain of a statement and each of the definition and operand domains is called *definition mapping* and *operand mapping*, respectively. The following examples better illustrate each of these definitions:

Example 5.2

```
int A[5][5], B[5][5];
int C[5];
for ( i = 0; i < 5; i++ )
    for ( j = 0; j < 5; j++ )
S1:     A[i][j] = (B[i][j] + C[i]);;
```

The variable domains of arrays A, B, and C are denoted by $\mathbf{D}_A^{\text{var}}$, $\mathbf{D}_B^{\text{var}}$, and $\mathbf{D}_C^{\text{var}}$, respectively. The boundaries of the domains can be extracted from the declarations:

$$\mathbf{D}_A^{\text{var}} = \{ \ [a_1, a_2] \mid 0 \le a_1 \le 4 \ \wedge \ 0 \le a_2 \le 4 \ \wedge \ [a_1, a_2] \in \mathbb{Z}^2 \ \}$$
$$\mathbf{D}_B^{\text{var}} = \{ \ [b_1, b_2] \mid 0 \le b_1 \le 4 \ \wedge \ 0 \le b_2 \le 4 \ \wedge \ [b_1, b_2] \in \mathbb{Z}^2 \ \}$$
$$\mathbf{D}_C^{\text{var}} = \{ \ c \mid 0 \le c \le 4 \ \wedge \ c \in \mathbb{Z} \ \}$$

Next, given the iteration domain of statement S1 and the index expressions of the array accesses, we can extract the descriptions of the definition and operand mappings (denoted by $\mathbf{M}_{11A}^{\text{def}}$ and $\mathbf{M}_{11B}^{\text{oper}}$, respectively) and definition and operand domains (denoted by $\mathbf{D}_{11A}^{\text{def}}$ and $\mathbf{D}_{11B}^{\text{oper}}$, respectively), which are the result of applying the respective mappings to the iteration domain:

$$\mathbf{D}_1^{\text{iter}} = \{ \ [i, j] \mid 0 \le i \le 4 \ \wedge \ 0 \le j \le 4 \ \wedge \ [i, j] \in \mathbb{Z}^2 \ \}$$
$$\mathbf{M}_{11A}^{\text{def}} = \{ \ [i, j] \rightarrow [a_1, a_2] \mid a_1 = i \ \wedge \ a_2 = j \ \wedge \ [a_1, a_2] \in \mathbb{Z}^2 \ \}$$
$$\begin{aligned}
\mathbf{D}_{11A}^{\text{def}} &= \mathbf{M}_{11A}^{\text{def}}(\mathbf{D}_1^{\text{iter}}) \\
&= \{ \ [a_1, a_2] \mid \exists \ [i, j] \in \mathbf{D}_1^{\text{iter}} \ \text{s.t.} \ a_1 = i \ \wedge \ a_2 = j \ \wedge \ [a_1, a_2] \in \mathbb{Z}^2 \ \} \\
&= \{ \ [a_1, a_2] \mid \exists \ [i, j] \in \mathbb{Z}^2 \ \text{s.t.} \ a_1 = i \ \wedge \ a_2 = j \ \wedge \\
&\quad\quad 0 \le i \le 4 \ \wedge \ 0 \le j \le 4 \ \wedge \ [a_1, a_2] \in \mathbb{Z}^2 \ \}
\end{aligned}$$
$$\mathbf{M}_{11B}^{\text{oper}} = \{ \ [i, j] \rightarrow [b_1, b_2] \mid b_1 = i \ \wedge \ b_2 = j \ \wedge \ [b_1, b_2] \in \mathbb{Z}^2 \ \}$$

$$\mathbf{D}^{\text{oper}}_{11B} = \mathbf{M}^{\text{oper}}_{11B}(\mathbf{D}^{\text{iter}}_1)$$
$$= \{\ [b_1, b_2] \mid \exists\, [i, j] \in \mathbf{D}^{\text{iter}}_1 \text{ s.t. } b_1 = i \ \wedge\ b_2 = j \ \wedge\ [b_1, b_2] \in \mathbb{Z}^2\ \}$$
$$= \{\ [b_1, b_2] \mid \exists\, [i, j] \in \mathbb{Z}^2 \text{ s.t. } b_1 = i \ \wedge\ b_2 = j \ \wedge$$
$$0 \le i \le 4 \ \wedge\ 0 \le j \le 4 \ \wedge\ [b_1, b_2] \in \mathbb{Z}^2\ \}$$
$$\mathbf{M}^{\text{oper}}_{12C} = \{\ [i, j] \to c \mid c = i \ \wedge\ c \in \mathbb{Z}\ \}$$
$$\mathbf{D}^{\text{oper}}_{12C} = \mathbf{M}^{\text{oper}}_{12C}(\mathbf{D}^{\text{iter}}_1)$$
$$= \{\ c \mid \exists\, [i, j] \in \mathbf{D}^{\text{iter}}_1 \text{ s.t. } c = i \ \wedge\ c \in \mathbb{Z}\ \}$$
$$= \{\ c \mid \exists\, [i, j] \in \mathbb{Z}^2 \text{ s.t. } c = i \ \wedge\ 0 \le i \le 4 \ \wedge\ 0 \le j \le 4 \ \wedge\ c \in \mathbb{Z}\ \}$$

The first index in the subscripts of $\mathbf{D}^{\text{def}}_{11A}$ and $\mathbf{D}^{\text{oper}}_{11B}$ refers to statement S1. The second index refers to the position of the definition or operand, respectively[4]), while the third index refers to the array of the domain to which it belongs.

5.2.1.3 Dealing with non-affine indices

The behavior of some programs is dependent on the data values of the program that are only known at runtime/execution. This is referred to in the compiler domain as non-manifest behavior, which implies that the data and/or control of the program does not manifest itself at compile time.

One of the simplest ways, to model the non-manifest behavior is the use of symbolic constants and symbolic expressions. [24,29–31] present such models and show how non-manifest behavior can be modeled. In this thesis, it is assumed that non-manifest behavior can be dealt with in a similar way. For dealing with non-affine cases that include modulo, division operation etc., piece-wise linear approximations can be made as shown in [32]. For simplicity's sake, all the examples used in the rest of the thesis are affine and manifest; however the modeling and compilation capability is not limited by this.

5.2.2 Order

Iteration, variable, definition, and operand domains form the basics of the model. However they do not represent the "time" and "space" notion that is needed for scheduling and allocation (and layout) of data in both the scratchpad and the VWR. For this purpose we introduce two concepts: execution order, which is the relative order of execution of statement instances; and storage order, which is the layout of the elements of an array.

---

[4]There can be multiple definitions and/or operands in the same statement; for example C is the second operand in statement S1.

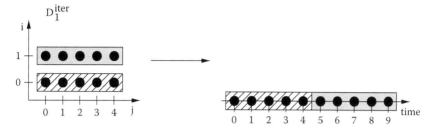

Figure 5.8 Graphical representation of iteration domain of statement S1 and its view in the time domain for Example 5.3

5.2.2.1 Execution order

To explain the term execution order in more detail consider the following example:

Example 5.3

```
int A[5][5], B[5][5];
int C[5];
for ( i = 0; i < 2; i++ )
    for ( j = 0; j < 5; j++ )
S1:     A[i][j] = (B[i][j] + C[i]);
```

Figure 5.8 shows both the $\mathbf{D}_1^{\text{iter}}$ and the corresponding view of the iteration domain in space. This view can also be represented in terms of the time or execution order equation for the statement S1 as follows:

$$O_1^{\text{time}}(i, j) = 5 * i + j$$

Another important fact to note is that the execution order can be a relative order instead of an absolute order. The absolute order is often not needed and it is also quite complex to compute accurately. As the execution order is relative and generic, it can be modeled the same way for both the VWR and the SRAM (L1 memory).

5.2.2.2 Storage order

Besides the representation of the iteration domain of a statement in time, the variable domain of each array can also be represented in space. This is known as the *storage order*. Storage order is needed to reason with the data layout of the variables in the storage elements (both memories as well as VWRs). The storage order usually has a linear (or piece-wise linear) relation between the storage address and the real memory address.

The following example illustrates both the storage order and the execution order in a formal way:

Example 5.4

```
int A[10][20], B[10][10];
int C[10];
for ( i = 0; i < 2; i++ )
    for ( j = 0; j < 5; j++ )
S1:     A[i][j] = (B[i][j] + C[i]);
```

As explained in the previous example, the execution order of statement S1 is as follows:

$$O_1^{\texttt{time}}(i,j) = 5 * i + j$$

For the arrays A, B, and C, the storage address can be written as functions of their dimensions of their respective variable domains:

$$O_A^{\texttt{addr}}(a_1, a_2, N) = N * a_1 + a_2$$
$$N = 10$$
$$O_B^{\texttt{addr}}(b_1, b_2, N) = N * b_1 + b_2$$
$$N = 10$$
$$O_C^{\texttt{addr}}(c_1, N) = c_1$$
$$N = 10 \tag{5.1}$$

Example 5.5 For the above example, the execution order has been explained with respect to any memory. In case of representing the storage order in a VWR, the order needs to be further annotated with the VWR number. For the example in Figure 5.5 and its corresponding layout in Figure 5.6, the VWR storage order can be given for the different arrays as a function of their respective dimensions of their variable domain:

$$Ovwr_A^{\texttt{addr}}(a_1, N) = \{ \ a_1 \% \mathbf{N}_{vwr3} \ | \ N = 64 \ \wedge \ \mathbf{N}_{vwr3} = 8 \ \wedge \ A \rightarrow VWR_3 \ \}$$
$$Ovwr_B^{\texttt{addr}}(b_1, N) = \{ \ b_1 \% \mathbf{N}_{vwr2} \ | \ N = 64 \ \wedge \ \mathbf{N}_{vwr2} = 8 \ \wedge \ B \rightarrow VWR_2 \ \}$$
$$Ovwr_C^{\texttt{addr}}(c_1, N) = \{ \ c_1 \% \mathbf{N}_{vwr1} \ | \ N = 64 \ \wedge \ \mathbf{N}_{vwr1} = 8 \ \wedge \ C \rightarrow VWR_1 \ \}$$
$$\tag{5.2}$$

In other words, this means that each of the arrays map in a piece-wise linear way to the corresponding VWR (for example array A is mapped to VWR3) and so forth. Also the size of the different VWRs is eight entries. For every VWR allocation, such a storage order gives the layout and mapping of each array to the corresponding VWR. The eventual goal of the compiler step that performs register allocation for VWR is to finalize the storage order of the different arrays of the application.

More small differences exist between the storage and execution order, and the reader is referred to [24] for more details.

5.2.3 Dependencies

The previous sections introduced the basic domains and order models. However, for the correct operation of the program and to perform data flow analysis, the concept of *data dependencies* needs to be introduced. This has been introduced in various previous research works [33–35], and they form a well established concept in compilers.

Dependencies are of three types: flow dependencies, output dependencies, and anti dependencies. However, only value-based flow dependencies form the real type of dependency. Other types of dependencies are purely storage related: for example due to sharing of storage locations etc. However, the other types of dependencies (output and anti) are only important for analysis of non-single assignment code. The goal of array *dependency analysis* is to find all the flow based dependencies and model them consistently such that none of these dependencies constraints are violated during any part of compiler.

5.2.3.1 SSA

To reduce the complexity of analysis often in many compilers, the intermediate code is converted to static single assignment (SSA) [33,36,37]. In static single assignment every variable is only assigned once. For modern code, however, static single assignment is not sufficient and dynamic single assignment (DSA) is required. For more details on how to convert the code to dynamic single assignment mode, the reader is referred to [33,38,39]. In the rest of the thesis, it is assumed that the code that is analyzed has already been converted to SSA or DSA form to allow maximal transformation freedom for the compiler.

Similar to iteration domains and operand domains, dependencies can also be modeled for complete groups instead of individual dependencies. The following example illustrates how dependencies are modeled:

Example 5.6

```
        int A[10][10];
        for ( i = 0; i < 10; i++ )
        {
           for ( j = 0; j < 10; j++ )
S1:          A[i][j] = ...;
        }
        for (m = 0; m < 10; m++ ) {
           for ( k = 0; k < 10; k++ )
S2:          ... = A[m][k];
        }
```

For the above code snippet, based on the various definitions introduced in this chapter, we can say the following:

$$\mathbf{D}_A^{\text{var}} = \{\ [a_1, a_2] \mid 0 \le a_1 \le 9 \ \wedge \ 0 \le a_2 \le 9 \ \wedge \ [a_1, a_2] \in \mathbb{Z}^2 \ \}$$

$$\mathbf{D}_1^{\text{iter}} = \{\ [i, j] \mid 0 \le i \le 9 \ \wedge \ 0 \le j \le 9 \ \wedge \ [i, j] \in \mathbb{Z}^2 \ \}$$

$$\mathbf{D}_{11A}^{\text{def}} = \{\ [a_1, a_2] \mid \exists\, [i, j] \in \mathbf{D}_1^{\text{iter}} \ \text{s.t.} \ a_1 = i \ \wedge \ a_2 = j \ \wedge \ [a_1, a_2] \in \mathbb{Z}^2 \ \}$$

$$\mathbf{D}_2^{\text{iter}} = \{\ [m, k] \mid 0 \le m \le 9 \ \wedge \ 0 \le k \le 9 \ \wedge \ [m, k] \in \mathbb{Z}^2 \ \}$$

$$\mathbf{D}_{21A}^{\text{oper}} = \{\ [a_1, a_2] \mid \exists\, [m, k] \in \mathbf{D}_2^{\text{iter}} \ \text{s.t.} \ a_1 = m \ \wedge \ a_2 = k \ \wedge \ [a_1, a_2] \in \mathbb{Z}^2 \ \}$$

From both the code as well as the $\mathbf{D}_{11A}^{\text{def}}$ and $\mathbf{D}_{21A}^{\text{oper}}$, we can say that there is a flow dependency from statement S1 to S2 over the entire variable domain of array A. This dependency as denoted by $\mathbf{M}_{1211A}^{\text{flow}}$ can be represented as:

$$
\begin{aligned}
\mathbf{M}_{1211A}^{\text{flow}} &= \{\ [i, j] \rightarrow [m, k] \mid \exists\, [a_1, a_2] \in \mathbf{D}_A^{\text{var}} \ \text{s.t.} \\
&\qquad \mathbf{M}_{11A}^{\text{def}}(i, j) = [a_1, a_2] = \mathbf{M}_{21A}^{\text{oper}}(m, k) \ \wedge \\
&\qquad [i, j] \in \mathbf{D}_1^{\text{iter}} \ \wedge \ [m, k] \in \mathbf{D}_2^{\text{iter}} \ \} \\
&= \{\ [i, j] \rightarrow [m, k] \mid i = m \ \wedge \ j = k \ \wedge \\
&\qquad 0 \le i \le 9 \ \wedge \ 0 \le j \le 9 \ \wedge \ 0 \le m \le 9 \ \wedge \ 0 \le k \le 9 \ \wedge \\
&\qquad [i, j] \in \mathbb{Z}^2 \ \wedge \ [m, k] \in \mathbb{Z}^2 \ \} \\
&= \{\ [i, j] \rightarrow [m, k] \mid 0 \le i = m \le 9 \ \wedge \ 0 \le j = k \le 9 \ \wedge \ [i, j, m, k] \in \mathbb{Z}^4 \ \}
\end{aligned}
$$

In other words we can say that whenever for every occurrence of statements S1 and S2 when i = m and j = k, a dependence exists. This happens when an element of A is produced in statement S1 and the same is consumed in S2.

5.2.3.2   Dependence weight of a variable in an iteration domain

For efficient mapping of data on the L1 memory and VWR, the different loops/ iteration domains need to be transformed. The dependencies give the true restriction on what the real constraints are for the transformations. Therefore this can be used as an estimate for the constraints on an iteration domain. This has also been used as a weight for performing transformations in other works like [40–43]. The dependence weight of a variable m in an iteration domain $\mathbf{D}_{ij}^{\text{iter}}$ can be defined as:

$$
\begin{aligned}
\mathbf{DVM}_{ijm}^{\text{flow}} &= \{\ \mathbf{M}_{ijklm}^{\text{flow}} \mid \forall \mathbf{M}_{klm}^{\text{oper}} \\
&\qquad \exists\, [a_1, a_2] \in \mathbf{D}_m^{\text{var}} \\
&\qquad \text{s.t.} \ \mathbf{M}_{ijm}^{\text{def}} = [a_1, a_2] = \mathbf{M}_{klm}^{\text{oper}} \ \} \\
&= \bigcup_{kl} \mathbf{M}_{ijklm}^{\text{flow}}
\end{aligned}
\tag{5.3}
$$

In other words, the dependence weight of variable m in iteration domain $\mathbf{D}_{ij}^{\text{iter}}$ is the union of all flow dependencies corresponding to variable m. The dependence weight of a variable is defined by the iteration domains where the variable is produced. Given that the code is SSA code, each variable is written only once, and therefore the dependence weight of each variable is unique and associated with a single iteration domain. This can be further extended to "dependence weight of a variable in all its iteration domains." For a variable m, this can be defined as:

$$\mathbf{VM}_m^{\texttt{allflow}} = \bigcup_{ij} \mathbf{DVM}_{ijm}^{\texttt{flow}} \tag{5.4}$$

5.2.4 Occupation domains

The basic domain models along with the storage and execution order allow us to model the program, and the dependency constraints ensure no real dependency is violated. To further perform data layout aware register allocation and schedule the load/store operations accurately, it is also required to model the lifetime of the variables accurately. Lifetime analysis is required both in the memory as well as in the registers. Therefore it is important to introduce the concept of *occupation domains*.

The lifetime of a variable can be defined as the time between the moment variable is produced until moment it is last consumed. Two statements have such a producer-consumer relation if a variable m is defined in statement Si and consumed in statement Sj if:

$$\mathbf{D}_{ijklm}^{\texttt{addr}} = \{\ a\ |\ \exists\ \mathbf{s} \in \mathbf{D}_m^{\texttt{var}}, w \text{ s.t. } a = O_m^{\texttt{addr}}(\mathbf{s}, w)\ \wedge$$
$$\mathbf{C}_m^{\texttt{addr}}(w) \geq 0\ \wedge\ \mathbf{s} \in \mathbf{D}_{ikm}^{\texttt{def}} \bigcap \mathbf{D}_{jlm}^{\texttt{oper}}\ \} \tag{5.5}$$

The constraint $\mathbf{C}_m^{\texttt{addr}}(w) \geq 0$ represents the constraint including all auxiliary dimensions constraints as symbolic constraints as mentioned in the previous sections for data dependent and non-manifest modeling. It can also further contain constraints from previous compilation steps. For example, various transformations could have been applied that improve the data locality in the higher layers of the memory. It is necessary that the compilation to the foreground memory does not undo these transformations.

5.2.4.1 BOAT/OAT domain

Given the execution order of the statements, it is also possible to precisely model from which time to which time different addresses are alive, or in other words which memory location is occupied during which time. This can be represented in a binary occupied address-time domain (as defined in [24]) as follows:

$$\mathbf{D}_{ijklm}^{\texttt{BOAT}} = \{\ [a, t]\ |\ \exists\ \mathbf{s} \in \mathbf{D}_m^{\texttt{var}}, \mathbf{i} \in \mathbf{D}_i^{\texttt{iter}}, \mathbf{j} \in \mathbf{D}_j^{\texttt{iter}}, x, y, w \text{ s.t.}$$
$$a = O_m^{\texttt{addr}}(\mathbf{s}, w)\ \wedge\ \mathbf{C}_m^{\texttt{addr}}(w) \geq 0\ \wedge$$
$$\mathbf{M}_{ikm}^{\texttt{def}}(\mathbf{i}) = \mathbf{s} = \mathbf{M}_{jlm}^{\texttt{oper}}(\mathbf{j})\ \wedge$$
$$t \geq O_{ikm}^{\texttt{wtime}}(O_i^{\texttt{time}}(\mathbf{i}, x))\ \wedge\ \mathbf{C}_{ikm}^{\texttt{time}}(x) \geq 0\ \wedge$$
$$t \leq O_{jlm}^{\texttt{rtime}}(O_j^{\texttt{time}}(\mathbf{j}, y))\ \wedge\ \mathbf{C}_{jlm}^{\texttt{time}}(y) \geq 0\ \} \tag{5.6}$$

This *binary occupied address-time domain* or *BOAT* domain is a two-dimensional domain where each point with integer coordinates represents an occupied address/time tuple or in other words an address that is possibly (worst case) being occupied at that time. In other words the BOAT domain can also be viewed as a mapping of a dependency in the time and space domain. The following example illustrates the BOAT domain in more detail:

Example 5.7

```
    int A[2][5];
    for ( i = 0; i < 2; i++ )
        for ( j = 0; j < 5; j++ )
S1:         A[i][4-j] = ...;

    for ( k = 0; k < 2; k ++ )
        for ( l = 0; l < 5; l++ )
S2:         ... = g(A[l][k]);
```

For this program, we have the following domain descriptions:

$$\mathbf{D}_A^{\mathtt{var}} = \{\ [a_1, a_2]\ |\ 0 \le a_1 \le 1\ \wedge\ 0 \le a_2 \le 4\ \wedge\ [a_1, a_2] \in \mathbb{Z}^2\ \}$$
$$\mathbf{D}_1^{\mathtt{iter}} = \{\ [i, j]\ |\ 0 \le i \le 4\ \wedge\ 0 \le j \le 1\ \wedge\ [i, j] \in \mathbb{Z}^2\ \}$$
$$\mathbf{D}_{11A}^{\mathtt{def}} = \{\ [d_1, d_2]\ |\ \exists\ [i, j] \in \mathbf{D}_1^{\mathtt{iter}}\ \text{s.t.}\ d_1 = i\ \wedge\ d_2 = 4 - j\ \}$$
$$\mathbf{D}_2^{\mathtt{iter}} = \{\ [k, l]\ |\ 0 \le k \le 1\ \wedge\ 0 \le l \le 4\ \wedge\ [k, l] \in \mathbb{Z}^2\ \}$$
$$\mathbf{D}_{21A}^{\mathtt{oper}} = \{\ [o_1, o_2]\ |\ \exists\ [k, l] \in \mathbf{D}_2^{\mathtt{iter}}\ \text{s.t.}\ o_1 = l\ \wedge\ o_2 = k\ \}$$

Assuming a row-major storage order function for array A, we can say:

$$O_A^{\mathtt{addr}}(a_1, a_2) = 5a_1 + a_2$$

If we assume that this program is executed sequentially, and that each of the statements S1 and S2 can be executed in one clock cycle (precise latency is not needed), we get the following time order functions:

$$O_1^{\mathtt{time}}(i, j) = 5i + j \tag{5.7}$$

$$O_2^{\mathtt{time}}(k, l) = 10 + 5k + l \tag{5.8}$$

This results in the following BOAT-domain description:

$$\mathbf{D}_{1211A}^{\mathtt{BOAT}} = \{\ [a, t]\ |\ \exists\ [a_1, a_2] \in \mathbb{Z}^2, [i, j] \in \mathbb{Z}^2, [k, l] \in \mathbb{Z}^2\ \text{s.t.}$$
$$a = 5a_1 + a_2\ \wedge\ t \ge 5i + j\ \wedge\ t \le 10 + 5k + l\ \wedge$$
$$a_1 = i\ \wedge\ a_2 = 4 - j\ \wedge\ a_1 = l\ \wedge\ a_2 = k\ \wedge$$
$$0 \le i \le 1\ \wedge\ 0 \le j \le 4\ \wedge$$
$$0 \le k \le 1\ \wedge\ 0 \le l \le 4\ \wedge$$
$$0 \le a_1 \le 1\ \wedge\ 0 \le a_2 \le 4\ \}$$

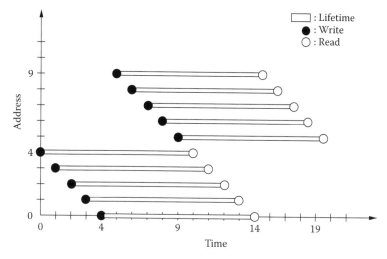

Figure 5.9 Graphical representation of the BOAT domain in Example 5.7.

This BOAT domain can also be graphically represented. This is shown in Figure 5.9.

Note that in this special case, where all constraints and ordering functions are manifest and affine, and all constraints are convex, the resulting BOAT domain is a linearly bounded lattice (LBL) [44].

Given the BOAT domain of an array variable per dependency, if we take a *union* across all the production-consumption chains (or flow dependencies) over all the iteration domains, we would get the memory occupation of an array. More formally, we can express the memory occupation (or OAT domain) of an array as follows:

$$\mathbf{D}_m^{\mathtt{OAT}} = \bigcup_{ijkl} \mathbf{D}_{ijklm}^{\mathtt{BOAT}} \qquad (5.9)$$

Furthermore we can extend the occupied address/time domain over all the array variables. If we take the union of the OAT domain of each variable, we can get the collective memory occupancy domain (COAT). This would tell us how the memory is utilized over the complete lifetime of the program. More formally we can express the COAT domain as follows:

$$\mathbf{D}^{\mathtt{COAT}} = \bigcup_m \mathbf{D}_m^{\mathtt{OAT}} \qquad (5.10)$$

5.2.4.2 L1 memory and VWR occupation domain

For the above concepts of memory occupation, one can express the memory occupation per level of the memory. This thesis assumes that by default all the above terms correspond to the L1 memory. The terms defined in the previous sections can be instantiated for either the L1 data memory or the VWR register file. If the terms are annotated with *vwr* and a *vwr number* then they correspond to the appropriate VWR. For example, the COAT domain for the very wide register 1 would be represented as $\mathbf{D}_{vwr1}^{\text{COAT}}$, and $\mathbf{D}_{vwr1}^{\text{COAT}}$ would represent how the contents of *VWR1* would evolve over the lifetime of the program.

5.2.5   Validity constraints

In the previous sections we have shown how the memory and register file occupancy, program's execution order etc. can be modeled. However, to optimize the program, both the execution order and the storage order can be transformed. Therefore it is important to come up with conditions under which a transformation would be valid or in other words a given storage and execution order of a program is valid. In short we can say that a combination of storage and execution order for a program is valid if it satisfies the following conditions:

- No value of any variable is ever read from a memory location before it has been written to that location
- No memory location might ever be overwritten when it might still contain a data value of a variable that might still have to be read (unless a copy of this has been made to the higher memory).

The formal proof for these valid execution and storage requirements is outside the scope of this thesis. For more detailed proof the reader is referred to [24].

A special validity constraint is also needed only for the VWR/register file, which is:

- Arrays (operands and result) required in one operation should be allocated to different VWRs (as each VWR has one ported cells and two data elements cannot be read from the VWR).

5.3   SARA: StreAm-Register-Allocation-Based Compilation

During typical compilation flow, register allocation is one of the intermediate steps in the compiler and various phases of the compiler exist before the register allocation. However, the target of the SARA is to only allocate arrays with spatial locality/streaming nature. A special note has to be taken into account for the preceding compiler steps that include vectorization and SPM locality optimization phases of the compilation. It is assumed for the following

---

Algorithm 5.1 SARA: Mapping arrays in SRAM to VWR and insert appropriate Load/Stores

---

Require: $\mathbf{D}_a^{\mathtt{OAT}}$ for all a
Ensure: Mapping from SRAM OAT Domain to VWR OAT Domain and insert required Load/Store
  {Phase 1: Enabling Transformations}
  Convert code to Static Single Assignment
  Convert all assignment to 3 operand format
  In case of multiple references to a single array in one operation, split the array into multiple arrays {Split array into multiple arrays}
  Apply access normalization (as in [47] and [48])
  Apply layout transformation for each $\mathbf{D}_i^{\mathtt{iter}}$: for each $\mathbf{D}_{ijkla}^{\mathtt{BOAT}}$: Find M s.t. spatial locality is optimized for array A as in [45] or [40]
  {Phase 2: Preparation}
 1: **for all** $\mathbf{D}_A^{\mathtt{OAT}}$ **do**
     Color $\mathbf{D}_A^{\mathtt{OAT}}$ with unique color
 2: **end for**
  Call Algorithm 5.2 to make prioritized list
  {Phase 3: Color Coalescing}
  Call Algorithm 5.3

---

discussions that the constraints from the SPM locality and the vectorization steps have to be respected. For each of the specific constraints that exist, they need to be added to constraint list during stream register allocation as shown in Section 5.2.4.

The proposed SARA algorithm for allocating arrays to the different VWRs is illustrated in Algorithm 5.1. The algorithm consists of four prominent phases. Each of the different phases are explained in the following subsections along with an example code for better understanding.

5.3.1 Phase 1: enabling transformations

The first phase of the SARA algorithm is transform the code such that the code is analyzable by the compiler. The first step of this phase is to Convert the code into static single assignment (or DSA) as explained in Section 5.2.3.1.

The second step of this phase is to convert the code into 3-operand format as a preparation for the next phases. Converting the code to 3-operand format may require intermediate scalar variables. However these scalar variables have a short lifetime. They can either be register allocated to a scalar register file or handled by a forwarding network inside the processor. In case of allocation to a scalar register file, a typical register allocation technique can be used.

The third step of the enabling transformation is "access normalization." The basic goal of access normalization is to ensure that the step between two successive accesses to an array is a step of the index. This transformation reduces the addressing overhead of the arrays, which in turn improves the addressing of the VWRs. However, note that "access normalization" is an optional step and is not needed for the proposed algorithm.

The last step of the enabling transformation is to exploit spatial locality. The goal of this step is to transform the data-layout of each of the different arrays in memory such that its spatial locality is optimized. Various techniques like [40,45] can be used to find the transformation required to exploit the spatial locality in the memory. For a detailed explanation of the trade-offs involved in this step, the reader is referred to [45], and for multiple loops to [40,46].

5.3.2 Phase 2: preparation

The phase 2 of the SARA algorithm consists of two basic steps: coloring of the different arrays and their prioritization. The prioritization algorithm is shown in Algorithm 5.2. Each of the individual OAT domains (e.g., $\mathbf{D}_a^{\mathtt{OAT}}$) in the application is given its own unique color. Note that dissimilar to the typical register allocator, complete arrays are given colors instead of all the scalar variables. This allows the proposed allocation technique to be scalable to realistic programs. The second part of phase 2 is to order the different arrays based on a priority function. The number of accesses and size are the two important issues that indicate the importance of an array. Therefore the

---

Algorithm 5.2 Rank the different arrays for prioritization for color coalescing and prioritized iteration domain for projection/folding

---

Require: $\mathbf{D}_a^{\mathtt{OAT}}$ for all a and $\mathbf{D}_i^{\mathtt{iter}}$ for all i
Ensure: Prioritized $\mathbf{D}_a^{\mathtt{OAT}}$ and Prioritized $\mathbf{D}_i^{\mathtt{iter}}$
   {Higher priority implies color coalescing would be done last}
  1: **for all $\mathbf{D}_a^{\mathtt{OAT}}$ do**
       Priority of $\mathbf{D}_a^{\mathtt{OAT}}$ = (No. of points in $\mathbf{D}_a^{\mathtt{OAT}}$) $\times$ log(size of array A)
  2: **end for**
  3: **for all $\mathbf{D}_a^{\mathtt{OAT}}$ with the same no. of points do**
       Reverse Order based on $\mathbf{VM}_a^{\mathtt{allflow}}$ {Since dependence vectors give an indication of freedom for a given array, it can be ordered in the descending order of freedom available}
  4: **end for**
  5: **for all $\mathbf{D}_i^{\mathtt{iter}}$ do**
       Priority of $\mathbf{D}_i^{\mathtt{iter}}$ = $\sum_{all\ arrays\ A}$ (No. of Points Used of $\mathbf{D}_a^{\mathtt{OAT}}$ used in $\mathbf{D}_i^{\mathtt{iter}}$)
  6: **end for**

---

product of the number of points on the OAT domain and log of its size can be used as its weight/priority. Note that this metric gives the focus on the access as it is the most important metric and secondary importance to the foot print (log of array size).

As mentioned in the previous section, the true constraints are the dependencies in the application. Given two arrays with approximately the same number of accesses, the array with more constraints needs to be optimized first. Therefore all dependencies to an array ($\mathbf{VM}_a^{\mathtt{allflow}}$) give a good idea of the flexibility of that array and it is used as a priority function. Other works like [42,46] have also used dependence angle as an estimate for performing transformations.

5.3.2.1 Phase 3: color coalescing for VWR allocation

The third phase of the algorithm is the color coalescing. In the preparation phase of the register allocation, all the different arrays in the application were given an initial color. Each color indicates a particular VWR register or port. Note that the number of VWRs and the effective number of ports may be different. In which case multiple words can be read from one VWR. Therefore, at the end of the coloring coalescing, each iteration domain should have at most as many colors as the effective number of ports from the different VWRs.

This phase of the algorithm is shown in Algorithm 5.3. In this phase of the algorithm, the different colors are "merged/coalesced" such that they can be allocated to the different VWRs. In case this is not feasible, there an may be a need to either move a array to the scalar register file or spill (based on the costs). Merging the colors of two arrays implies that they share the same VWR. This implies that the number of load/stores would be higher (implying a higher cost). Therefore colors of arrays with the lowest priorities must be merged. The algorithm steps in the reverse order of the different arrays (using the ordering as done in Phase 2) and identifies which are the best candidates for color coalescing based on their usage together (see Algorithm 5.3) and based on their access pattern.

The condition that "arrays with same access pattern are merged" is present as this allows lowered data layout transformation to match the two arrays. If the arrays are accessed in different patterns, their data layout may need to be transformed for efficient VWR usage. However, the access normalization and spatial locality optimization transformations form enablers for this condition.

5.3.2.2 Phase 4: allocation

At some point during phase 3, the number of colors in each of the different iteration domains meets the architecture constraints (number of colors ≤ number of effective ports from the VWRs). At this point the different arrays can be allocated to the VWRs. This allocation algorithm is shown in Algorithm 5.4. For clearly understanding the allocation phase of the algorithm, it is

Algorithm 5.3 Color coalescing of all iteration domains till register allocation can be done

---

Require: Initial Colored list of OAT Domains
Ensure: Colored list of OAT Domains that are register allocatable and call the register allocator
　　{Phase 3: Color Coalescing}
　　Call Algorithm 5.4
1: **for all** $\mathbf{D}_A^{\mathtt{OAT}}$ in reverse order of priorities **do**
2: 　　**for all** $\mathbf{D}_B^{\mathtt{OAT}}$ in order of number of usages with $\mathbf{D}_A^{\mathtt{OAT}}$ in the same iteration domains $\mathbf{D}_i^{\mathtt{iter}}$ **do**
3: 　　　　**if** Array A and B used together later in same $\mathbf{D}_i^{\mathtt{iter}}$ **and** Not in same statement **and** Same access pattern (for e.g. $\mathbf{D}_{ijA}^{\mathtt{oper}} = \mathbf{D}_{ijB}^{\mathtt{oper}}$) **then**
　　　　　　Merge Colors of $\mathbf{D}_A^{\mathtt{OAT}}$ and $\mathbf{D}_B^{\mathtt{OAT}}$
　　　　　　Call Algorithm 5.4
4: 　　　　**end if**
5: 　　　　**if** Array A and B used together later in same $\mathbf{D}_i^{\mathtt{iter}}$ **and** Not in same statement **then**
　　　　　　Merge Colors of $\mathbf{D}_A^{\mathtt{OAT}}$ and $\mathbf{D}_B^{\mathtt{OAT}}$
　　　　　　Call Algorithm 5.4
6: 　　　　**end if**
7: 　　**end for**
8: **end for**
　　{Need to look beyond the same iteration domain for merging candidates}
9: **for all** $\mathbf{D}_A^{\mathtt{OAT}}$ in reverse order of priorities **do**
10: 　　**for all** $\mathbf{D}_B^{\mathtt{OAT}}$ in reverse order of priorities not in the same iteration domains $\mathbf{D}_i^{\mathtt{iter}}$ **do**
11: 　　　　**if** Array A and B have the same access pattern (for e.g. $\mathbf{D}_{ijA}^{\mathtt{oper}} = \mathbf{D}_{ijB}^{\mathtt{oper}}$) **then**
　　　　　　Merge Colors of $\mathbf{D}_A^{\mathtt{OAT}}$ and $\mathbf{D}_B^{\mathtt{OAT}}$
　　　　　　Call Algorithm 5.4
12: 　　　　**end if**
13: 　　**end for**
14: **end for**
　　{Need to look beyond the same iteration domain and different access patterns for merging candidates}
15: **for all** $\mathbf{D}_A^{\mathtt{OAT}}$ in reverse order of priorities **do**
16: 　　**for all** $\mathbf{D}_B^{\mathtt{OAT}}$ in reverse order of priorities not in the same iteration domains $\mathbf{D}_i^{\mathtt{iter}}$ **do**
　　　　　Merge Colors of $\mathbf{D}_A^{\mathtt{OAT}}$ and $\mathbf{D}_B^{\mathtt{OAT}}$
　　　　　Call Algorithm 5.4
17: 　　**end for**
18: **end for**
　　{Need to spill inside one iteration domain}
　　Move the array with lowest priority to scalar register file {This implies that this array need not be considered for allocation to the VWR.}
　　Call Algorithm 5.3

Algorithm 5.4 Check if allocation can be done and allocate

Require: Colored set of $\mathbf{D}_a^{\text{OAT}}$
Ensure: State of VWR allocation
 1: **for all** $\mathbf{D}_i^{\text{iter}}$ **do**
 2:   **if** No. of Colors in $\mathbf{D}_i^{\text{iter}} \geq$ No. of effective VWR ports **then**
       return {Register allocation will fail}
 3:   **end if**
 4: **end for**
     {Phase 4: Allocation} {Fold the OAT Domains to allocate onto VWRs}
 5: **for all** $\mathbf{D}_i^{\text{iter}}$ in order of priority **do**
     Projection on invariant dimension
     Fold inside variant dimension on non-overlapping lifetimes
 6:   **for all** Color in $\mathbf{D}_i^{\text{iter}}$ **do**
 7:     **for all** $\mathbf{D}_m^{\text{OAT}}$ belonging to this color in $\mathbf{D}_i^{\text{iter}}$ **do**
         Fold $\mathbf{D}_{ijklm}^{\text{BOAT}}$ by a factor = (VWR size/No. of arrays in this color)
         For each fold insert corresponding load(s)/store(s) at the appropriate region boundaries
 8:     **end for**
 9:   **end for**
10: **end for**

---

important to understand a few more concepts and transformations in the BOAT domain.

Projection and folding in the OAT domain: To ensure that the different elements fit on the VWR and they are efficiently laid out in the VWR to reduce the load/stores, it is important that the appropriate transformations are made in the OAT domain. For performing this, a few basic transformation techniques are required. This includes folding and projection in the OAT domain. This can be better explained using the following example:

Example 5.8

```
    int A[3][4];
    for ( i = 0; i < 3; ++i )
    {
        for ( j = 0; j < 7; ++j )
        {
            if ( j < 4 ) 0 1 2 3
S1:         A[i][j] = ...
            ...
            if ( j >= 3 ) 3 4 5 6
S2:         ... = A[i][j-3];
        }
    }
```

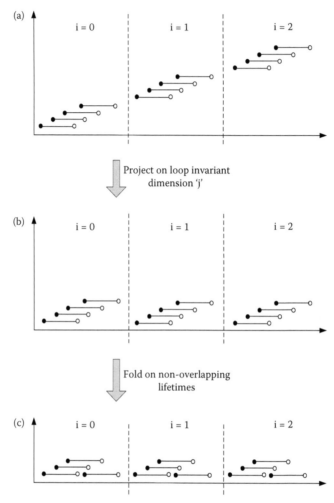

Figure 5.10 Graphical representation of OAT domain of array A in Example 5.8 before projection and after projection.

Note in the above code no dependencies exist across the i dimension. This is also referred to as an "invariant" dimension. Note that often parallelism is available across an "invariant" dimension. However in this case the parallelism is exploited as spatial locality. Part (A) of Figure 5.10 shows the OAT domain corresponding to the above example code. It can be seen that along the i dimension the address space can be projected down. This implies that at a given point of time only the elements for a given value of i needs to kept alive. Part (B) of Figure 5.10 shows the projection along the loop invariant dimension. Such a transformation reduces the amount of memory footprint that is required. Therefore this transformation (also called as performing "inplace" inside an array or intra-array inplace) is both beneficial to reduce the memory footprint as well as for placement of data inside the very wide register.

Note that besides projecting on the loop invariant dimension, the address space can be folded in a more aggressive way. Even inside a single dimension, arrays elements can be folded provided their lifetimes do not overlap. Part (C) of Figure 5.10 shows the case when along the "j" dimension the last element is folded to the same location as the first element as their lifetimes do not overlap. A more detailed discussion on intra-array inplace can be found in [24].

Folding with discontinuities in OAT domain: All data in a single loop would not always fit in the VWR. Therefore a need exists to bring parts of the array to the VWR. This would cause a discontinuity in the OAT domain of the array in the VWR. This can be better explained using the following example:

Example 5.9

```
      int A[6], B[6], C[6];
BB1: for ( i = 0; i < 6; ++i )
      {
S1:       C[i] = f(A[i], B[i]);
      }
BB2: for ( j = 0; j < 6; ++j )
      {
S2:       .. = g(A[j], C[j]);
      }
```

Consider the above example that consists of three arrays and two iteration domains. Assume that in the architecture there are three VWR registers and each VWR has three words that it can store. In this case each array can be allocated one VWR. However the VWR cannot store the complete array, and therefore the SRAM OAT domain has to be transformed to match the VWR OAT domain.

This procedure requires to demarcate in the SRAM OAT domain where to insert the load/stores to the VWR. The top part of Figure 5.11 shows the SRAM OAT domain and also the two basic blocks to which the statement corresponds. The figure also shows the places where load/stores have to be inserted when each VWR has three locations. These are shown with solid vertical lines in the SRAM OAT domain. The bottom part of Figure 5.11 also shows the VWR OAT domain that would be obtained on folding the SRAM OAT domain at the boundaries obtained by the demarcations. Based on a read or write operation between two demarcated regions, a load can be inserted at the beginning or a store at the end of the region. On insertion of the load/stores, the final VWR OAT domain is shown at the bottom of Figure 5.11.

VWR allocation with projection and folding: The previous two subsections have introduced the concepts of projection and folding, which are crucial to perform VWR allocation. The previous phase (phase 3) would have

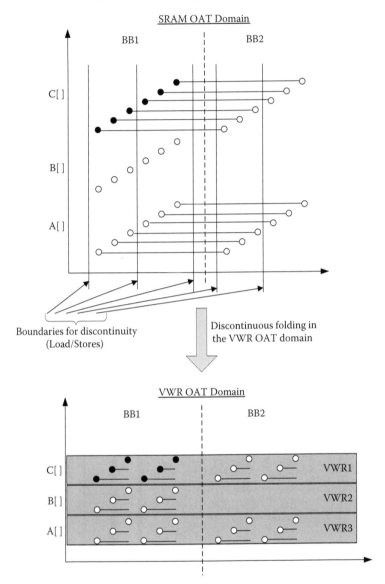

Figure 5.11   Graphical representation of OAT domain with discontinuous folding for arrays in Example 5.9.

decided which arrays need to be allocated together. In phase 4, there needs to be projection and folding to ensure that for each of the different colors (in turn VWRs), all the arrays can be put together efficiently. This would reduce the size required in the VWR per array. However without folding/projection, it would always be "feasible" to fit data on the VWR, but perhaps in a poor way.

The first step of Phase 4 is to perform projection on the invariant dimension for all the individual array's OAT domains. This is followed by folding the individual OAT domains along the variant dimension in case of non-overlapping lifetimes. This transformation can be done in the SRAM OAT domain itself as it reduces the memory footprint of each of the arrays. These steps make the VWR allocation process more efficient. However they are not required.

For each iteration domain in order of its priority, all the arrays in one color can be considered for folding with the insertion of load stores. Based on the number of arrays in one color, the number of locations can be distributed to each of the individual arrays. Note that in the proposed algorithm this distribution is done equally among the arrays in one color. However this may be suboptimal, and a better heuristic for unequal allocation on a VWR may be required. The folding with multiple arrays is illustrated better in Section 5.3.3.

In case the number of colors for a particular iteration domain is greater than the effective number of ports from VWRs, then there is a need for spilling. This can be either introduced by additional load/stores or by reducing the number of arrays that are considered for allocation to the VWR. The latter is considered by removing the array with the lowest priority from the list.

5.3.3 Example illustration of SARA

Below is a code snippet that will be used for illustrating the SARA algorithm. Assume that the proposed architecture has three VWR registers and each VWR register in turn has one read/write port to the datapath. Each VWR register also has the capacity to store six words.

Example 5.10

```
BB1: for (i = 0; i<6; i++) {
S1:      a[i] = x(b[i],c[i]);
S2:      d[i] = y(f[i]);
     }

BB2: for (j = 0; j<6; j++)
S3:      g[j] = z(a[j],b[j]);

BB3: for (k = 0; k<6; k++) {
S4:      h[k] = p(a[k],d[k]);
```

Figure 5.12 gives the original OAT domain for the code shown in Example 5.10. It is obvious from the example that the code cannot be immediately mapped onto the very wide registers as is.

Phase 1: Since the code is already in static single assignment, no need exists to convert it. The example code is also in 3 operand format where each of the operations are embedded inside the code as "functions" to abstract away from them. This is similar to the methodology followed in other data

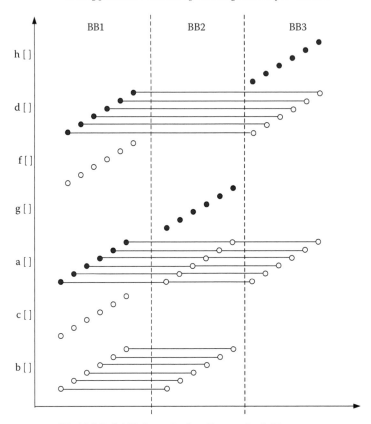

Figure 5.12   SRAM BOAT domain for Example 5.10.

transformations works like [41,49] etc. Given that the access to the arrays are also normal in all the cases and the original data is trivial, no "access normalization" or spatial locality optimization is required.

Phase 2: The initial coloring would have given seven colors in total. The ordered list of OAT domains in order of priority based on algorithm 5.2 would be as follows:

1. A

2. B, D

3. F, G, H, C

This list can now be used for color coalescing such that the number of colors per iteration domain is less than three (= total number of very wide registers for this example).

Phase 3: Register allocation cannot be done yet as the number of colors for basic block 1 is greater 3. Therefore color coalescing needs to be performed first. On executing algorithm 5.3, the first color merging that would be done is between arrays F and C as they have the lowest priority. This will be followed by the merging of the colors of arrays B and D. At this point the number of colors for all the iteration domains is equal to three, and therefore the register allocation can be done. The list of colors for each of the iteration domains would be:

- BB1: {F,C}, {B,D}, {A}
- BB2: {G}, {B}, {A}
- BB3: {D}, {H}, {A}

Phase 4: Based on the above groupings, we can now allocate the arrays to the different very wide registers starting from the basic block 1. For all the different arrays, no loop invariant dimensions exist. Therefore no projection can be done along the loop invariant dimension. Also all the elements in each array have overlapping lifetimes. Therefore no projection exists along the non-overlapping lifetime elements either.

For the basic block (iteration domain) BB1, the two arrays F and C can be allocated three elements each in (say) VWR1. In a similar fashion arrays B and D can be allocated three elements each in (say) VWR2. However array A is allocated one complete very wide register (say) VWR3. This is because array A is used often and it makes sense to minimize the number of load/stores "more" compared to other arrays. These constraints from allocation of registers in BB1 can be propagated to the rest of the iteration domains/basic blocks (as shown in Algorithm 5.4). And a similar allocation can be followed for remaining basic blocks BB2 and BB3 for the un-allocated arrays as well. Finally the VWR OAT domain that would be obtained is shown in Figure 5.13.

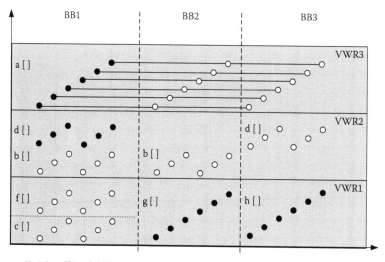

Figure 5.13 Final VWR COAT domain for Example 5.10.

Note that by ordering the iteration domains and propagating constraints of allocation from one to another domain (basic block), an efficient allocation can be performed. In the above example, this ordering allows array A to get a full VWR in all the iteration domains as it is the most important array. Furthermore allocation is performed first on the iteration domain BB1 and the constraints are passed onto the other iteration domains. For example array B gets only three locations based on iteration domain BB1, and this is also the case for BB2. This also makes sure the "low priority" arrays do not need any re-layouting between two iteration domains.

5.3.4   Comparison to state of the art

The related work to the area of stream register allocation can be obtained from various different classes of work:

Transformations: A large part of work exists in the space of loop transformations. Various transformation work like IMEC's DTSE [49], [28,40,50–54] target improving spatial and temporal locality in memories. Some research works like [24] target reducing the memory foot print by performing in-place mapping inside and between different arrays in the application. Other research works like [21,34,55,56] present efficient modeling techniques for modeling application in a polyhedral/geometrical space. They also discuss in detail how transformations can be performed in the geometrical space while maintaining correctness. These research works are complementary to the proposed register allocation algorithm as the modeling work forms the basis for register allocation and the transformation work is an enabler for efficient stream register allocation.

Register file allocation: Another domain of related work is typical register allocation techniques. Research in graph coloring based register allocation has been active since the early 80s. One of the first research work on graph coloring based register allocation was started in IBM by Chaitin [57,58]. More recent works on register allocation and different coloring heuristics for register allocation also exist [59–63]. These works also deal with iterative methods for coloring for efficient register allocation. However, all these techniques purely deal with all the different variables in the application in the same way viz. as scalars. Traditionally register allocation has always tried to solve/improve the temporal locality of the different variables by analyzing only the lifetime of variables.

Stream register: Another class of related work in register allocation is in the space of stream register files. However, stream register files as defined in [64] do not share the asymmetrical interface as presented in the proposed VWR architecture. Therefore, the register allocation problem is different between streaming register as defined in [64] and the proposed VWR architecture. [65,66] propose two different register allocation techniques for the stream register file [64]. Both these techniques use StreamC language instead of generic ANSI C as input code, which heavily simplifies the compilation problem as the different streams have already been identified by the programmer. [66] does consider both space as well as time while performing analysis. Given that the stream register file in their case is a L1 like memory, the problem they solve is similar to [24], namely, placement of the data in the SRF memory. Also the work does not include intra-in-place, which overlaps lifetimes of different elements of the same array in the same memory space.

The closest related work in the space of stream register file allocation is [67], which performs register allocation for the IBM's iVMX architecture [15]. The iVMX architecture also has asymmetrical interfaces: a wide interface to the L2 cache and a narrower interface toward the datapath. However the management of the access to the wide register is done via a map management register. Therefore the compiler technique for the iVMX architecture described in [67] cannot be directly utilized. Furthermore [67], like most register allocation work, it does not consider arrays as arrays but as scalars. Therefore it misses out on various data locality optimization possibilities. A more quantitative comparison between an adapted version of [67] and the proposed SARA technique is presented in the next subsection.

5.3.5 Results/comparison

As explained in the previous section, a very limited literature exists for compiling arrays to streaming registers, so quantitative comparison is difficult. The closest compilation technique present in state of the art is [67]. However, it targets another architecture with different constraints. Section 5.3.5.1 presents a version of the algorithm proposed in [67] adapted for the VWR based architecture. Section 5.3.6 then compares quantitatively the adapted iVMX algorithm to the SARA algorithm for various benchmarks.

5.3.5.1 iVMX-based compilation technique adaptation for VWR architecture

The iVMX architecture is relatively different with respect to the VWR based architecture as already described in Section 5.1.2. Therefore the compilation technique of iVMX [67] cannot be reused as is for compiling code to the VWR architecture. This section describes an adapted version of the iVMX compilation technique to compile for the VWR architecture.

While re-implementing the iVMX compilation technique to the VWR architecture a few assumptions/adaptions have been made. These assumptions are listed below:

1. The same primitive models and scopes as described in [67] are used.

2. The architecture constraints of the VWR architecture (single port, no map management system etc.) are enforced.

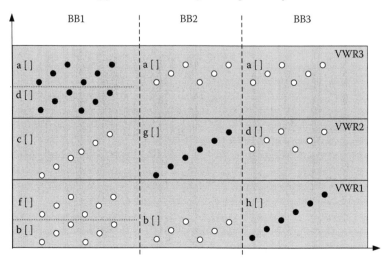

Figure 5.14   VWR COAT domain for Example 5.10 using adapted iVMX Algorithm 5.5.

3. Indirection logic of iVMX and its corresponding overhead in the algorithm has been removed.

4. Only array variables are considered for VWR allocation (unlike all variables in [67]).

5. An extra enabling transformation phase has been added to enable efficient mapping on the VWRs.

Algorithm 5.5 shows the adapted version of the iVMX compilation technique for the VWR architecture. Note that the adapted algorithm is quite optimized[5] already within these assumptions and therefore it is better than the direct implementation as described in [67].

5.3.6   Comparison of adapted iVMX compilation with the proposed SARA technique

Figure 5.14 shows the register allocation based on iVMX compilation for a running example represented in the OAT domain for better visibility and comparison. The architecture considered is the same as in the example with three SRs each having six locations.

Comparing the iVMX compilation shown in Figure 5.14 with the SARA-based result shown in Figure 5.13, it is clear that the SARA technique can use the stream registers more efficiently compared to the iVMX based technique.

---

[5]Note that additional enabling transformations, restriction of variables for register allocation etc. optimize the iVMX algorithm substantially.

Algorithm 5.5 Adapted version of [67] compilation technique for very wide register based compilation

---

Require: C source code
Ensure: Mapping from all array variables to VWR
    {Phase 1: Initial preparation/transformation phase}
    Convert code to Static Single Assignment
    Convert all assignment to 3 operand format
1: **for all** Basic-blocks in order of appearance **do**
        Unroll basic block till the number of array variables = number of locations in all the VWRs {Note that the number of array variables should be as close to number of locations in all VWRs and need not be equal}
2: **end for**
    {Phase 2: Preparation}
3: **for all** Basic-blocks in order of appearance **do**
        Perform initial register allocation based on coloring with virtual registers and compute for each register r0 its live-range lr_0.
4: **end for**
5: **for all** Basic-blocks in order of appearance **do**
        Calculate sequence of live-ranges lr_0, lr_1, ..., lr_{n-1} called a *chain* such that for every i: (a) all live ranges have the same operand types; (b) All live ranges interfere (cannot be assigned same VWR); and (c) two instructions with same operand are ordered based on live range of sequence.
6: **end for**
    {Phase 3: Register allocation}
7: **for all** Basic-blocks in order of appearance **do**
8:     **for all** *Chains* in the the basic-block **do**
            Assign all elements (virtual registers of the chain) to same VWR
9:     **end for**
10: **end for**

Furthermore the number of loads/stores that would be required for the iVMX based technique are higher than those needed by the SARA-based technique. This is tabulated in Table 5.1.

Another key observation that can be made from the two OAT domains is that the average occupancy or usage of the iVMX based technique is poor. Also the gains of using the SARA algorithm can be greater if there is higher reuse. It is clear (from Figures 5.13 and 5.14) that for array A[] the allocation performed by SARA is superior. Table 5.1 also shows the results for a set of wireless kernels that are used in software defined radio. The kernels consist of filters, FFTs, and and power estimation of the signal. Note that for such realistic applications the gains are even higher. Note that only a ratio of load/stores as presented in Table 5.1 corresponds to a limited set of samples.

Table 5.1   Comparison of number of loads and stores and SR footprint between iVMX and SARA-based compilation techniques on different benchmarks

Benchmarks	Total number of Loads/Stores		Total memory accesses	
	iVMX	**SARA**	**iVMX**	**SARA**
Example 5	11/4	8/5	15	13
SDR Kernel sequence	48/17	36/5	65	41
SDR Ker. 1/Filter	32/8	32/2	40	34
SDR Ker. 2/FFT	8/8	2/2	16	4
SDR Ker. 3/Power Est.	8/1	2/1	9	3

It can also be seen from the example that the gains of the mapping increase with the size of the application, which is an interesting property of the SARA technique. This is because the larger the application, the more globally optimal job the technique can do.

The key reasons on why the iVMX algorithm has lower performance with respect to the proposed algorithm are:

1. Treatment of all variables as the same (array, non-array) with the same weight

2. Register allocation per basic block without passing constraints from one to another

3. Inability to modify the data layout of the array in memory. However this is relaxed for the re-instantiation of the iVMX algorithm presented in this chapter.

The above example showed that the proposed SARA technique can efficiently map code on the SR even for a small example. Based on the above reasoning, it is clear that for a larger realistic example the proposed technique should be able to map code very well.

5.4   Conclusion

This chapter introduced an asymmetrical foreground memory architecture called very wide register (VWR), which can reduce the power consumption in low power embedded processors. Since this architecture uses spatial locality to gain in energy consumption, an efficient compilation technique was also introduced that allocates arrays to the VWRs. The proposed technique was shown to work across different basic blocks and to produce an efficient data layout and allocation to the VWRs using various benchmarks.

References

1. Praveen Raghavan. *Low Energy Architecture Extensions for Embedded Processors (working title)*. PhD thesis, IMEC vzw, ESAT, KULeuven, March/April 2009 (in preparation).

2. Andy Lambrechts. *Energy-Aware Datapath Optimizations at the Architecture-Compiler Interface*. PhD thesis, IMEC vzw, ESAT, KULeuven, February 2009 (in preparation).

3. Francky Catthoor, Praveen Raghavan, Andy Lambrechts, Murali Jayapala, Angeliki Kritikakou, and Javed Absar. *Ultra Low Power Application Specific Processor Design*. Springer, June 2010.

4. Andy Lambrechts, Praveen Raghavan, Anthony Leroy, Guillermo Talavera, Tom VanderAa, Murali Jayapala, Francky Catthoor, Diederik Verkest, Geert Deconinck, Henk Coporaal, Frédéric Robert, and Jordi Carrabina. Power breakdown analysis for a heterogeneous NoC platform running a video application. In *Proc. of IEEE 16th International Conference on Application-Specific Systems, Architectures and Processors (ASAP)*, pages 179–184, July 2005.

5. Scott Rixner, William J. Dally, Brucek Khailany, Peter R. Mattson, Ujval J. Kapasi, and John D. Owens. Register organization for media processing. In *HPCA*, pages 375–386, January 2000.

6. Jesús Sánchez and Antonio González. Modulo scheduling for a fully-distributed clustered VLIW architecture. In *Proc. of 29th International Symposium on Microarchitecture (MICRO)*, December 2001.

7. Viktor Lapinskii, Margarida F. Jacome, and Gustavo de Veciana. Application-specific clustered VLIW datapaths: Early exploration on a parameterized design space. *IEEE Transactions on Computer Aided Design of Integrated Circuits and Systems*, 21(8):889–903, August 2002.

8. J. Janssen and H. Corporaal. Partitioned register file for TTAs. In *Proc. of Micro*, pages 303–312, 1995.

9. Paul R. Wilson, Mark S. Johnstone, Michael Neely, and David Boles. Dynamic storage allocation: A survey and critical review. In *In 1995 International Workshop on Memory Management*, pages 1–116. Springer-Verlag, 1995.

10. Hugo DeMan. Ambient intelligence: Giga-scale dreams and nano-scale realities. In *Proc. of ISSCC, Keynote Speech*, February 2005.

11. M. Joshi, N.S. Nagaraj, and A. Hill. Impact of interconnect scaling and process variations on performance. In *Proceedings of CMOS Emerging Technologies*, Vancouver, Canada, Published online at www.cmoset.com 2006.

12. Dennis Sylvester and Kurt Keutzer. Getting to the bottom of deep submicron ii: A global wiring paradigm. In *ISPD '99: Proceedings of the 1999 International Symposium on Physical Design*, pages 193–200, ACM Press, New York, NY, 1999.

13. ITRS. International technology roadmap for semiconductors 2007 edition: Interconnect. Technical report, ITRS, http://www.itrs.net/Links/2007ITRS/2007_Chapters/2007_Interconnect.pdf, 2007.

14. William J. Dally, Ujval J. Kapasi, Brucek Khailany, Jung Ho Ahn, and Abhishek Das. Stream processors: Programmability with efficiency. *ACM Queue*, 2(1), March 2004.

15. J.H. Derby, R.K. Montoye, and J. Moreira. Victoria — vmx indirect compute technology oriented towards in-line acceleration. In *Proc of CF*, pages 303–311, May 2006.

16. B. Amrutur and M. Horowitz. Speed and power scaling of SRAM's. In *IEEE Journal of Solid-State Circuits*, volume 35, February 2000.

17. P.D. Evans and R.J. Franzon. Energy consumption modeling and optimization for SRAM's. In *IEEE Journal of Solid-State Circuits*, volume 30, pages 571–579, May 1995.

18. P. Raghavan and F. Catthoor. Register file exploration for a multi-standard wireless forward error correction asip. In *Proc. of SIPS*, 2009.

19. P. Raghavan, A. Lambrechts, M. Jayapala, F. Catthoor, D. Verkest, and H. Corporaal. Very wide register: An asymmetric register file organization for low power embedded processors. In *DATE '07: Proceedings of the Conference on Design, Automation and Test in Europe*, 2007.

20. Jean Baka Domelevo. Working on the design of a customizable ultra-low power processor: A few experiments. Master's thesis, ENS Cachan Bretage and IMEC, September 2005.

21. P. Feautrier. Dataflow analysis of array and scalar references. *International Journal of Parallel Programming*, 20(1):23–53, 1991.

22. W. Pugh. The omega test: A fast and practical integer programming algorithm for dependence analysis. *Proceedings of Supercomputing'91*, November 1991.

23. F. Franssen, M. van Swaaij, F. Catthoor, and H. De Man. Modeling piece-wise linear and data dependent signal indexing for multidimensional signal processing. In *Proceedings of the 6th ACM/IEEE International Workshop on High-Level Synthesis*, pages 245–255, Dana Point, CA, November 1992.

24. Eddy De Greef. *Storage Size Reduction for Multimedia Applications*. PhD thesis, Department of Electrical Engineering (ESAT), KULeuven, Belgium, 1998.

25. F. Catthoor, F. Balasa, E.D. Greef, and L. Nachtergaele. *Custom Memory Management Methodology: Exploration of Memory Organization for Embedded Multimedia System Design*. Kluwer Academic Publisher, 1998.

26. M. van Swaaij, F. Franssen, F. Catthoor, and H. De Man. Modeling data flow and control flow for high level memory management. In *Proceedings of EDAC '92*, pages 8–13, Brussels, Belgium, March 1992.

27. Michael E. Wolf and Monica S. Lam. A data locality optimizing algorithm. In *PLDI '91: Proceedings of the ACM SIGPLAN 1991 Conference on Programming Language Design and Implementation*, pages 30–44, ACM Press, New York, NY, 1991.

28. Preeti Ranjan Panda, Alexandru Nicolau, and Nikil Dutt. *Memory Issues in Embedded Systems-on-Chip: Optimizations and Exploration*. Kluwer Academic Publishers, Norwell, MA, 1998.

29. W. Pugh and D. Wonnacott. Static analysis of upper and lower bounds on dependences and parallelism. *ACM Transactions on Programming Languages and Systems*, 16(4):1248–1278, July 1994.

30. V. Maslov. Lazy array data-flow dependence analysis. In *Proceedings of the 21st Annual ACM SIGPLAN-SIGACT Symposium on Principles of Programming Languages*, pages 311–325, January 1994.

31. J.F. Collard, D. Barthou, and P. Feautrier. Fuzzy array dataflow analysis. In *5th ACM SIGPLAN Conference on Principles and Practice of Parallel Programming*, Santa Barbara, CA, July 1995.

32. F. Franssen, M. van Swaaij, F. Catthoor, and H.De Man. Modelling piece-wise linear and data dependent signal indexing for multidimensional signal processing. In *Proc. 6th Intnl. Wsh. on High-Level Synthesis*, Dana Point, California, IEEE, ACM SIGDA, November 1992.

33. Paul Feautrier. Dataflow analysis of array and scalar references. *International Journal of Parallel Programming*, 20(1):23–51, February 1991.

34. U. Banerjee. *Dependence Analysis for Supercomputing*. Kluwer Aacdemic Publishers, 1988.

35. Randy Allen and Ken Kennedy. *Optimizing Compilers for Modern Architectures*. Morgan Kaufmann, 2002.

36. *GCC, the GNU Compiler Collection*. http://gcc.gnu.org, 2007.

37. Kathleen Knobe and Vivek Sarkar. Array SSA form and its use in parallelization. In *Proceedings of the 25th ACM SIGPLAN-SIGACT Symposium on Principles of Programming Languages*, pages 107–120, 1998.

38. Carl Offner and Kathleen Knobe. Weak dynamic single assignment form. Technical Report TR-HPL-2003-169, HP Labs, November 2003.

39. Peter Vanbroekhoven, Gerda Janssens, Maurice Bruynooghe, Henk Corporaal, and Francky Catthoor. A step towards a scalable dynamic single assignment conversion. Technical Report CW 360, Department of Computer Science, Katholieke Universiteit Leuven, April 2003.

40. Javed Absar, Praveen Raghavan, Andy Lambrechts, Min Li, Murali Jayapala, and Francky Catthoor. Locality optimizations in a compiler for wireless applications. *Design Automation of Embedded Systems (DAEM)*, vol. 13, numbers 1–2, pp. 53–72, April 2008.

41. P. R. Panda, F. Catthoor, N. D. Dutt, K. Danckaert, E. Brockmeyer, C. Kulkarni, A. Vandercappelle, and P. G. Kjeldsberg. Data and memory optimization techniques for embedded systems. *ACM Trans. Des. Autom. Electron. Syst.*, 6(2):149–206, 2001.

42. Koen Danckaert. *Loop Transformations for Data Transfer and Storage Reduction on Multiprocessor Systems*. PhD thesis, K.U. Leuven, May 2001.

43. Yi qing Yang, Corinne Ancourt, and Francois Irigoin. Minimal data dependence abstractions for loop transformations. In *International Journal of Parallel Programming*, pages 359–388, 1994.

44. L. Thiele. Compiler techniques for massive parallel architectures. In P. Dewilde, editor, *State of the Art in Computer Science*, chapter 1, pages 19–50. Kluwer Academic Publishers, May 1992.

45. Mahmut Taylan Kandemir and J. Ramanujam. *Data Relation Vectors: A New Abstraction for Data Optimizations*. IEEE Trans. on Computers, volume 50, no. 8, pages 798–810, August 2001.

46. Javed Absar. *Locality Optimization in a Compiler for Embedded Systems*. PhD thesis, IMEC vzw, ESAT, KU Leuven, July 2007.

47. Wei Li and Keshav Pingali. Access normalization: Loop restructuring for numa computers. *ACM Trans. Comput. Syst.*, 11(4):353–375, 1993.

48. Dattatraya Kulkarni and Michael Stumm. *Loop and Data Transformations: A Tutorial*, University of Toronto, Toronto, 1993.

49. E. Brockmeyer, C. Ghez, W. Baetens, and F. Catthoor. *Unified Low-Power Design Flow for Data-Dominated Multi-Media and Telecom Applications*. Kluwer Academic Publishers, Boston, 2000.

50. Peter Marwedel. *Embedded System Design*. Kluwer Academic Publishers (Springer), Norwell, MA, 2003.

51. Michael E. Wolf and Monica S. Lam. A data locality optimizing algorithm. In *PLDI '91: Proceedings of the ACM SIGPLAN 1991 Conference on Programming Language Design and Implementation*, pages 30–44, ACM Press, New York, NY, 1991.

52. Janis Sermulins, William Thies, Rodric Rabbah, and Saman Amarasinghe. Cache aware optimization of stream programs. In *LCTES'05: Proceedings of the 2005 ACM SIGPLAN/SIGBED Conference on Languages, Compilers, and Tools for Embedded Systems*, pages 115–126, ACM Press, New York, NY, 2005.

53. M. Bruynooghe, S. Verdoolaege, G. Janssens, and F. Catthoor. Multidimensional incremental loop fusion for data locality. In *Proc. of ASAP*, pages 17–27, 2003.

54. Dattatraya Kulkarni. Transformations for improving data access locality in non-perfectly nested loops. In *Proc. of Seventh International Conference on Parallel Architectures and Compilation Techniques*, pages 314–321, 1998.

55. Sylvain Girbal, Nicolas Vasilache, Cedric Bastoul, Albert Cohen, David Parello, March Sigler, and Olivier Temam. Semi-automatic composition of loop transformations for deep parallelism and memory hierarchies. In *International Journal of Parallel Programming*, ACM Conference: Proceedings of the 1982 SIGPLAN Compiler Construction, Boston, MA, pages 261–317, October 2006.

56. C. Bastoul. Code generation in the polyhedral model is easier than you think. In *PACT'13 IEEE International Conference on Parallel Architecture and Compilation Techniques*, ACM Conference: Proceedings of the 1982 SIGPLAN Compiler Construction, Boston, MA, pages 7–16, September 2004.

57. G Chaitin. Register allocation and spilling via graph coloring. In *Proc. of Compiler Construction*, ACM 1982.

58. Gregory Chaitin. Register allocation and spilling via graph coloring. *SIGPLAN Not.*, 39(4):66–74, 2004.

59. Yumin Zhang and Danny Z. Chen. Efficient global register allocation for minimizing energy consumption. *SIGPLAN Not.*, 37(4):42–53, 2002.

60. Fernando Magno Quintao Pereira and Jens Palsberg. Register allocation by puzzle solving. In *PLDI '08: Proceedings of the 2008 ACM SIGPLAN Conference on Programming Language Design and Implementation*, pages 216–226, ACM Press, New York, NY, 2008.

61. Guei-Yuan Lueh, Thomas Gross, and Ali-Reza Adl-Tabatabai. Fusion-based register allocation. *ACM Trans. Program. Lang. Syst.*, 22(3):431–470, 2000.

62. Michael D. Smith, Norman Ramsey, and Glenn Holloway. A generalized algorithm for graph-coloring register allocation. In *PLDI '04: Proceedings of the ACM SIGPLAN 2004 Conference on Programming Language Design and Implementation*, pages 277–288, ACM Press, New York, NY, 2004.

63. Preston Briggs, Keith D. Cooper, Ken Kennedy, and Linda Torczon. Coloring heuristics for register allocation. *SIGPLAN Not.*, 39(4):283–294, 2004.

64. N. Jayasena, M. Erez, J.H. Anh, and W.J. Dally. Stream register files with indexed access. In *HPCA*, pages 60–72, February 2004.

65. Li Wang, Xuejun Yang, Jingling Xue, Yu Deng, Xiaobo Yan, Tao Tang, and Quan Hoang Nguyen. Optimizing scientific application loops on stream processors. In *LCTES '08: Proceedings of the 2008 ACM SIGPLAN-SIGBED Conference on Languages, Compilers, and Tools for Embedded Systems*, pages 161–170, ACM Press, New York, NY, 2008.

66. Abhishek Das, William J. Dally, and Peter Mattson. Compiling for stream processing. In *PACT '06: Proceedings of the 15th International Conference on Parallel Architectures and Compilation Techniques*, pages 33–42, ACM Press, New York, NY, 2006.

67. D. Nuzman, M. Namolaru, A. Zaks, and J.H. Derby. Compiling for an indirect vector register architecture. In *Proc of CF*, pages 199–205, May 2008.

Chapter 6

Optimization of the Dynamic Energy Consumption and Signal Mapping in Hierarchical Memory Organizations

Florin Balasa
American University in Cairo, New Cairo, Egypt

Ilie I. Luican
Microsoft Corp., Redmond, Washington

Hongwei Zhu
ARM, Inc., Sunnyvale, California

Doru V. Nasui
American International Radio, Inc., Rolling Meadows, Illinois

Contents

6.1 Introduction

Many signal processing systems are synthesized to execute data-dominated applications in various domains including video and image processing, artificial vision and medical imaging, real-time 3D rendering, advanced audio and speech coding. The behavior of many of these systems is described in a high-level programming language, where the code is typically organized in sequences of loop nests and the main data structures are multidimensional arrays whose references have as indices linear functions of the loop iterators.

The problem of memory allocation is central to any computer-aided design tool focusing on memory management. Since data transfer and storage have a significant impact on both the system performance and the major cost parameters—power consumption and chip area, the designer must spend a significant effort during the system development process on the exploration of the possible memory organizations in order to achieve a cost-optimized design [1,2].

In particular, the memory subsystem is, typically, a major contributor to the overall energy budget of the entire system [3]. The *dynamic* energy consumption is caused by memory accesses, whereas the *static* energy consumption is due to leakage currents. Savings of dynamic energy at the level of the whole memory subsystem can be potentially obtained by accessing frequently used data from smaller on-chip memories rather than from the large off-chip main memory, the problem being how to optimally assign the data to the memory layers.

As on-chip storage, the scratch-pad memories (SPMs)—compiler-controlled static random-access memories, more energy-efficient than the hardware-managed caches—are widely used in embedded systems, where caches incur a significant penalty (in aspects like area cost, energy consumption, hit latency) and where the flexibility of caches in terms of workload adaptability is often unnecessary, whereas power consumption and cost play a much more critical role [4]. Different from caches, the SPM occupies a distinct part of the virtual address space, with the rest of the address space occupied by the main memory. The consequence is that there is no need to check for the availability of the data in the SPM. Another consequence is that in cache memory systems, the mapping of data to the cache is done during the code execution, whereas in SPM-based systems this can be done either manually by the designer, or automatically—by a compiler, using a suitable algorithm, as this chapter will show.

Several approaches for the energy-efficient memory hierarchy design are intrinsically explorative, exploiting the fact that the memory design space can be discretized to allow near-exhaustive search. Such works assume a certain memory hierarchy with one or more levels of caching and an off-chip memory. A finite number of cache sizes and organization options are considered, as well as different off-chip memory alternatives (e.g., number of ports, number of banks). The best memory organization is obtained by simulating the workload for the possible alternative architectures. The various approaches mainly differ in the number of hierarchy levels, or the number of dimensions in the design space. Research works based on explorative techniques include [5–9]—that focus on cache memories, and [10]—that analyze embedded SRAMs. An advantage of the explorative techniques is that they allow concurrent evaluation of multiple cost metrics—such as performance and area—that can be used as constraints during exploration. The main limitation is that they require extensive data collection and only a small set of architectures are actually compared.

Other research works focused on the energy-efficient assignment of signals to the on- and off-chip memories. Their guiding strategy was to partition the signals from the application code into so-called *copy candidates* (since the on-chip memories were usually caches), and on the optimal selection and assignment of these to different layers into the memory hierarchy [11–14]. Their general idea was to identify the most frequently accessed data in each loop nest. Copying these heavily accessed data from the large off-chip memory to a smaller on-chip memory can potentially save energy (since most accesses will take place on the smaller copy and not on the large, more energy consuming, original array) and also improve performance. Many different possibilities exist for deciding which parts of the arrays should be copy candidates and, also, for selecting among the candidates those which will be instantiated as *copies* and their assignment to the different memory layers. For instance, Kandemir and Choudhary analyzed and exploited the temporal locality by inserting local copies [12]. Their layer assignment built a separate hierarchy per loop nest and then combined them into a single global hierarchy. However, the approach lacks a global view on the lifetimes of array elements in applications having imperfect nested loops. Brockmeyer et al. used the steering heuristic of assigning the arrays having the lowest access number over size ratio to the lowest memory layer first, followed by incremental reassignments [13]. They take into account the relative lifetime differences between arrays and between the scalars covered by each array. However, it is not clear whether the copy candidates can be also *parts* of arrays instead of entire arrays (and if so, how they identify these parts) since the access patterns are, in general, not uniform. Hu et al. used *parts* of arrays as copies, but they typically were cuts along the array dimensions [14] (like rows and columns of matrices).

Udayakumaran and Barua proposed a dynamic allocation model for SPM-based embedded systems [15], but the focus was global and stack data rather than multidimensional signals. Issenin et al. performed a data reuse analysis in a multilayer memory organization [16], but the mapping of the signals into the hierarchical data storage was not considered.

Within a given memory hierarchy level, power consumption can be reduced by memory partitioning techniques. The principle is to subdivide the address space in several smaller blocks and to map these blocks to different physical memory banks that can be independently enabled and disabled. Memory partitioning is a low-power approach, providing the opportunity of selectively shutting down the memory banks that are not accessed.[1] Arbitrary

---

[1] Memory partitioning can be also used as performance-oriented approach because of the reduced latency when accessing smaller memory blocks.

fine partitioning is typically prevented: an excessive number of small banks is area inefficient, imposing a severe wire overhead (which also tends to increase communication power and performance).

Techniques of an exploratory nature analyze possible partitions, matching them against the access patterns of the application [9,10]. Other approaches exploit the properties of the dynamic energy cost and the resulting structure of the partitioning space to come up with algorithms able to derive the optimal partition for a given access pattern [17,18]. Leakage-aware partitioning of memory structures is addressed at circuit-level—especially for caches. The cache-decay architecture turns off the cache lines during the time they are not used [19]. The drowsy cache architecture puts the cache lines into state-preserving low-power modes based on usage statistics [20]. These techniques, together with dynamic resizing, require the modification of the internal structure of caches, which are normally highly optimized designs. On a higher level of abstraction, Kandemir et al. exploit bank locality for maximizing the idleness, thus ensuring maximal amortization of the energy spent on memory re-activation [21]. Golubeva et al. proposed a trace-based architectural approach, considering both the dynamic and static energy consumption [22].

This chapter presents a memory allocation methodology for embedded data-intensive signal processing applications [23]. Starting from the high-level behavioral specification of a given application, where the code is organized in sequences of loop nests and the main data structures are multidimensional arrays, this framework performs the assignment of the multidimensional signals to the memory layers—the on-chip scratch-pad memory and the off-chip main memory—the goal being the reduction of the dynamic energy consumption in the memory subsystem [24,25]. The previous works do not take into account the nonuniform access patterns within the same arrays; they make a distinction only between the access intensity of *entire* arrays [13] (e.g., array A is more accessed than array B), or they try to heuristically identify the more accessed parts by imposing constraints on their shape and/or size (e.g., rows 1 and 2 of array A are more accessed than rows 3 and 4). In contrast to the previous works, our analysis is more refined, allowing to formally identify those intensely accessed areas of the array space—independent of their shape, size, or array dimensions. Assigning the most heavily accessed parts of the arrays into the scratch-pad layer (requiring less energy consumed per access) entails important savings of the dynamic energy consumption in the memory subsystem. This is why this framework is *energy-aware*.

Based on the assignment results, the framework subsequently performs the mapping of signals into both memory layers such that the overall amount of data storage be significantly reduced. Different from the previous works, this mapping technique is designed to work in hierarchical memory organizations [26,27], operating with parts of the arrays that can be assigned to different physical memories. The polyhedral framework, common to both design phases (the signal assignment to the memory layers and the signal mapping into the data memories), entails a high computation efficiency since both phases rely on similar polyhedral operations.

This software system yields a complete allocation solution [23]: the exact storage amount on each memory layer, the mapping functions that determine the exact locations for any array element (scalar signal) in the specification, metrics of quality for the allocation solution, and an estimation of the dynamic energy consumption in the memory subsystem using the CACTI power model [28,29].

The rest of the chapter is organized as follows. Section 6.2 presents an algorithm that assigns the signals to the memory layers, aiming to minimize the dynamic energy consumption in the hierarchical memory subsystem subject to SPM size constraints. Section 6.3 discusses several signal-to-memory mapping techniques. Section 6.4 presents the basic ideas of a novel signal mapping algorithm—able to cope with hierarchical memory organizations—used in a memory allocation framework. Section 6.5 addresses implementation aspects and discusses experimental results. Section 6.6 summarizes the main conclusions of this chapter.

6.2 Energy-Aware Signal Assignment to the Memory Layers

The algorithms describing the functionality of real-time multimedia and telecom applications are typically specified in a high-level programming language, where the code is organized in sequences of loop nests having as boundaries linear functions of the outer loop iterators. Conditional instructions are very common as well, and the multidimensional array references have (possibly complex) linear indexes (the variables being the loop iterators). As already mentioned in Chapter 3 (Section 3.3), this class of specifications is often referred to as *affine*.

Figure 6.1 shows an illustrative example whose structure and computation flow is similar to the kernel of a motion detection algorithm[2] [30]. The problem is to automatically identify those parts of arrays from the given application code that are more intensely accessed, in order to steer their assignment to the energy-efficient data storage layer (the on-chip scratch-pad memory) such that the dynamic energy consumption in the hierarchical memory subsystem be reduced. The number of storage accesses for each array element can certainly be computed by the simulated execution of the code.

The result of such a simulation is displayed in the maps from Figure 6.2. The top map represents the *index space* (that is, the set of vectors whose elements are all the possible values of the signal's indexes) of the 2D signal A

---

[2]The actual code contains also a *delay* operator, irrelevant in this context.

```
optDelta[0] = 0 ;                    // int A[81][81]: input;
for ( i=16; i<=64; i++ )
  for ( j=16; j<=64; j++ )
  { Delta[i][j][0] = 0 ;
    for ( k=i-16; k<=i+16; k++ )
      for ( l=j-16; l<=j+16; l++ )
        Delta[i][j][33*k-33*i+l-j+545] = A[i][j] - A[k][l]
                       + Delta[i][j][33*k-33*i+l-j+544] ;
      optDelta[49*i+j-799] = Delta[i][j][1089] + optDelta[49*i+j-800];
  }
opt[0] = optDelta[2401];
```

Figure 6.1 *Example 6.1*, whose code structure and computation flow is similar to a motion detection kernel [30] ($m = n = 16, M = N = 64$) [25]. (© 2009 IEEE).

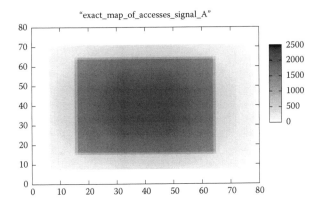

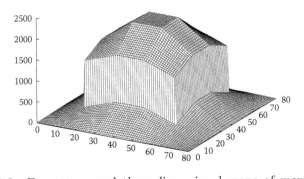

Figure 6.2 Exact two- and three-dimensional maps of memory *read* accesses, obtained by simulation, for the 2D signal A from *Example 6.1* [25]. (© 2009 IEEE).

from the illustrative code in Figure 6.1. For each pair of possible indexes (between 0 and 80), the number of accesses was counted and the level of grey in the map area depends on the intensity with which the array elements are accessed (the darker the color, the higher the number of accesses). In the bottom map, signal A's index space is in the horizontal plane xOy; the numbers of memory accesses for every A-element are on the vertical axis Oz. One can see the array elements near the center of the index space are accessed with high intensity (for instance, $A[40][40]$ is accessed 2178 times; $A[16][40]$ is accessed 1650 times), whereas the array elements at the periphery are accessed with a significantly lower intensity (for instance, $A[0][40]$ is accessed 33 times and $A[0][0]$ only once).

The drawbacks of such an approach are twofold. First, the simulated execution may be computationally ineffective when the number of array elements is very significant, or when the application code contains deep loop nests. Second, even if the simulated execution were feasible, such a scalar-oriented technique would not be helpful since the addressing hardware of the data memories would result very complex. An address generation unit (AGU) is typically implemented to compute arithmetic expressions in order to generate sequences of addresses [31] (see also Chapter 10); a set of array elements is not a good input for the design of an efficient AGU.

The energy-aware signal assignment to the memory layers is described below. The mathematical concepts have already been presented in Chapter 3, Sections 3.3 and 3.4.

Algorithm 6.1 *Given an affine behavioral specification, determine the assignment of various parts of arrays to the on-chip SPM memory layer (whose size is constrained) and to the off-chip memory layer such that the dynamic energy consumption in the memory subsystem be optimized.*

Step 1 *Let M be an indexed signal from the algorithmic specification. Decompose the array references $M[x_1(i_1, \ldots, i_n)] \cdots [x_m(i_1, \ldots, i_n)]$ into disjoint linearly bounded lattices (LBLs).*

The motivation of the decomposition of the array references in our context relies on the following intuitive idea: the elements belonging to a lattice included in many array references are likely to be more heavily accessed during the code execution.

This operation was explained in detail in Chapter 3, Section 3.4.3. The decomposition is performed analytically, by recursively intersecting the array references of every indexed (multidimensional) signal in the code.

Let M be an indexed signal in the algorithmic specification and let $\mathcal{L}_M$ be the initial set of M's lattices (these are the lattices of the array references of M). An inclusion graph is gradually built; this is a directed acyclic graph whose nodes are lattices of M and whose arcs denote inclusion relations between the respective sets. For instance, an arc from node X to node Y shows that the lattice X in included in Y.

Figure 6.3b shows such an inclusion graph and the result of the decomposition of the six ("bold") array references of the 2-D signal A in the illustrative example from Figure 6.3a. The graph displays the inclusion relations (*arcs*) between the lattices of A (*nodes*). The six "bold" nodes represent the six array references of signal A. For instance, the node $A1$ represents the lattice of $A[k][i]$ from the first loop nest. The nodes are labeled with the size of the corresponding lattice, that is, the number of array elements covered by it. In this example, $A1 \cap A2 = A4$ and $A1 - A4$ is also a bounded lattice—denoted $A7$ in Figure 6.3b. However, the difference $A6 - A17$ is not a lattice due to the non-convexity of this set, so the resulting set had to be decomposed further.

At the end of the decomposition, $\mathcal{L}_A$ contains the 17 lattices $A1, \ldots, A17$; each of the 11 nodes without incident arcs $A7, \ldots, A17$ represents a disjoint lattice in the partition $\mathcal{P}_A$ of the array space of A. Every array reference in the code is now either a disjoint lattice itself (like $A15$, $A16$, and $A17$), or a union of disjoint lattices (for instance, $A1 = A7 \cup A4 = A7 \cup A9 \cup A6 = A7 \cup A9 \cup \bigcup_{i=11}^{17} A_i$).

```
optDelta[0] = 0 ;                    // A[67][161] : input
for ( j=32 ; j<=128 ; j++)           // The first loop nest
{ Delta[32][j][0] = 0 ;
  for ( k=0 ; k<=64 ; k++)
    for ( i=j-32 ; i<=j+32 ; i++)
      Delta[32][j][65*k+i-j+33] = A[32][j] -A[k][i] // (*)
                         + Delta[32][j][65*k+i-j+32] ;
    optDelta[j-31] = Delta[32][j][4225] + optDelta[j-32] ;
}
for( j=32 ; j<=128 ; j++)            // The second loop nest
{ Delta[33][j][0] = 0 ;
  for( k=1 ; k<=65 ; k++)
    for( i=j-32 ; i<=j+32 ; i++)
      Delta[33][j][65*k+i-j-32] = A[33][j] -A[k][i]
                         + Delta[33][j][65*k+i-j-33] ;
    optDelta[j+66] = Delta[33][j][4225] + optDelta[j+65] ;
}
for( j=32 ; j<=128 ; j++)            // The third loop nest
{ Delta[34][j][0] = 0 ;
  for( k=2 ; k<=66 ; k++)
    for( i=j-32 ; i<=j+32 ; i++)
      Delta[34][j][65*k+i-j-97] = A[34][j] -A[k][i]
                         + Delta[34][j][65*k+i-j-98] ;
    optDelta[j+163] = Delta[34][j][4225] + optDelta[j+162] ;
}
opt = optDelta[291];                 // opt : output
                                     (a)
```

Figure 6.3   (a) *Example 6.2.* (b) Decomposition of the index space of the signal $A[x][y]$ into disjoint linearly-bounded lattices; the arcs in the graph show inclusion relations [27]. (© 2009 IEEE).

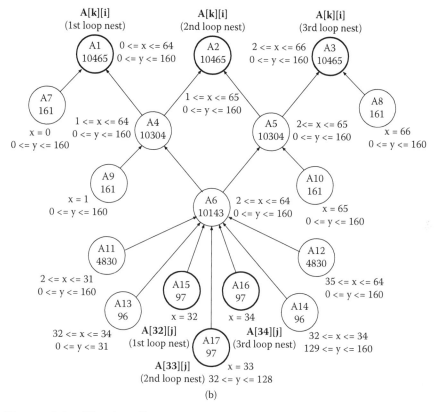

Figure 6.3 (Continued)

Step 2 *Compute the average number of memory accesses for each disjoint lattice of signal M.*

The total number of memory accesses to a certain linearly bounded lattice in the collection of disjoint lattices $\mathcal{P}_M$ is computed as follows:

Step 2.1 Select an array reference of M and intersect the given lattice with it. If the intersection result is not an empty set, it follows that the selected array reference and the given lattice have M-elements in common. The intersection is done in order to determine the expressions of the loop iterators for these common M-elements.

Step 2.2 Compute the number of points in the (nonempty) intersection. The number of accesses to the elements of the given lattice, as part of the selected array reference, is actually the size of a LBL—computed as explained in Chapter 3, Section 3.4.4.

Step 2.3 Repeat steps 2.1 and 2.2 for all the signal's array references in the code, cumulating the numbers of accesses to the given lattice.

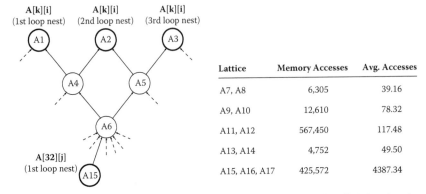

Lattice	Memory Accesses	Avg. Accesses
A7, A8	6,305	39.16
A9, A10	12,610	78.32
A11, A12	567,450	117.48
A13, A14	4,752	49.50
A15, A16, A17	425,572	4387.34

Figure 6.4 The computation of memory accesses for the disjoint lattices. *A*15 is included in four array references: *A*1, *A*2, *A*3, and itself. The number of accesses from each array reference is 5249, 5249, 5249, and 409,825. The total numbers of accesses are listed together with the average numbers of accesses per *A*-element [27]. (© 2009 IEEE).

Consider, for instance, the lattice $A15 = \{x = 32, y = j \mid 32 \le j \le 128\}$ derived from the illustrative example in Figure 6.3a. As *A*15 is included in the array references $A[k][i]$ from all the three loop nests and it also coincides with the operand $A[32][j]$ from the first loop nest (see Figure 6.4), the contributions of memory accesses for all these four array references must be computed. Since the lattice of the first array reference is $\{x = k, y = i \mid 128 \ge j \ge 32, 64 \ge k \ge 0, j + 32 \ge i \ge j - 32\}$, the expressions of the iterators mapping the array elements into *A*15 are $\{j = t_1, k = 32, i = t_2 \mid 128 \ge t_1, t_2 \ge 32, t_1 + 32 \ge t_2 \ge t_1 - 32\}$. The size of this lattice is 5249 (computed as explained in Chapter 3, Section 3.4.4), representing the number of memory accesses to *A*15 due to the first array reference $A[k][i]$. Performing similar computations, the distribution of accesses in the whole array space is computed.

The lattices *A*15, *A*16, and *A*17 covering roughly the center of the array space of signal *A* are the most accessed array parts (see Figure 6.4): 425,572 memory accesses (the total number of accesses being 2,458,950) for each of the three lattices *A*15, *A*16, and *A*17; therefore, an average of 4,387.34 accesses for each *A*-element covered by them.[3] Approximate maps of memory accesses can be thus computed: their advantage is that the (usually time-expensive) simulation is not needed any more, being replaced by algebraic computations.

Figure 6.5 displays in 3-D the distribution of memory accesses for the whole array space $A[67][161]$ of the 2D signal *A* from *Example 2*. The area covered by the lattices *A*15, *A*16, and *A*17 is intensely accessed: it appears like a steep mountain crest in the first 3D graph of the figure. The second graph

---

[3] Although the number of memory accesses is higher for *A*11 and *A*12, these lattices are significantly larger (4830 *A*-elements each) and, hence, the average number of accesses per element is much lower.

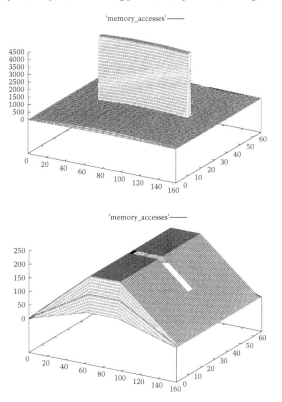

Figure 6.5 The distribution of memory accesses in the array space of the 2D signal *A* from *Example 6.2*.

displays the same memory access distribution, but with a scale limit of only 250 accesses: it can be seen better than in the first graph that the area around the three lattices is not flat (that is, the other array elements are not uniformly accessed), but still the access intensity is much lower.

Step 3 *Select the lattices having the highest average access numbers, whose total size does not exceed the maximum SPM size (assumed to be a design constraint), and assign them to the SPM layer. The other lattices will be assigned to the main memory.*

Storing on-chip all the signals is, obviously, the most desirable scenario in point of view of dynamic energy consumption. This is usually not possible since the SPM size is, typically, limited to smaller values than the overall data storage requirement of the algorithmic specification. In our tests (Section 6.5), we compute the ratio between the expected dynamic energy reduction and the SPM size after mapping (see Section 6.3); the value of the SPM size maximizing this ratio is selected, the goal being to obtain the maximum benefit in energy point of view for the smallest SPM size.

An optimally customized storage allocation solution comprises two memory layers: besides the background memory of 10,787,291 locations, an on-chip SPM of $3 \times 97 = 291$ locations (the total sizes of the lattices $A15$, $A16$, and $A17$), closer to the processor. The hierarchical data memory organization is compensated here by savings of dynamic energy of about 51.5%, according to the CACTI power model [28,29]. Indeed, assuming typical SPM and off-chip memory energy values for *read* accesses of 0.048 nJ and 3.57 nJ (for memory sizes as in this illustrative example), the dynamic energy consumption caused by accesses to the array A would decrease from 8.75 to 4.25 mJ.

6.3   Signal-to-Memory Mapping Techniques

Since these (typically large) arrays from the algorithmic specification must be stored during the execution of the application code, the memory allocation solution is significantly influenced on how these data structures from the specification are mapped into the physical memory. The goals of the *mapping* operation are the following:

(a) To assign the arrays from the behavioral specification into an amount of data storage as small as possible; moreover, to be able to compute this amount of storage (after mapping) and be able to determine the memory location of any array element from the specification;

(b) To use mapping functions simple enough in order to ensure an address generation hardware of a reasonable complexity;

(c) To ascertain that any distinct scalar signals (array elements) *simultaneously alive* are mapped to distinct storage locations. The lifetime of a scalar signal is the time interval between the clock cycles when the scalar is *produced* or written, and it is read for the last time, i.e., *consumed*, during the code execution. Two scalars are simultaneously alive if their lifetimes do overlap. Obviously, in such a case, they must occupy different memory locations; otherwise, they can share the same location.

Several mapping models have been proposed in the past, trading off between the first two goals (that is, accepting a certain excess of storage to ensure a less complex address generation hardware) while ascertaining that the third goal be strictly satisfied.

To reduce the size of a multidimensional array mapped to memory, De Greef et al. consider all the possible *canonical* linearizations of the array; for any linearization, they compute the largest distance at any time between two live elements in the linearized array [32]. This distance plus 1 is then the storage "window" required for the mapping of the array into the data memory.

More formally, $|W_A| = \min \max \{dist(A_i, A_j)\} + 1$, where $|W_A|$ is the size of the storage window of a signal A, the minimum is taken over all the canonical linearizations, while the maximum is taken over all the pairs of A-elements (A_i, A_j) simultaneously alive.

Note that for an m-dimensional array there are $m!$ orderings of the indices. For instance, a two-dimensional array can be typically linearized concatenating the rows, or concatenating the columns. In addition, the elements in a given dimension can be mapped in the increasing or decreasing order of the respective index. De Greef et al. consider in their model all these $2^m \cdot m!$ possible linearizations, called *canonical*. (For $m \geq 6$, an heuristic is proposed to limit the search.)

In order to avoid the inconvenience of analyzing different linearization schemes, Tronçon et al. proposed [33] to reduce the size of an m-dimensional array A mapped to the data memory, computing a window $W_A = (w_1, \ldots, w_m)$, whose elements can be used as operands in modulo operations that redirect all accesses to A. An access to the element $A[index_1] \ldots [index_m]$ is redirected to $W_A[index_1 \bmod w_1] \ldots [index_m \bmod w_m]$; in its turn, this bounding window is mapped (relative to a base address) into the physical memory by a typical canonical linearization, like row or column concatenation for 2D arrays. Signal A's window is denoted as W_A in any of the two models [32] and [33] (in spite of being one-dimensional in the case of the former model and m-dimensional for the latter) since the meaning will be clear enough from the context. Each window element w_k is computed as the maximum difference (in absolute value) between the kth indices of any two A-elements (A_i, A_j) simultaneously alive, plus 1. More formally, $w_k = \max \{|x_k(A_i) - x_k(A_j)|\} + 1$, for $k = 1, \ldots, m$. This ensures that any two array elements simultaneously alive are mapped to distinct memory locations. The bounding window is identical to the storage window from model [32] for 1D arrays. The amount of data memory required for storing (after mapping) the array A is the volume of the window W_A, that is, $|W_A| = \Pi_{k=1}^{m} w_k$.

Lefebvre and Feautrier, addressing parallelization of static control programs, developed in [34] an approach based on modular mapping as well. They first compute the lexicographically maximal "time delay" between the write and the last read operations, which is a super-approximation of the distance between conflicting index vectors (i.e., whose corresponding array elements are simultaneously alive). Then, the modulo operands are computed successively as follows: the modulo operand b_1, applied on the first array index, is set to 1 plus the maximal difference between the first indices over the conflicting index vectors; the modulo operand b_2 of the second index is set to 1 plus the maximal difference between the second indices over the conflicting index vector, when the first indices are equal; and so on.

Quilleré and Rajopadhye perform first an affine mapping into a linear space of *smallest* dimension (what they call a "projection") before modulo operations are applied to the array indices [35].

Darte et al. proposed a very refined mathematical framework, establishing a correspondence between valid linear storage allocations and integer lattices [36] called *strictly admissible* relative to the set of differences of the conflicting indices [37]. Heuristic techniques for building strictly admissible integer lattices, hence building valid storage allocations, are proposed.

All these signal-to-memory assignment approaches treat *separately* the arrays from the algorithmic specification, computing windows in the physical memory for each individual array. They exploit the possibility of memory sharing only between the elements of a same array. However, since the arrays are handled separately, the possibility of memory sharing between elements of different arrays is inherently ignored. This can lead to an excessive data storage, as Section 6.3.2 will illustrate. The implementation of some mapping techniques may detect simpler cases when, for instance, two entire arrays have disjoint lifetimes and, consequently, the two arrays may share the same (largest) window in the physical memory. Here, we are referring to the more general and typical situation when elements of different arrays are simultaneously alive.

Interestingly, the possibility of memory sharing between elements of different arrays with disjoint lifetimes was observed a long time ago (e.g., [38]) and it has been taken into account by several approaches for memory size evaluation (e.g., [39,40]). The memory sharing is sometimes called "inter-array in-place mapping" when the elements belong to different arrays, and "intra-array in-place mapping" when the elements belong to the same array [2]. It must be emphasized that these terms can create confusion since they do not necessarily refer to signal-to-memory mapping techniques, where an explicit correspondence between the array elements and their addresses in the physical memory is indicated; they rather refer to the possible reuse of data storage by array elements having disjoint lifetimes.

A novel signal-to-memory assignment approach exploiting the possibility of memory sharing between elements of different arrays [41] will be presented. Section 6.4 will explain the advantages of the novel assignment model, will describe the computation methodology, and will illustrate the algorithm flow. To the best of our knowledge, it is the only mapping approach with the capability of inter-array memory sharing even when the arrays do not have disjoint lifetimes, thus producing better savings of data storage than all the previous techniques.

The first part of this discussion will focus on the previous mapping models [32] and [33], since they are relevant for the rest of the chapter. The second part will illustrate the advantage of mapping with inter-array memory sharing.

6.3.1 Previous mapping models exemplified

The mapping model [32] will be first illustrated for *Example 6.3*—the loop nest from Figure 6.6a. The graph above the code represents the array (index) space of signal A. The points represent the A-elements $A[index_1][index_2]$ that are

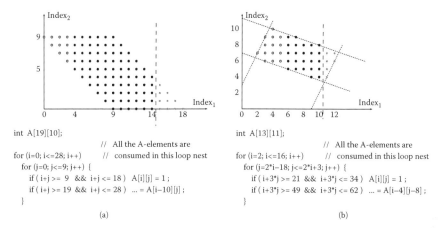

```
int A[19][10];
                          // All the A-elements are
for (i=0; i<=28; i++)     // consumed in this loop nest
  for (j=0; j<=9; j++) {
    if ( i+j >= 9  && i+j <= 18 )  A[i][j] = 1;
    if ( i+j >= 19 && i+j <= 28 )  ... = A[i−10][j];
  }

                          (a)
```

```
int A[13][11];
                          // All the A-elements are
for (i=2; i<=16; i++)     // consumed in this loop nest
  for (j=2*i−18; j<=2*i+3; j++) {
    if ( i+3*j >= 21 && i+3*j <= 34 )  A[i][j] = 1;
    if ( i+3*j >= 49 && i+3*j <= 62 )  ... = A[i−4][j−8];
  }

                          (b)
```

Figure 6.6 (a) *Example 6.3* whose memory allocation is better in storage point of view when the mapping model [32] is used. The graph shows the array (index) space of the 2D signal A at the end of the iteration ($i = 14, j = 4$). (b) *Example 6.4* has a similar code structure, but its allocation solution is better when the mapping model [33] is used. The graph shows the index space of the 2D signal A at the end of the iteration ($i = 10, j = 8$) [41]. (© 2011 Springer. With kind permission from Springer Science and Business Media).

produced (and also consumed) in the loop nest. In the figure, the A-elements are produced column by column from left to right. (Note that a column of points in the figure is actually a row of the 2D signal.) Inside each column, the A-elements are produced bottom-up; they are consumed in the same order, but with a "delay" of 10 columns. The points to the left of the dashed line represent the elements produced till the end of the iteration ($i = 14$, $j = 4$), the black points being the elements still alive (i.e., produced and still used as operands in the next iterations), while the circles represent A-elements already 'dead' (i.e., not needed as operands any more). The light grey points to the right of the dashed line are A-elements still unborn (to be produced in the next iterations). If we consider the array linearization by column concatenation in the increasing order of the columns [($A[index_1][index_2]$, $index_1 = 0, 18$), $index_2 = 0, 9$), the two elements simultaneously alive and placed the farthest apart from each other are $A[9][0]$ and $A[9][9]$. The distance between their positions in the linearization is $9 \times 19 = 171$. If the columns are concatenated decreasingly [($A[index_1][index_2]$, $index_1 = 0, 18$), $index_2 = 9, 0$], there are nine pairs of elements simultaneously alive, mapped at a maximum distance. Two such elements are, for instance, $A[14][0]$—produced in the iteration ($i = 14, j = 0$) and consumed in ($i = 24, j = 0$), and $A[4][9]$—produced in the iteration ($i = 4, j = 9$) and consumed in ($i = 14, j = 9$), the distance between them in the linearization being $9 \times 19 + 10 = 181$.

Now, if we consider the array linearization by row concatenation in the increasing order of the rows [($A[index_1][index_2]$, $index_2 = 0, 9$), $index_1 = 0, 18$], there are nine pairs of elements simultaneously alive, maximally distanced from each other. Two such elements are, for instance, $A[4][5]$—produced in the iteration $(i = 4, j = 5)$ and consumed in $(i = 14, j = 5)$, and $A[14][4]$—produced in the iteration $(i = 14, j = 4)$ and consumed in $(i = 24, j = 4)$. The distance between them in the linearization is $10 \times 10 - 1 = 99$. Finally, if the rows are concatenated decreasingly $[(A[index_1][index_2], index_2 = 0, 9)$, $index_1 = 18, 0]$, there are nine pairs of elements simultaneously alive as well, (e.g., $A[14][0]$ and $A[4][9]$), placed at the maximum distance $10 \times 10 + 9 = 109$.

For the other four linearizations, the maximum distances obtained have the same values (i.e., 171, 181, 99, 109), since the array elements are stored in reverse order relative to one of the four linearizations analyzed above.

According to the mapping model [32], the best linearization (among those considered above) for the array A is the concatenation row by row, increasingly. A memory window W_A of $99 + 1 = 100$ successive locations (relative to a certain base address) is sufficient to store the array without mapping conflicts: it is sufficient that any access to $A[index_1][index_2]$ be redirected to $W_A[(10 * index_1 + index_2) \bmod 100]$.

The assignment model proposed by Tronçon et al. [33] circumvents the need of analyzing different linearization schemes by computing a minimal bounding window, having the same dimension as the signal, large enough to cover all the array elements simultaneously alive in any moment of the computation. In *Example 6.3* shown in Figure 6.6a, the window corresponding to the signal A is $W_A = (11, 10)$. Indeed, the maximum horizontal and vertical distances between black points is $d_1 = 10$ and $d_2 = 9$, respectively. A 2D window whose elements are $w_1 = d_1 + 1$ and $w_2 = d_2 + 1$ is large enough to store the A-elements without any mapping conflict if any access to $A[index_1][index_2]$ is redirected to $W_A[index_1 \bmod 11][index_2 \bmod 10]$.

From the above discussion, it follows that the storage allocation for signal A is 100 locations if the mapping model [32] is used, whereas it is $w_1 \times w_2 = 110$ locations if the model [33] is applied. The better allocation result of the mapping model [32] for the illustrative code in Figure 6.6a should not be construed as proof that the strategy based on canonical linearizations is generally advantageous in storage point of view. In *Example 6.4* shown in Figure 6.6b, where the code has a similar structure as *Example 6.3*, the mapping model [33] yields a better allocation result.

Indeed, the 2D window corresponding to the array A is $W_A = (5, 6)$: it can be easily verified that the maximum distance between the first (respectively, second) indices of two alive elements cannot exceed 4 (respectively, 5). For instance, the simultaneously alive elements $A[6][9]$ and $A[10][4]$ (see the black points in the graph from Figure 6.6b) have both indices the farthest apart from each other. Therefore, a memory access to the element $A[index_1][index_2]$ can be safely redirected to $W_A[index_1 \bmod 5][index_2 \bmod 6]$, the data storage requirement after mapping being 30 locations.

Taking into account that the ranges of the two indexes are $12 - 2 + 1 = 11$ and, respectively, $10 - 4 + 1 = 7$, the maximum distance in the canonical linearization by column concatenation, in the increasing order of the columns, is 53 (e.g., between the elements $A[9][4]$ and $A[7][9]$), whereas in the decreasing order of the columns, it is 59. In the canonical linearization by row concatenation, the maximum distances are 31 (e.g., between the elements $A[6][5]$ and $A[10][8]$) and 33, respectively. It follows that the memory requirement using the mapping model [32] is 32 locations, slightly worse than the allocation result using the m-dimensional window model [33]. For many benchmarks, the latter model finds smaller window sizes: this happens especially when the array space contains *holes* and, also, when the size of the array can be reduced in every dimension, since, in such cases, any linearization will contain a number of unused array elements. Anyway, the two illustrative examples in Figure 6.6 show that, given an algorithmic specification, one cannot decide in advance which of the assignment techniques [32] and [33] yields a better solution.

It should be noticed that, actually, only 80 locations are really needed for the execution of *Example 6.3*, since no more than 80 A-elements can be simultaneously alive: see again the graph in Figure 6.6a, where the 80 black dots represent the live elements at the end of the iteration ($i = 14, j = 4$). Similarly, the minimum data storage required for the execution of *Example 6.4* is 24 locations: see the graph in Figure 6.6b, where the 24 black dots represent the live elements at the end of the iteration ($i = 10, j = 8$). Figure 6.7 shows the variation of the storage requirement for the signal A during the execution of both codes. These memory traces are generated using the tool described in [40], allowing to evaluate the minimum data storage. A *minimum* array window is not only difficult to compute but, typically, difficult to use in practical allocation problems as, in most of the cases, it requires a significantly more complex addressing hardware. A signal-to-memory mapping model like the ones described above trades-off an excess of storage for a less complex address generation hardware.

6.3.2   Exploiting inter-array memory sharing

Let us analyze now *Example 6.5* in Figure 6.8a, where the A-elements produced in the first loop nest are consumed in the second loop nest, and the B-elements produced in the second loop nest are consumed in the second one as well. The variation of the storage requirements for each of the signals A and B, as well as for the entire code, are shown in Figure 6.8b. If we assume that the A-elements and B-elements are stored in separate windows of the physical memory, the minimum data memory (maximum storage requirement) for A is 16 locations, while for B it is 23 locations (therefore, a total of 39 locations). Otherwise, since part of the B-elements can be stored in memory locations previously occupied by consumed A-elements, the minimum data storage ensuring the code execution is only 23 locations, as shown by the second graph in Figure 6.8b.

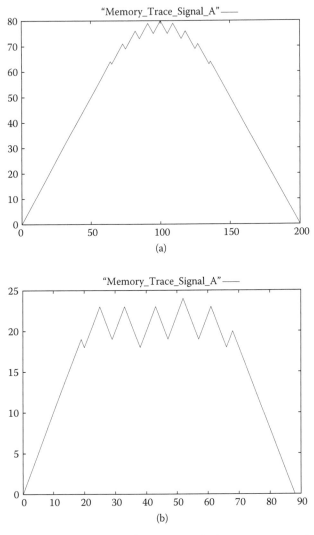

Figure 6.7 (a) Memory trace (variation of the storage requirement) for the signal A in *Example 6.3*. The abscissae are the number of assignment operations and the ordinates are the number of occupied memory locations. (b) Memory trace for the signal A in *Example 6.4* [41]. (© 2011 Springer. With kind permission from Springer Science and Business Media).

An analysis—similar to the one in Section 6.3.1—of the canonical linearizations of the signals A and B yields mapping windows of sizes $|W_A| = 22$ and $|W_B| = 25$ locations, with assignment functions

$$A[index_1][index_2] \mapsto W_A[(4 * index_1 + index_2) \bmod 22]$$

```
int A[7][4], B[11][6] ;

for ( i=0; i<=6; i++ )
    for ( j=0; j<=3; j++ )
        if ( 3<=i+j && i+j<=6 ) A[i][j] = ... ;

for ( i=0; i<=14; i++ )
    for ( j=0; j<=5; j++ )
    { if ( 5<=i+j )
        if ( i<=3 ) B[i][j] = A[i][j-2] + A[6-i][5-j] ;
        else if ( i+j<=10 ) B[i][j] = ... ;
        if ( 9<=i+j && i+j<=14 ) ... = B[i-4][j];
    }
```

(a)

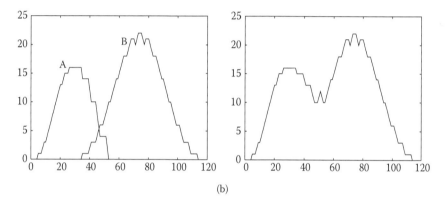

(b)

Figure 6.8 (a) *Example 6.5* with two arrays. (b) The variation of the storage requirements during the execution of the code: for the signals *A* and *B* (the first graph), and for the whole code (the second graph) [41]. (© 2011 Springer. With kind permission from Springer Science and Business Media).

and

$$B[index_1][index_2] \mapsto W_B[(6 * index_1 + index_2) \bmod 25],$$

relative to some base address. Hence, the model [32] yields a data memory of $|W_A| + |W_B| = 22 + 25 = 47$ locations.

Applied to *Example 6.5*, the mapping model [33] computes 2D windows for the signals *A* and *B*, large enough along each dimension to bound their elements simultaneously alive. Without further details, these bounding windows result to be $W_A = (7, 4)$ and $W_B = (5, 6)$. The assignment functions mapping the array spaces into these windows are $A[index_1][index_2] \mapsto W_A[index_1 \bmod 7][index_2 \bmod 4]$ and, respectively, $B[index_1][index_2] \mapsto W_B[index_1 \bmod 5][index_2 \bmod 6]$. Hence, the model [33] yields a data memory of $|W_A| + |W_B| = 28 + 30 = 58$ locations.

Both these previous signal assignment techniques yield data memory allocation solutions quite poor for this illustrative example, at least twice the size of the minimum data memory (of 23 locations). The main cause of this relatively poor behavior is the separate handling of the arrays. Since the mapping windows are all disjoint, no assignment technique adopting such a strategy could yield a better result than the sum of the minimum storage requirement per signal, that is, $16 + 23 = 39$ locations in this case.

Now, suppose the A-elements and B-elements are mapped first as in De Greef's model [32]: $A[index_1][index_2] \mapsto W_A[(4 * index_1 + index_2) \bmod 22]$ and $B[index_1][index_2] \mapsto W_B[(6 * index_1 + index_2) \bmod 25]$. If the two memory windows W_A and W_B are contiguous, the maximum distance between locations simultaneously occupied by live elements is 32. This is the distance between the locations occupied by $A[3][0]$ and $B[3][4]$: $dist(W_A[12], W_B[22]) = 32$, if W_B follows W_A in the physical memory. This means that a common window W_{AB} of size 33 is actually sufficient for both signals. Indeed, it can be verified that mapping A and B in the same window W_{AB} such that $A[index_1][index_2] \mapsto W_{AB}[(4 * index_1 + index_2) \bmod 22]$ and $B[index_1][index_2] \mapsto W_{AB}[(22 + (6 * index_1 + index_2) \bmod 25) \bmod 33]$ does not create any conflict between simultaneously alive elements. Note that an amount of 33 storage locations after mapping is better than the lower bound of 39 locations when the arrays are handled separately. Actually, there is an even better result of only 25 locations, very close to the absolute minimum of 23 locations when the assignment function for the B-elements is $B[index_1][index_2] \mapsto W_{AB}[(6 * index_1 + index_2 + 14) \bmod 25]$, the mapping of A being the same.

There is a price to be paid though: the cost of the address generation hardware may increase (here, the mapping function for B is more complex), which is not unexpected. However, in the system-level exploration phase, the designer must be offered as many possible meaningful options.

This illustrative example was used to convincingly prove that developing signal-to-memory assignment techniques allowing memory sharing between different arrays can be very beneficial in terms of data storage.

The next section will present a signal-to-memory mapping algorithm very efficient in terms of amount of data storage. In a first phase, the two mapping models [32] and [33] are implemented within our polyhedral framework [42], which is advantageous both from the point of view of computational efficiency and the amount of allocated data storage—since the allocation solution for each signal is the best of the two models. In a second phase, the assignment algorithm contains a mechanism of investigating and exploiting the possibility of memory sharing between elements of different arrays [41].

Different from all the previous works that do not provide realistic metrics of quality for their memory allocation solutions, this memory management software tool also computes the minimum storage requirement of each multidimensional signal in the specification [40] (therefore the optimal memory sharing between the elements of the same array), as well as the minimum data storage for the entire algorithmic specification (therefore the optimal memory sharing between all the array elements and scalars in the code), as explained in Chapter 3.

6.4 The Signal-to-Memory Mapping Model

The idea of the signal assignment algorithm is to start from the mapping solution of either the model based on canonical linearizations [32], or the model based on a multidimensional bounding window [33]—which one is better for the given algorithmic specification (see the discussion in Section 6.3.1). Afterward, search for a pairwise grouping of the array windows that will yield the maximum benefit in terms of data storage reduction by mutual memory sharing.

First, Section 6.4.1 will present in detail the computation of the mapping windows of a linearly bounded lattice (LBL) according to the two mapping models. The flow of the entire algorithm will be described in Section 6.4.2. Afterward, in Section 6.4.3, an example will illustrate its main steps.

6.4.1 The mapping window of an array reference

The computation method of an array window employed by De Greef et al. consists of a sequence of integer linear programming (ILP) optimizations for each array linearization [32]. To compute the minimal bounding window for the live elements of an array, Tronçon et al. perform first a "liveness analysis." During this phase, a set of program points is firstly decided. For instance, if the code contains n nested loops, $2n + 2$ program points are considered: one at the beginning of each loop body and one at the exit from each loop body, together with points at the start and at the end of the code. Every program point is annotated with a set of $\mathbf{Z}$-polyhedra, whose integer solutions specify sets of live array elements. In this way, all the live array elements in the chosen program points can be determined. Afterward, an evaluation of the lower-bounds of window sides w_k is performed. Starting with $w_k = 1$, all the integer solutions differing in the kth coordinate ($x'_k \neq x''_k$) of the computed $\mathbf{Z}$-polyhedra are tested whether $x'_k \bmod w_k \neq x''_k \bmod w_k$ or not. The equality implies that a potential mapping conflict was found and, consequently, the window side w_k should be increased. Subsequent incremental upward adjustments of w_k are performed for all the chosen program points till no mapping conflict occurs.

In contrast, our methodology is developed within a polyhedral framework operating with ($\mathbf{Z}$-) polytopes and linearly bounded lattices (see Chapter 3, Sections 3.3.3 and 3.3.4). The key idea of this approach is the reduction of the computation of memory windows for entire *arrays* to the computation of windows for *lattices*. Note that the way an array window is defined depends on the mapping model employed, as exemplified in Section 6.3.1 for the two previous models [32] and [33]. Consequently, the computation of windows for lattices will be also dependent on the mapping model employed. Leaving this difference aside, the rest of the computation is practically independent of the mapping model.

In the model [32], the problem is to compute a maximal one-dimensional window for every canonical linearization of the array. Let us consider the array reference $A[i][j]$ from the code in *Example 6.6*, assuming that its elements remain all alive at the end of the loop nest.

Example 6.1

```
for (i=2; i<=12; i++)
        for (j=2*i-18; j<=2*i+3; j++)
            if (i+3*j ≥21  &&  i+3*j ≤ 34)  A[i][j] =  ...
```

The index space of the 2D signal A is than shown in Figure 6.9a, all the points (A-elements) being black (alive) after the execution of the loop nest.

Now take, for instance, the linearization by row concatenation in the increasing order of the rows. Denoting x and y in the two indexes, the distance between two A-elements $A_1(x_1, y_1)$ and $A_2(x_2, y_2)$ is: $dist(A_1, A_2) = (x_2 - x_1)\Delta y + (y_2 - y_1)$, where $\Delta y = \max\{y\} - \min\{y\} + 1$ is the range of the second index in the array space. To ensure that the distance is a nonnegative number, we shall assume that $[x_2\ y_2]^T \succ [x_1\ y_1]^T$ is relative to the lexicographic order.[4] Then, it can be easily observed that the maximum distance is reached when the index vectors of the A-elements in the array reference are the minimum and, respectively, the maximum relative to the lexicographic order. These array A-elements are represented by the points $M = A[2][7]$ and $N = A[12][7]$ in Figure 6.9(a), and $dist(M, N) = (12 - 2) \times 7 + (0 - 0) = 70$.

Similarly, the linearization by column concatenation in the increasing order of the columns, the distance between two A-elements $A_1(x_1, y_1)$ and $A_2(x_2, y_2)$ is: $dist(A_1, A_2) = (y_2 - y_1)\Delta x + (x_2 - x_1)$, where $\Delta x = \max\{x\} - \min\{x\} + 1$ is the range of the first index in the array space. To ensure that the distance is a nonnegative number, we shall assume that $[y_2\ x_2]^T \succ [y_1\ x_1]^T$ is relative to the lexicographic order. Then, the array elements at the maximum distance from each other are still the elements with (lexicographically) minimum and maximum index vectors, provided an interchange of the indices is applied first. In Figure 6.9b, the elements $A[4][9]$ and $A[10][4]$ (represented by the points M and N) are the farthest away from each other; they correspond to the elements $A[9][4]$ and $A[4][10]$ in the initial figure, and the distance between them is $(10 - 4) \times 11 + (4 - 9) = 61$.

It follows that finding the points in a lattice having the (lexicographically) minimum and maximum index vectors is crucial for De Greef's model [32].

---

[4]Let $\mathbf{x} = [x_1, \ldots, x_m]^T$ and $\mathbf{y} = [y_1, \ldots, y_m]^T$ be two m-dimensional vectors. The vector $\mathbf{y}$ is larger lexicographically than $\mathbf{x}$ (written $\mathbf{y} \succ \mathbf{x}$) if $(y_1 > x_1)$, or $(y_1 = x_1$ and $y_2 > x_2)$, $\ldots$, or $(y_1 = x_1, \ldots, y_{m-1} = x_{m-1}$, and $y_m > x_m)$.

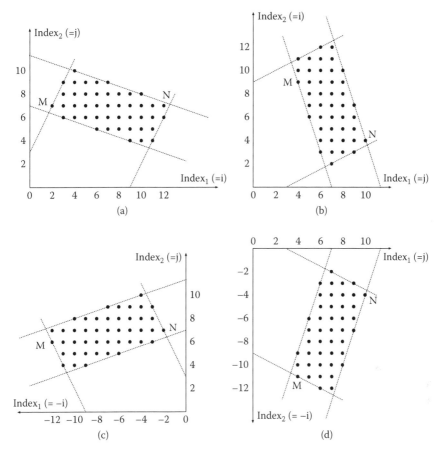

Figure 6.9 (a) The index space of the array reference $A[i][j]$ from *Example 6.6.* (b) The index space of the array reference $A[j][i]$ (with indexes interchanged) assuming the same iterator space. (c and d) The index spaces of the array references $A[-i][j]$ and $A[j][-i]$ [23]. (© 2009 IEICE).

For any canonical linearization, it is sufficient to apply an index permutation first, followed at the end by the inverse permutation of the resulting index vectors. The distance between the array elements $A_i(x_1^i, x_2^i, \ldots, x_m^i)$ and $A_j(x_1^j, x_2^j, \ldots, x_m^j)$ is:

$$dist(A_i, A_j) = \left(x_1^j - x_1^i\right)\Delta x_2 \cdots \Delta x_m + \left(x_2^j - x_2^i\right)\Delta x_3 \cdots \Delta x_m$$
$$+ \cdots + \left(x_{m-1}^j - x_{m-1}^i\right)\Delta x_m + \left(x_m^j - x_m^i\right)$$

where the index vector of A_j is (lexicographically) larger than that of A_i.

If in the linearization some dimension is traversed backward, then a simple transformation reversing the index variation must be also applied as shown in Figure 6.9c—row concatenation in the decreasing order of the columns, where $dist(M, N) = 71$—and in Figure 6.9d—column concatenation in the decreasing order of the columns, where $dist(M, N) = 73$.

Computing the (lexicographically) minimum and maximum index vectors of the array reference $A[i][j]$ in *Example 6.6* can be done easier since the iterator space of the array reference is identical to its index space (the indexes being equal to the iterators). In general, it is desirable to reduce this problem to the computation of the minimum and maximum *iterator* vectors since the iterator space is a ($\mathbf{Z}$-) polytope (or decomposable into disjoint polytopes), whereas the index space is not always a ($\mathbf{Z}$-) polytope: see, for instance, the holes in the quadrilateral from the right-hand side of Figure 3.4 of Chapter 3. Moreover, note that the maximum/minimum *iterator* vector is not always mapped into the maximum/minimum *index* vector. For instance, the (lexicographically) maximum iterator vector $[i \ j \ k]^T = [2 \ 0 \ 0]^T$ from *Example 3.1* of Chapter 3 is mapped to the index vector $[x \ y]^T = [5 \ 0]^T$, which is not the largest index vector. (As Figure 3.4 of Chapter 3 clearly shows, $[3, 9]^T$ is the largest index vector.)

However, if the mapping matrix $\mathbf{T}$ of the array reference is brought first to the Hermite Normal Form (HNF) [36] (assuming for simplicity that it has full rank) with all the diagonal elements strictly positive,[5] then the maximum/minimum iterator vectors of the transformed polytope (by the same unimodular matrix) will be mapped to the maximum/minimum index vectors: indeed, it easy to prove that $\mathbf{i_2} \succ \mathbf{i_1}$ entails $\mathbf{x_2} \succ \mathbf{x_1}$, when $\mathbf{x} = \mathbf{T} \cdot \mathbf{i} + \mathbf{u}$ and $\mathbf{T}$ is lower-triangular with strictly positive diagonal elements.

Algorithm 6.2 *Computation of the storage window of a given array reference according to the mapping model based on canonical linearizations [32].*

Step 1 Let $\mathbf{x} = \mathbf{T} \cdot \mathbf{i} + \mathbf{u}$ be the mapping of the array reference, and its iterator space be $P = \{\mathbf{i} \in \mathbf{Z}^n \mid \mathbf{A} \cdot \mathbf{i} \geq \mathbf{b}\}$. (It can be assumed that the index space of the given array reference can be represented by one LBL.) Let $\mathbf{S}$ be an n-order unimodular matrix (a square matrix whose determinant is ± 1) bringing $\mathbf{T}$ to the Hermite Normal Form (HNF) [36]: $\mathbf{H} = \mathbf{T} \cdot \mathbf{S}$ (see, e.g., [36] for building $\mathbf{S}$).

Step 2 After applying the unimodular transformation $\mathbf{S}$, the new iterator polytope becomes:

$$\bar{P} = \{\bar{\mathbf{i}} \in \mathbf{Z}^n \mid \mathbf{A} \cdot \mathbf{S} \cdot \bar{\mathbf{i}} \geq \mathbf{b}\}.$$

Step 3 Compute the maximum (minimum) iterator vector in the ($\mathbf{Z}$-) polytope $\bar{P}$ as follows. Compute the maximum (minimum) value of $\bar{i}_1$ (the first element of $\bar{\mathbf{i}}$) by projecting the polytope $\bar{P}$ on the first axis [44]. Verify that $\bar{i}_1$ is in the "dark shadow" [44], otherwise decrement (increment) it till it is. Then, replacing this value in $\bar{P}$, compute the maximum (minimum) value of $\bar{i}_2$ by projection on the second axis, and so on. The iterator vector whose elements

---

[5]If $\mathbf{T}$ does not have full rank, a *reduced* Hermite form [43] will suffice.

are determined as explained above is the maximum (minimum) iterator vector in lexicographic order, denoted $\bar{\mathbf{i}}^{max}$ (respectively, $\bar{\mathbf{i}}^{min}$).

Step 4 Compute the maximum and minimum index vectors of the array reference:

$$\mathbf{x}^{min} = \mathbf{H} \cdot \bar{\mathbf{i}}^{min} + \mathbf{u} \text{ and } \mathbf{x}^{max} = \mathbf{H} \cdot \bar{\mathbf{i}}^{max} + \mathbf{u}.$$

The mapping window of the array reference has the size: $|W| = dist(\mathbf{x}^{min}, \mathbf{x}^{max}) + 1$.

The algorithm will be illustrated for the array reference A[i+2*j+3][j+2*k] from *Example 3.1* of Chapter 3, assuming first the linearization by row concatenation. Since the unimodular matrix $\mathbf{S} = \begin{bmatrix} 1 & -2 & 4 \\ 0 & 1 & -2 \\ 0 & 0 & 1 \end{bmatrix}$ brings matrix $\mathbf{T} = \begin{bmatrix} 1 & 2 & 0 \\ 0 & 1 & 2 \end{bmatrix}$ to the HNF:

$$\mathbf{H} = \mathbf{T} \cdot \mathbf{S} = \begin{bmatrix} 1 & 0 & 0 \\ 0 & 1 & 0 \end{bmatrix},$$

the new iterator polytope (*Step 2*) is computed:

$$\bar{P} = \{\bar{\mathbf{i}} \in \mathbf{Z}^n \,|\, \mathbf{A} \cdot \mathbf{S} \cdot \bar{\mathbf{i}} \geq \mathbf{b}\}$$

$$= \left\{ \begin{bmatrix} \bar{i} \\ \bar{j} \\ \bar{k} \end{bmatrix} \in \mathbf{Z}^3 \,\middle|\, \begin{bmatrix} 1 & -2 & 4 \\ 0 & 1 & -2 \\ 0 & 0 & 1 \\ -6 & 8 & -19 \end{bmatrix} \begin{bmatrix} \bar{i} \\ \bar{j} \\ \bar{k} \end{bmatrix} \geq \begin{bmatrix} 0 \\ 0 \\ 0 \\ -12 \end{bmatrix} \right\}.$$

By successive projections on the axes (*Step 3*), the (lexicographically) maximum iterator vector in $\bar{P}$ results

$$\begin{bmatrix} \bar{i} \\ \bar{j} \\ \bar{k} \end{bmatrix}^{max} = \begin{bmatrix} 6 \\ 3 \\ 0 \end{bmatrix},$$

which yields the maximum index vector

$$\begin{bmatrix} x \\ y \end{bmatrix}^{max} = \mathbf{H} \begin{bmatrix} \bar{i} \\ \bar{j} \\ \bar{k} \end{bmatrix}^{max} + \mathbf{u} = \begin{bmatrix} 9 \\ 3 \end{bmatrix}.$$

The (lexicographically) minimum iterator vector in $\bar{P}$ is

$$\begin{bmatrix} \bar{i} \\ \bar{j} \\ \bar{k} \end{bmatrix}^{min} = \begin{bmatrix} 0 \\ 0 \\ 0 \end{bmatrix},$$

which yields the minimum index vector

$$\begin{bmatrix} x \\ y \end{bmatrix}^{min} = \mathbf{H} \begin{bmatrix} \bar{i} \\ \bar{j} \\ \bar{k} \end{bmatrix}^{min} + \mathbf{u} = \begin{bmatrix} 3 \\ 0 \end{bmatrix};$$

the distance between the A-elements $A[3][0]$ and $A[9][3]$ is 57.

In the linearization by column concatenation, the array indices are reversed. Matrix $\mathbf{T}$ is thus $\begin{bmatrix} 0 & 1 & 2 \\ 1 & 2 & 0 \end{bmatrix}$ and $\mathbf{u} = \begin{bmatrix} 0 \\ 3 \end{bmatrix}$. The unimodular matrix $\mathbf{S} = \begin{bmatrix} -2 & 1 & 4 \\ 1 & 0 & -2 \\ 0 & 0 & 1 \end{bmatrix}$ and the Hermite Normal Form is:

$$\mathbf{H} = \mathbf{T} \cdot \mathbf{S} = \begin{bmatrix} 1 & 0 & 0 \\ 0 & 1 & 0 \end{bmatrix}.$$

The new iterator polytope is:

$$\bar{P} = \left\{ \begin{bmatrix} \bar{i} \\ \bar{j} \\ \bar{k} \end{bmatrix} \in \mathbf{Z}^3 \ \left| \ \begin{bmatrix} -2 & 1 & 4 \\ 1 & 0 & -2 \\ 0 & 0 & 1 \\ 8 & -6 & -19 \end{bmatrix} \begin{bmatrix} \bar{i} \\ \bar{j} \\ \bar{k} \end{bmatrix} \geq \begin{bmatrix} 0 \\ 0 \\ 0 \\ -12 \end{bmatrix} \right. \right\}.$$

The (lexicographically) maximum iterator vector in $\bar{P}$ is

$$\begin{bmatrix} \bar{i} \\ \bar{j} \\ \bar{k} \end{bmatrix}^{max} = \begin{bmatrix} 8 \\ 0 \\ 4 \end{bmatrix},$$

which yields the maximum index vector

$$\begin{bmatrix} x \\ y \end{bmatrix}^{max} = \mathbf{H} \begin{bmatrix} \bar{i} \\ \bar{j} \\ \bar{k} \end{bmatrix}^{max} + \mathbf{u} = \begin{bmatrix} 8 \\ 3 \end{bmatrix}.$$

The (lexicographically) minimum iterator vector in $\bar{P}$ is

$$\begin{bmatrix} \bar{i} \\ \bar{j} \\ \bar{k} \end{bmatrix}^{min} = \begin{bmatrix} 0 \\ 0 \\ 0 \end{bmatrix},$$

which yields the minimum index vector

$$\begin{bmatrix} x \\ y \end{bmatrix}^{min} = \mathbf{H} \begin{bmatrix} \bar{i} \\ \bar{j} \\ \bar{k} \end{bmatrix}^{min} + \mathbf{u} = \begin{bmatrix} 0 \\ 3 \end{bmatrix}.$$

Since the array indices were interchanged, these index vectors correspond to the array elements $A[3][0]$ and $A[3][8]$, the distance between them being 56.

In the mapping model [33], the problem is to compute the projection spans of the index space of the array reference on all the m coordinate axes, the elements of the m-dimensional bounding window being the sizes of these m projection spans. Computing the projections of the array reference $A[i][j]$ in *Example 6.6* can be done easier since the iterator space of the array reference is identical to its index space (the indexes being equal to the iterators). Here, the bounding window of the array reference is $W = (12 - 2 + 1, 10 - 4 + 1) = (11, 7)$.

In general, it is desirable to reduce this problem to the computation of the projections of the iterator space since this is a (**Z**-) polytope (or is reducible to disjoint polytopes), which is a well-studied problem [44,45]. The general idea for solving this problem is to find a transformation **S** such that the extreme values of one of the iterators (here, we shall consider the first iterator) correspond to the extreme values of, say, the kth index, for any value of k.

Algorithm 6.3 *Computation of the storage window of a given array reference according to the mapping model based on the minimal bounding window* [33].

Step 1 Let $\mathbf{x} = \mathbf{T} \cdot \mathbf{i} + \mathbf{u}$ be the mapping of the array reference, and its iterator space be $P = \{\mathbf{i} \in \mathbf{Z}^n \mid \mathbf{A} \cdot \mathbf{i} \geq \mathbf{b}\}$. The kth index has the expression: $x_k = \mathbf{t}_k \cdot \mathbf{i} + u_k$, where $\mathbf{t}_k$ is the kth row of the matrix **T**. Let **S** be a n-order unimodular matrix bringing $\mathbf{t}_k$ to the Hermite Normal Form [36]: $[h_1 \ 0 \cdots 0] = \mathbf{t}_k \cdot \mathbf{S}$, where $h_1 > 0$. (If the row $\mathbf{t}_k$ is null, then the window reduces to one point: $x_k^{min} = x_k^{max} = u_k$.)

Step 2 After applying the unimodular transformation **S**, the new iterator polytope becomes:

$$\bar{P} = \{\bar{\mathbf{i}} \in \mathbf{Z}^n \mid \mathbf{A} \cdot \mathbf{S} \cdot \bar{\mathbf{i}} \geq \mathbf{b}\}.$$

Step 3 Compute the extreme values of $\bar{i}_1$ (denoted $\bar{i}_1^{min}$ and $\bar{i}_1^{max}$) by projecting the (**Z**-) polytope $\bar{P}$ on the first axis [44]. Then, $x_k^{min} = h_1 \bar{i}_1^{min} + u_k$ and $x_k^{max} = h_1 \bar{i}_1^{max} + u_k$. The kth window element $w_k = x_k^{max} - x_k^{min} + 1$.

Performing this 3-step algorithm for $k = 1, \ldots, m$, the mapping window $(w_1, \ldots, w_m)$ of the array reference is obtained. Its size is: $|W| = \Pi_{k=1}^m w_k$.

The algorithm will be exemplified projecting the array reference from *Example 3.1* of Chapter 3 on the first axis and finding the extreme values of the first index x, where $[x] = \begin{bmatrix} 1 & 2 & 0 \end{bmatrix} \begin{bmatrix} i \\ j \\ k \end{bmatrix} + [3]$. The unimodular matrix $\mathbf{S} = \begin{bmatrix} 1 & -2 & 0 \\ 0 & 1 & 0 \\ 0 & 0 & 1 \end{bmatrix}$ brings $\mathbf{t}_1 = \begin{bmatrix} 1 & 2 & 0 \end{bmatrix}$ to the Hermite Normal Form:

$\mathbf{t}_1 \cdot \mathbf{S} = \begin{bmatrix} 1 & 0 & 0 \end{bmatrix}$. The new iterator polytope (*Step 2*) is:

$$\bar{P} = \left\{ \begin{bmatrix} \bar{i} \\ \bar{j} \\ \bar{k} \end{bmatrix} \in \mathbf{Z}^3 \,\middle|\, \begin{bmatrix} 1 & -2 & 0 \\ 0 & 1 & 0 \\ 0 & 0 & 1 \\ -6 & 8 & -3 \end{bmatrix} \begin{bmatrix} \bar{i} \\ \bar{j} \\ \bar{k} \end{bmatrix} \geq \begin{bmatrix} 0 \\ 0 \\ 0 \\ -12 \end{bmatrix} \right\}.$$

Eliminating $\bar{j}$ and $\bar{k}$ from the inequalities of $\bar{P}$ with a Fourier-Motzkin technique [46], the extreme values of the *exact shadow* [44] of $\bar{P}$ on the first axis are $\bar{i}^{min} = 0$ and $\bar{i}^{max} = 6$, and these extreme points on the first axis in the iterator space are valid projections. Since $h_1 = 1$ and $u_1 = 3$, it follows immediately that $x^{min} = 3$ and $x^{max} = 9$, which can be easily observed from Figure 3.4 in Chapter 3. With a similar computation, it follows that $y^{min} = 0$ and $y^{max} = 8$. Therefore, the bounding window of the array reference $A[i + 2 * j + 3][j + 2 * k]$ is $W_A = (7, 9)$.

6.4.2   The flow of the signal-to-memory mapping algorithm

Algorithm 6.4 *Computation of a global mapping solution for all the multidimensional signals from a given affine behavioral specification.*

Step 1 For every array A in the algorithmic specification, compute the size of the memory window $|W_A|$ as the minimum window of the two assignment models [32] and [33].

Keeping in mind the distinct computation methodologies for the memory windows of lattices discussed in Section 6.4.1, the computation flow of *Step 1* is the following:

Step 1a *Extract the array references from the given algorithmic specification and decompose the array references for every multidimensional signal into disjoint linearly-bounded lattices.*

This operation was thoroughly presented in Chapter 3, Section 3.4.3. It was also used in *Algorithm 6.1* at *Step 1*. (Obviously, this operation must be done only once in the framework if several memory management algorithms are using it.) In this way, the next substeps will have to deal only with windows of lattices in order to eventually obtain the windows of entire arrays.

Step 1b *Compute underestimations of the window sizes for each indexed signal taking into account the live elements at the boundaries between the loop nests.*

Let A be an m-dimensional signal in the algorithmic specification, and let $\mathcal{P}_A$ be the set of disjoint lattices partitioning the index space of A. A high-level pseudo-code of the computation of A's preliminary windows is given below. Preliminary window sizes for each canonical linearization according to DeGreef's model [32] are computed first, followed by the computation of the window size underestimate according to Tronçon's model [33] in the same framework operating with lattices. The meaning of the variables are explained as comments.

> **for** (*each canonical linearization* $\mathcal{C}$) {
>   **for** (*each disjoint lattice* $L \in \mathcal{P}_A$)
>   // compute the (lexicographically) minimum and maximum ...
>   compute $x^{min}(L)$ and $x^{max}(L)$ using *Algorithm 6.2*;
>   // ... index vectors of L relative to $\mathcal{C}$
>   **for** (*each boundary* n *between the loop nests* n *and* $n+1$) {
>   // the start of the code is boundary 0
>   let $\mathcal{P}_A(n)$ be the collection of disjoint lattices of A, which are
>   alive at the boundary n ;
>   // these are disjoint lattices produced *before* the
>   boundary and consumed *after* it
>   let $X_n^{min} = \min_{L \in \mathcal{P}_A(n)} \{x^{min}(L)\}$ and $X_n^{max} = \max_{L \in \mathcal{P}_A(n)} \{x^{max}(L)\}$;
> $|W_{\mathcal{C}}(n)| = dist(X_n^{min}, X_n^{max}) + 1$;
>   // The distance is computed in the canonical linearization $\mathcal{C}$
>   }
> $|W_{\mathcal{C}}| = \max_n \{ |W_{\mathcal{C}}(n)| \}$;
>   // the window size according to [32] for the canonical linearization $\mathcal{C}$
> }   // (possibly, an underestimate)
> **for** (*each disjoint lattice* $L \in \mathcal{P}_A$)
>   **for** (*each dimension* k *of signal* A)
>   // compute the extremes of the integer projection of L ...
>   compute $x_k^{min}(L)$ and $x_k^{max}(L)$ using *Algorithm 6.3*;
>   // ... on the kth axis
> **for** (*each boundary* n *between the loop nests* n *and* $n+1$) {
>   // the start of the code is boundary 0
>   let $\mathcal{P}_A(n)$ be the collection of disjoint lattices of A, which are alive
>   at the boundary n ;
>   **for** (*each dimension* k *of signal* A) {
>   let $X_k^{min} = \min_{L \in \mathcal{P}_A(n)} \{x_k^{min}(L)\}$ and $X_k^{max} = \max_{L \in \mathcal{P}_A(n)} \{x_k^{max}(L)\}$;
> $w_k(n) = X_k^{max} - X_k^{min} + 1$;
>   // The kth side of A's bounding window at boundary n
>   }
> }
> **for** (*each dimension* k *of signal* A) $w_k = \max_n \{w_k(n)\}$;
> // kth side of A's window over all boundaries $|W| = \Pi_{k=1}^m w_k$;
> // the window size according to [33] (possibly, an underestimate)

Steps 1a and *1b* find the exact values of the window sizes for both models when every loop nest either produces or consumes (but not both!) the signal's elements. When elements of the signal are both produced and consumed in the same loop nest (like in the illustrative examples from Figure 6.6), then the window sizes obtained after *Step 1b* may be only underestimates since an increase of the storage requirement can happen *inside* such a loop nest. In order to determine the exact values of the window sizes, an additional step is required.

Step 1c *Update the mapping windows for each indexed signal in every loop nest producing and consuming elements of the signal.*

Let $L \in \mathcal{P}_A$ be a disjoint lattice of an indexed signal A, which is consumed in a certain loop nest where A-elements are produced as well. Then,

> **for** (*each A-element covered by the lattice L*) {
>   compute the iteration vector **i** when the A-element is consumed
>   (accessed for the last time) ;
>   // this requires the computation of the maximum iterator vector
>   relative to the lexicographic order [40]
>   // if L is included in several array references, the overall maximum
>   iterator vector is considered
>   determine the disjoint lattices $\mathcal{L}_p$ of A partially produced until the
> A-element is consumed in iteration **i** ;
>   determine the disjoint lattices $\mathcal{L}_c$ of A partially consumed until the
> A-element is consumed in iteration **i** ;
>
>   **for** (*each canonical linearization $\mathcal{C}$*)
>   **for** (*each lattice L' in $\mathcal{L}_p \cup \mathcal{L}_c$*) {
>   compute $x^{min}(L')$ and $x^{max}(L')$;
> $|W_\mathcal{C}|$ = max { $|W_\mathcal{C}|$, $dist(x^{min}(L'), x^{max}(L')) + 1$ } ;
>   // $|W_\mathcal{C}|$ may increase
>   }
>   **for** (*each dimension k of signal A*)
>   **for** (*each lattice L' in $\mathcal{L}_p \cup \mathcal{L}_c$*) {
>   compute $x_k^{min}(L')$ and $x_k^{max}(L')$;
> w_k = max { w_k, $x_k^{max}(L') - x_k^{min}(L') + 1$ } ;
>   // w_k may increase
>   update $|W|$ if w_k increased its value ;
>   // $|W|$ may increase
>   }
> }

The guiding idea is that local or global maxima of w_k are reached immediately before the consumption of an A-element, which may entail a shrinkage of some side of the bounding window encompassing the live elements. Similarly, the local or global maxima of $|W_\mathcal{C}|$ are reached immediately before the consumption of an A-element, which may entail a decrease of the maximum distance between live elements.

Finally, the memory window for signal A has the minimum size over the two models:

$|W_A| = \min\{|W|, \min_C\{|W_C|\}\}$. The mapping functions are also available (as exemplified in Section 6.3.1).

The next steps investigate the possibility of memory sharing between the elements of different arrays.

Step 2 Build a complete graph G, where each vertex represents an array in the application code. Compute weights for every edge (A,B) in the following way:

(a) If the two arrays A and B have disjoint lifetimes, the weight is $\min\{|W_A|, |W_B|\}$.

(b) When the lifetimes of the two arrays overlap, compute the maximum distance between the locations occupied by simultaneously alive A- and B-elements, taking into account the mapping functions found at *Step 1* and assuming the two memory windows are contiguous; the size of the common window W_{AB} is this maximum distance plus 1; the weight of the edge (A,B) is $|W_A| + |W_B| - |W_{AB}|$. This weight represents the data storage saved when the two arrays A and B share the same memory space versus the situation when the two arrays would be stored separately (in disjoint memory windows).

Step 3 Find the maximum weighted matching in the graph G. A matching in graph is a set of edges, no two of which meet at a common vertex. The weight of the matching is the sum of the weights of its edges. A maximum weighted matching represents a matching of maximum weight, as shown in Figure 6.10 for a graph with 18 vertices. In this case, the matching will produce the most beneficial pairwise grouping of the arrays in terms of storage reduction. The matching solution will maximize the overall savings of data storage when the arrays are sharing pairwise the memory space. Note that even larger savings could be achieved, in principle, if more complex array groups (larger than two) shared the same memory space, but the computation effort would become prohibitive.

Maximum matching has been a subject of interest in graph theory for many years. Matching algorithms were first developed for bipartite graphs. For non-bipartite graphs, most of the best matching algorithms are based on the work of Claude Berge [47], who proposed searching for augmenting paths as a general strategy for maximum matching. Based on Berge's theorems, Edmonds proposed an efficient algorithm whose computation time is proportional to V^4, where V is the number of vertices [48]. The algorithm works in $O(V)$ stages (of cubic complexity each) by finding augmenting paths by a tree search combined with a process of shrinking certain subgraphs called *blossoms* into single nodes of a reduced graph (most often Edmond's algorithm is called the "blossom shrinking algorithm"). Edmond's algorithm has been refined by

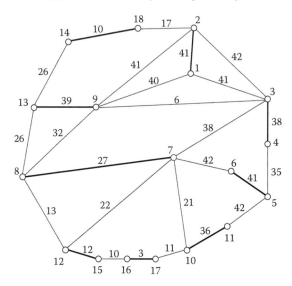

Figure 6.10 The maximum weighted matching in a weighted graph with 18 vertices (bold edges) [41]. (© 2011 Springer. With kind permission from Springer Science and Business Media).

Gabow [49], who obtained an overall complexity of $O(V^3)$. The fastest known algorithm (under the assumption of integral weights that are not particularly high) was developed by Gabow and Tarjan [50]. Further theoretical and practical improvements for large-scale matching problems were proposed by Applegate and Cook [51]. We are using an implementation of Gabow's algorithm [49] due to Ed Rothberg, available online [52]. The running times of this implementation is very suitable for our needs since the graphs built by the assignment algorithm have rather small numbers of vertices (each representing an array in the application code).

Step 4 Compute the overall data storage corresponding to the maximum weighted matching in the graph. In principle, the overall amount of data memory after mapping is the sum of the window sizes resulted after matching. This step takes into account the possibility of disjoint lifetimes between the (pairs of) arrays in this matching, in which case the (common) windows can overlap and share the largest of the two windows. Finally, determine the mapping functions for each array.

6.4.3 Example illustrating the flow of the algorithm

The algorithm will be illustrated on *Example 6.7* from Figure 6.11a.

Step 1 computes the sizes of the memory windows for each array in the code, according to the assignment models [32] and [33], implemented as described

```
int A[7][4], B[9][5], C[11][6], D[13][7], E[15][8] ;

for ( i=0; i<=6; i++ )
  for ( j=0; j<=3; j++ )
    if ( 3<=i+j && i+j<=6 ) A[i][j] = 1 ;
for ( i=0; i<=8; i++ )
  for ( j=0; j<=4; j++ )
    if ( 4<=i+j )
      if ( i<=3 ) B[i][j] = A[i][j-1] + A[6-i][4-j] ;
      else if ( i+j<=8 ) B[i][j] = 2 ;
for ( i=0; i<=10; i++ )
  for ( j=0; j<=5; j++ )
    if ( 5<=i+j )
      if ( i<=4 ) C[i][j] = B[i][j-1] + B[8-i][5-j] ;
      else if ( i+j<=10 ) C[i][j] = 3 ;
for ( i=0; i<=12; i++ )
  for ( j=0; j<=6; j++ )
    if ( 6<=i+j )
      if ( i<=5 ) D[i][j] = C[i][j-1] + C[10-i][6-j] ;
      else if ( i+j<=12 ) D[i][j] = 4 ;
for ( i=0; i<=14; i++ )
  for ( j=0; j<=7; j++ )
    if ( 7<=i+j )
      if ( i<=6 ) E[i][j] = D[i][j-1] + D[12-i][7-j] ;
      else if ( i+j<=14 ) E[i][j] = 5 ;
for ( i=8; i<=22; i++ )
  for ( j=0; j<=7; j++ )
    if ( 15<=i+j && i+j<=22 ) ... = E[i-8][j] ;
```
(a)

Memory windows	Array A	Array B	Array C	Array D	Array E	Array total
Model[4]	22	37	56	79	106	300
Model[5]	28	45	66	91	120	350
Min|W|	16	25	36	49	64	190

(b)

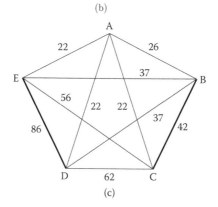

(c)

Figure 6.11 (a) *Example 6.7* illustrating the flow of the algorithm. (b) Table with the sizes of the mapping memory windows for the signal assignment models [32] and [33], as computed at *Step 1*. The last row of the table displays the minimum window sizes, computed as in [40]. (c) The complete graph built at *Step 2* and the maximum weighted matching computed at *Step 3* [41]. (© 2011 Springer. With kind permission from Springer Science and Business Media).

in Section 6.4.1. These window sizes are displayed in the first and the second rows of the table in Figure 6.11b. For comparison, the table in Figure 6.11b displays also the minimum storage requirement separately for each array [40]. The best result for each array is selected.

Step 2 builds a complete weighted graph with five vertices, one for each array in the code. Afterward, it computes the weights of its edges. This graph is shown in Figure 6.11c. For instance, the arrays A and E have disjoint lifetimes and, consequently, the weight of the edge (A,E) is $\min\{|W_A|, |W_E|\} = \min\{22, 106\} = 22$. Since A can share the memory space of E, the storage saving is 22 locations—the window size of A (see the first row of the table in Figure 6.11). On the other hand, since D and E have overlapping lifetimes, the size of the common memory window of the two signals results to be $|W_{DE}| = 99$ locations. The weight of the edge (D,E) is $|W_D| + |W_E| - |W_{DE}| = 79 + 106 - 99 = 86$ locations, which is the memory saving when D and E share the same storage space.

The size of the common memory window for signals whose lifetimes are not disjoint, like the signals D and E, is determined as follows. Considering the window W_D followed contiguously in the memory by the window W_E, we compute the minimum distance between locations occupied by live D- and E-elements. This computation involves the *mapped* disjoint lattices of D and E, whose lifetime information is known. If the minimum distance results larger than zero, than the two windows can be overlapped by this amount. For the signals D, E the minimum distance is 86, hence yielding a common window of $|W_{DE}| = 79 + 106 - 86 = 99$ locations. Note that the weight of the edge (D,E) is actually this minimum distance. The computation is repeated assuming that the window W_E is followed contiguously by W_D and the better of the two results is chosen.

Step 3 computes the maximum weighted matching in the graph. The edges in this matching are (D,E) and (B,C). It follows that the arrays D and E will share a common memory space (W_{DE}) of 99 locations. Similarly, B and C will share a common window (W_{BC}) of size 51, while A will be stored by itself in a memory window of size 22. The amount of data memory after mapping is, for the time being, $|W_{DE}| + |W_{BC}| + |W_A| = 99 + 51 + 22 = 172$ locations.

Step 4 finds that, actually, A has a disjoint lifetime from both arrays in the pair (D,E), so it can share their window. Therefore, the final result is $|W_{DE}| + |W_{BC}| = 99 + 51 = 150$ locations, whereas the previous models [32] and [33] yield 300 and 350 locations, respectively.

The mapping functions for each array are:

$$A[i][j] \;\mapsto\; W_{DE}[(4i + j) \bmod 22]$$
$$B[i][j] \;\mapsto\; W_{BC}[(5i + j) \bmod 37]$$
$$C[i][j] \;\mapsto\; W_{BC}[(37 + (6i + j) \bmod 56) \bmod 51]$$
$$D[i][j] \;\mapsto\; W_{DE}[(7i + j) \bmod 79]$$
$$E[i][j] \;\mapsto\; W_{DE}[(79 + (8i + j) \bmod 106) \bmod 99]$$

Note that the assignment algorithm can be stopped after *Step 1*, if desired: after the first phase, this algorithm will already offer an allocation solution at least as good as the best of the two models [32] and [33], since the algorithm picks the smaller window for each of the arrays. *Step 1* can even work with only one of the previous models, rather than with both of them. But, since any of the models could yield the best result (see the examples in Section 6.3.1) and since the computation overhead is not significant (the implementation of both models in a common polyhedral framework being advantageous both from the point of view of computational efficiency and the amount of data storage), the execution of *Step 1* with two models seems to be the right choice.

Optionally, *Steps 2 to 4* can be run in a second phase, to investigate possibilities of memory sharing between distinct arrays. More reduction of the data storage can thus be obtained, the price being some additional computation time and, possibly, a more complex address generation unit (ADU), as suggested by the mapping functions for the signals C and E. This unit will need to compute additions, multiplications, and modulo operations, like many typical ADUs [31]. Having more than one allocation solution should be a benefit for the designer since it offers the possibility of trading-off the amount of data storage against the complexity of the ADU.

6.5   Experimental Results

The polyhedral framework for the memory management of multidimensional signal processing applications has been implemented in C++, incorporating the algorithms described in this chapter. The algorithmic specifications of the applications are expressed in a subset of the C language, illustrated in the examples used in the chapter.

Table 6.1 summarizes our experiments carried out on a PC with an Intel Core 2 Duo 1.8 GHz processor and 512 MB RAM running Ubuntu 6.06. The benchmark tests (column 1) are signal processing applications and typical algebraic kernels used in this domain: (1) a motion detection algorithm used in the transmission of real-time video signals on data networks [2]; (2) the kernel of a motion estimation algorithm for moving objects (MPEG-4); (3) Durbin's algorithm for solving Toeplitz systems with N unknowns; (4) a singular value decomposition (SVD) updating algorithm [53] used in spatial division multiplex access (SDMA) modulation in mobile communication receivers, in beamforming, and Kalman filtering; (5) the kernel of a voice coding application—essential component of a mobile radio terminal.

Table 6.1 Experimental results on signal mapping. Column 6 displays the data storage requirements obtained after executing the mapping algorithm; the last column shows the corresponding running times. In order to assess the quality of the mapping solutions, column 7 displays the sum of the minimum storage requirement of each signal, together with the absolute minimum storage requirement (the optimal memory sharing) of the whole code—obtained with the algorithm described in Chapter 3. (© 2011 Springer. With kind permission form Springer Science and Business Media).

Application parameters	Num. array references	Num. array elements	Mem. size after mapping			$\sum min\vert W\vert / min(Mem_{code})$	CPU [sec]
			$\sum\vert W\vert_{[32]}$	$\sum\vert W\vert_{[33]}$	$\sum\vert W\vert$		
Motion detection	11						
M = N = 32, m = n = 4		72,543	2,741	2,741	2,740	2,741 / 2,740	3
M = N = 64, m = n = 4		318,367	9,525	9,525	9,524	9,525 / 9,524	13
M = N = 120, m = n = 8		3,749,063	33,285	33,285	33,284	33,285 / 33,284	196
Motion estimation	17						
M = 32, N = 16		265,633	4,513	4,513	3,624	4,513 / 2,465	22
M = 48, N = 32		2,368,865	11,873	11,873	10,054	11,873 / 7,265	182
M = 64, N = 32		4,208,449	18,241	18,241	14,670	18,241 / 10,049	464
Durbin algorithm	27						
N = 50		2,749	201	203	164	176 / 124	2
N = 100		10,499	401	403	327	351 / 249	4
SVD updating	86						
n = 20		25,908	2,464	3,243	2,120	1,664 / 1,430	55
n = 30		85,213	5,494	7,263	4,751	3,694 / 3,195	162
n = 40		199,418	9,724	12,883	8,356	6,524 / 5,660	285
Voice coder	251	33,751	14,498	14,525	12,690	12,965 / 11,899	130

Columns 2 and 3 display the numbers of array references and array elements in the specification code. The next group of three columns summarizes the allocation results: the sum of the memory windows computed according to the [32] model, the sum of the memory windows computed according to the [33] model, and the final allocation result—the overall data storage after mapping. (For *Example 6.5* in Figure 6.8, the values in these columns would be 47, 58, and 25—as explained in Section 6.3.2.) The seventh column displays two metrics of quality for the memory allocation solutions: (a) the sum of the *minimum* array windows (that is, the optimum memory sharing between elements of same arrays), and (b) the minimum storage requirement for the execution of the application code, that is, the optimum memory sharing between all the scalar signals or array elements in the code [40]. (Again, for *Example 6.5* in Figure 6.8, these two values are 39 and 23.) The last column displays the total running times for the results shown in columns 4 to 7.

The data memory sizes after mapping yielded according to the models [32] and [33] (columns 4 and 5) are always lower-bounded by the first values in the seventh column, since the memory window of an array cannot be smaller than the minimum storage requirement of the array. However, our allocation results (column 6) can be better than the first values in the seventh column, since the array windows may overlap significantly (the effect of memory sharing between arrays).

The two metrics of performance from column 7 help the designer to assess the quality of the allocation solution. For example, for the motion detection, the data memory after mapping results equal to the absolute lower bound—the minimum storage requirement for the code execution. It follows that the allocation solution is optimal and cannot be improved. For the motion estimation the individual array windows are optimal as well, but the memory allocation solution could still be improved, enhancing the memory sharing between elements of different arrays.

For these benchmark tests, the memory windows computed according to the model [32] resulted in equal or smaller than the windows computed according to the model [33], although this is not a general conclusion—as exemplified in Section 6.3.1. (The largest difference is observed for the SVD updating.) However, the largest part of the overall running times (column 8) is spent computing the windows in column 4. This is due in part to the number of canonical linearizations that have to be analyzed.

Our polyhedral framework entails significantly faster computations of the initial mapping windows. The running times reported in [32] and [33] are typically of the order of minutes or even tens of minutes, whereas our implementation runs for the same examples, or of similar complexity, in tens of seconds at most (only for columns 4 and 5). For instance, the *voice coding* application was processed by [33] in over 25 minutes using a 300 MHz

Pentium II; in contrast, our running time is significantly shorter—only 12 seconds for *Step 1*—in spite of the different computation platforms. (For evaluating the difference of performance between the two computation platforms [54], a scaling factor of at most 20 between CPU times is fairly accurate.) In another common benchmark—the SVD updating algorithm [53]—their test has only 6038 array elements, which corresponds to matrices of order $n = 12$ in the application code. Their reported run time is 87 seconds [33], whereas ours was about 1.5 seconds (only for column 5). In spite of the difference between the computation platforms, we can safely state that our implementation is several times faster. The investigation of inter-array memory sharing increases our running times, but the additional savings of data memory are quite significant.

Table 6.2 displays in columns 2 and 3 the total memory accesses and the dynamic energy consumption in the case of a single (off-chip) memory layer. Using the CACTI power model, we chose as parameters a technology of 65 nm, 4 memory banks, and a line size of 16 bytes. Assuming a two-layer memory organization, columns 4 to 7 display the SPM sizes and the savings of dynamic energy applying, respectively, a previous assignment model steered by the total number of accesses for whole arrays [13], another model steered by the most accessed array rows/columns, and the current assignment model (Section 6.2), versus the single-layer memory scenario (column 3). Finally, column 8 shows the CPU times.

The SPM sizes (column 4) are computed as follows: the lattices of all the arrays in the application are ordered decreasingly based on the average number of accesses per array element; in this order (such that the most accessed lattices come first), the lattices are gradually assigned to the SPM, increasing the SPM size with discrete amounts; for each new SPM size, the CACTI model computes the energy per access. Afterward, we determine the reduction of dynamic energy versus the scenario when all the signals are stored off chip, choosing the SPM size that maximizes the ratio between the dynamic energy reduction and the size.

Our results are regularly better than those of the other models since, building the map of memory accesses for each array, our framework identifies with accuracy those parts of arrays intensely accessed, whose assignment to the SPM layer yields the highest benefit in terms of dynamic energy consumption. For instance, the energy consumptions for the motion estimation benchmark were, respectively, 1894, 1832, and 1522 μJ; the saved energies relative to the energy in column 3 are displayed as percentages in columns 5 to 7.

Besides the memory allocation solution, the signal mapping algorithm computes the mapping functions for all the arrays, so we can determine the exact locations for any array element in the specification. This provides the necessary information for the automated design of the address generation unit.

Table 6.2 Experimental results on signal assignment to memory layers. Column 5 displays the savings of dynamic energy consumption if the assignment is steered by the most accessed arrays. Column 6 displays the savings of dynamic energy consumption if the assignment is steered by the most accessed cuts (e.g., rows/columns) of arrays. Column 7 displays the savings of dynamic energy consumption of our model (Section 6.2) that identifies the most accessed parts of the arrays [23]. (© 2009 IEICE).

Application parameters	#Memory accesses	Dyn. energy 1-layer [μJ]	SPM size	Dyn. energy saved (1)	Dyn. energy saved (2)	Dyn. energy saved	CPU [sec]
Motion detection $M = N = 32$, $m = n = 4$	136,242	486	841	30.2%	44.5%	49.2%	4
Motion estimation $M = 32$, $N = 16$	864,900	3,088	1,416	38.7%	40.7%	50.7%	23
Durbin algorithm $N = 500$	1,004,993	3,588	764	55.2%	58.5%	73.2%	28
SVD updating $n = 100$	6,227,124	22,231	12,672	35.9%	38.4%	46.0%	37
Vocoder	200,000	714	3,879	30.8%	32.5%	39.5%	8

6.6   Conclusions

This chapter has presented an integrated CAD methodology for power-aware memory allocation, targeting embedded data-intensive signal processing applications. The memory management tasks—the signal assignment to the memory layers and their mapping to the physical memories—are efficiently addressed within a common polyhedral framework. Starting from the high-level behavioral specification of a given application, this framework performs the assignment of the multidimensional signals to the memory layers—the on-chip scratch-pad memory and the off-chip main memory—the goal being the reduction of the dynamic energy consumption in the memory subsystem. Based on the assignment results, the framework subsequently performs the mapping of signals into both memory layers such that the overall amount of data storage be reduced. This software system yields a complete allocation solution: the exact storage amount on each memory layer, the mapping functions that determine the exact locations for any array element (scalar signal) in the specification, metrics of quality for the allocation solution, and an estimation of the dynamic energy consumption in the memory subsystem.

References

1. P.R. Panda, F. Catthoor, N. Dutt, K. Dankaert, E. Brockmeyer, C. Kulkarni, and P.G. Kjeldsberg, "Data and memory optimization techniques for embedded systems," *ACM Trans. Design Automation of Electronic Syst.*, vol. 6, no. 2, pp. 149–206, April 2001.

2. F. Catthoor, K. Danckaert, C. Kulkarni, E. Brockmeyer, P.G. Kjeldsberg, T. Van Achteren, and T. Omnes, *Data Access and Storage Management for Embedded Programmable Processors*, Boston, Kluwer Academic Publishers, 2002.

3. A. Macii, L. Benini, and M. Poncino, *Memory Design Techniques for Low Energy Embedded Systems*, Boston, Kluwer Academic Publishers, 2002.

4. R. Banakar, S. Steinke, B.-S. Lee, M. Balakrishnan, and P. Marwedel, "Scratchpad memory: A design alternative for cache on-chip memory in embedded systems," *Proc. 10th Int. Workshop on Hardware/Software Codesign*, Estes Park, CO, May 2002.

5. C.L. Su and A.M. Despain, "Cache design trade-offs for power and performance optimization: A case study," *Proc. ACM/IEEE Int. Symposium on Low-Power Design*, Dana Point, CA, pp. 63–68, April 1995.

6. M.B. Kamble and K. Ghose, "Analytical dissipation models for low-power caches," *Proc. ACM/IEEE Int. Symposium on Low-Power Design*, Monterey, CA, pp. 143–148, August 1997.

7. U. Ko, P.T. Balsara, A.K. Nanda, "Energy optimization of multilevel cache architectures for RISC and CISC processors," *IEEE Trans. on VLSI Syst.*, vol. 6, no. 2, pp. 299–308, June 1998.

8. R.I. Bahar, G. Albera, and S. Manne, "Power and performance trade-offs using cache strategies," *Proc. ACM/IEEE Int. Symposium on Low-Power Design*, Monterey, CA, pp. 64–69, August 1998.

9. W. Shiue and C. Chakrabarti, "Memory exploration for low power embedded systems," *Proc. 36th ACM/IEEE Design Automation Conf.*, New Orleans, LA, pp. 140–145, June 1999.

10. S. Coumeri and D.E. Thomas, "Memory modeling for system synthesis," *IEEE Trans. VLSI Systems*, vol. 8, no. 3, pp. 327–334, June 2000.

11. S. Wuytack, J.-P. Diguet, F. Catthoor, and H. De Man, "Formalized methodology for data reuse exploration for low-power hierarchical memory mappings," *IEEE Trans. VLSI Syst.*, vol. 6, no. 4, pp. 529–537, December 1998.

12. M. Kandemir and A. Choudhary, "Compiler-directed scratch-pad memory hierarchy design and management," *Proc. 39th ACM/IEEE Design Automation Conf.*, Las Vegas, NV, pp. 690–695, June 2002.

13. E. Brockmeyer, M. Miranda, H. Corporaal, and F. Catthoor, "Layer assignment techniques for low energy in multi-layered memory organisations," *Proc. ACM/IEEE Design Aut. & Test in Europe*, Munich, Germany, pp. 1070–1075, March 2003.

14. Q. Hu, A. Vandecapelle, M. Palkovic, P.G. Kjeldsberg, E. Brockmeyer, and F. Catthoor, "Hierarchical memory size estimation for loop fusion and loop shifting in data-dominated applications," *Proc. Asia & South-Pacific Design Automation Conf.*, Yokohama, Japan, pp. 606–611, January 2006.

15. S. Udayakumaran and R. Barua, "Compiler-decided dynamic memory allocation for scratch-pad based embedded systems," *Proc. Int. Conf. on Compilers, Architecture, and Synthesis for Embedded Systems*, New York, NY, pp. 276–286, Oct. 2003.

16. I. Issenin, E. Brockmeyer, M. Miranda, and N. Dutt, "Data reuse analysis technique for software-controlled memory hierarchies," *Proc. ACM/IEEE Design, Automation and Test in Europe*, Paris, France, pp. 202–207, February 2004.

17. L. Benini, L. Macchiarulo, A. Macii, E. Macii, and M. Poncino, "Layout-driven memory synthesis for embedded systems-on-chip," *IEEE Trans. VLSI Syst.*, vol. 10, no. 2, pp. 96–105, April 2002.

18. F. Angiolini, L. Benini, and A. Caprara, "An efficient profile-based algorithm for scratchpad memory partitioning," *IEEE Trans. CAD*, vol. 24, no. 11, pp. 1660–1676, November 2005.

19. S. Kaxiras, Z. Hu, and M. Martonosi, "Cache decay: Exploiting generational behavior to reduce cache leakage power," *Proc. Symp. Computer Architecture*, pp. 240–251, June 2001.

20. K. Flautner, N. Kim, S. Martin, D. Blaauw, and T. Mudge, "Drowsy caches: Simple techniques for reducing leakage power," *Proc. Symp. Computer Architecture*, pp. 148–157, May 2002.

21. M. Kandemir, M.J. Irwin, G. Chen, and I. Kolcu, "Compiler-guided leakage optimization for banked scratch-pad memories," *IEEE Trans. VLSI Systems*, vol.13, no. 10, pp. 1136–1146, 2005.

22. O. Golubeva, M. Loghi, M. Poncino, and E. Macii, "Architectural leakage-aware management of partitioned scratchpad memories," *Proc. Design, Automation and Test in Europe*, Nice, France, pp. 1665–1670, 2007.

23. F. Balasa, I.I. Luican, H. Zhu, and D.V. Nasui, "Energy-aware memory allocation framework for embedded data-intensive signal processing applications," *IEICE Trans. on Fundamentals of Electronics, Communications and Computer Sciences* (Special section on *VLSI Design and CAD Algorithms*), Japan, vol. E92-A, no. 12, pp. 3160–3168, December 2009.

24. I.I. Luican, H. Zhu, and F. Balasa, "Formal model of data reuse analysis for hierarchical memory organizations," *Proc. IEEE/ACM Int. Conf. on Comp.-Aided Design*, San Jose CA, pp. 595–600, November 2006.

25. F. Balasa, I.I. Luican, H. Zhu, and D.V. Nasui, "Automatic generation of maps of memory accesses for energy-aware memory management," *Proc. IEEE Int. Conf. Acoustics, Speech, and Signal Processing*, pp. 629–632, Taipei, Taiwan, April 2009.

26. H. Zhu, I.I. Luican, and F. Balasa, "Mapping multi-dimensional signals into hierarchical memory organizations," *Proc. ACM/IEEE Design, Automation and Test in Europe*, Nice, France, pp. 385–390, April 2007.

27. F. Balasa, H. Zhu, and I.I. Luican, "Signal assignment to hierarchical memory organizations for embedded multidimensional signal processing systems," *IEEE Trans. on VLSI Systems*, vol. 17, no. 9, pp. 1304–1317, September 2009.

28. S. Wilton and N. Jouppi, "CACTI: An enhanced access and cycle time model," *IEEE J. Solid-State Circ.*, vol. 31, pp. 677–688, May 1996.

29. G. Reinman and N.P. Jouppi, "CACTI2.0: An integrated cache timing and power model," COMPAQ Western Research Lab, 1999.

30. E. Chan and S. Panchanathan, "Motion estimation architecture for video compression," *IEEE Trans. Consumer Electronics*, vol. 39, pp. 292–297, August 1993.

31. G. Talavera, M. Jayapala, J. Carrabina, and F. Catthoor, "Address generation optimization for embedded high-performance processors: A survey," *J. Signal Processing Systems*, Springer, vol. 53, no. 3, pp. 271–284, December 2008.

32. E. De Greef, F. Catthoor, and H. De Man, "Memory size reduction through storage order optimization for embedded parallel multimedia applications," special issue on "Parallel Processing and Multimedia," A. Krikelis (ed.), in *Parallel Computing*, Elsevier, vol. 23, no. 12, pp. 1811–1837, December 1997.

33. R. Tronçon, M. Bruynooghe, G. Janssens, and F. Catthoor, "Storage size reduction by in-place mapping of arrays," in *Verification, Model Checking and Abstract Interpretation*, A. Coresi (ed.), pp. 167–181, 2002.

34. V. Lefebvre and P. Feautrier, "Automatic storage management for parallel programs," *Parallel Computing*, vol. 24, pp. 649–671, 1998.

35. F. Quilleré and S. Rajopadhye, "Optimizing memory usage in the polyhedral model," *ACM Trans. Programming Languages and Syst.*, vol. 22, no. 5, pp. 773–815, 2000.

36. A. Schrijver, *Theory of Linear and Integer Programming*, New York, John Wiley, 1986.

37. A. Darte, R. Schreiber, and G. Villard, "Lattice-based memory allocation," *IEEE Trans. Computers*, vol. 54, pp. 1242–1257, October 2005.

38. I. Verbauwhede, F. Catthoor, J. Vandewalle, and H. De Man, "In-place memory management of algebraic algorithms on application specific processors," in *Algorithms and Parallel VLSI Architectures*, E. Deprettere et al. (eds.), Elsevier Sc., 1991.

39. P.G. Kjeldsberg, F. Catthoor, and E.J. Aas, "Data dependency size estimation for use in memory optimization," *IEEE Trans. Comp.-Aided Design of ICs and Syst.*, vol. 22, no. 7, pp. 908–921, July 2003.

40. F. Balasa, H. Zhu, and I.I. Luican, "Computation of storage requirements for multi-dimensional signal processing applications," *IEEE Trans. VLSI Systems*, vol. 15, no. 4, pp. 447–460, April 2007.

41. F. Balasa, I.I. Luican, H. Zhu, and D.V. Nasui, "Signal assessment model for the memory management of multidimensional signal processing applications," *J. Signal Processing Systems* (Special Issue on *Design and Implementation of Signal Processing Systems*), Springer, vol. 63, no. 1, pp. 51–65, March 2011.

42. I.I. Luican, H. Zhu, and F. Balasa, "Signal-to-memory mapping analysis for multimedia signal processing," *Proc. Asia & South-Pacific Design Automation Conf.*, Yokohama, Japan, pp. 486–491, 2007.

43. M. Minoux, *Mathematical Programming—Theory and Algorithms*, New York, John Wiley, 1986.

44. W. Pugh, "A practical algorithm for exact array dependence analysis," *Comm. of the ACM*, vol. 35, no. 8, pp. 102–114, August 1992.

45. S. Verdoolaege, K. Beyls, M. Bruynooghe, and F. Catthoor, "Experiences with enumeration of integer projections of parametric polytopes," in *Compiler Construction: 14th Int. Conf.*, R. Bodik (ed.), vol. 3443, Springer, pp. 91–105, 2005.

46. G.B. Dantzig and B.C. Eaves, "Fourier-Motzkin elimination and its dual," *J. Combinatorial Theory (A)*, vol. 14, pp. 288–297, 1973.

47. C. Berge, "Two theorems in graph theory," in *Proc. Nat. Acad. Science USA*, vol. 43, pp. 842–844, September 1957.

48. J. Edmonds, "Paths, trees, and flowers," *Canadian J. of Mathematics*, vol. 17, pp. 449–467, 1965.

49. H.N. Gabow, *Implementation of Algorithms for Maximum Matching on Non-Bipartite Graphs*, Ph.D. Thesis, Stanford University, 1973.

50. H.N. Gabow and R.E. Tarjan, "Faster scaling algorithms for general graph-matching problems," *J. of the ACM*, vol. 38, no. 4, pp. 815–853, 1991.

51. D. Applegate and W. Cook, "Solving large-scale matching problems," in *Network Flows and Matchings*, D. Johnson and C.C. McGeoch (eds.), *DIMACS Series in Discrete Mathematics and Theoretical Computer Science*, vol. 12, pp. 557–576, American Mathematical Society, 1993.

52. Ed Rothberg [Online]. Available: ftp://dimacs.rutgers.edu/pub/netflow/

53. M. Moonen, P.V. Dooren, and J. Vandewalle, "An SVD updating algorithm for subspace tracking," *SIAM J. Matrix Anal. Appl.*, vol. 13, no. 4, pp. 1015–1038, 1992.

54. F. Volkel, "Benchmark Marathon: 65 CPUs from 100 MHz to 3066 MHz [Online]. Available: http://www.tomshardware.com/reviews/benchmark-marathon.590.html

Chapter 7

Leakage Current Mechanisms and Estimation in Memories and Logic

Ashoka Sathanur
IMEC, Eindhoven, The Netherlands

Praveen Raghavan
IMEC, Heverlee, Belgium

Stefan Cosemans
K.U. Leuven, Leuven, Belgium

Wim Dahaene
K.U. Leuven, Leuven, Belgium

Contents

7.1   Introduction

Leakage power is becoming a dominant component of the total power consumption of a digital system as technology scale toward sub-90 nm. Following Moore's law, the supply voltage is scaled at every technology node. But to maintain sufficient drive strength, and hence performance, the threshold voltage has to be scaled, thus increasing the subthreshold leakage exponentially. Similarly, gate-oxide t_{ox} scaling increases gate-oxide tunneling leakage current. As CMOS technology scaled beyond 100 nm, transistor leakage has become one of the major show stoppers for further device scaling and integration.

Embedded memories (predominantly SRAMs) constitute a major chunk of the total system power consumption in present-day SoCs. During long idle times, the stand-by leakage contribution of the SRAMs can be substantial and hence requires careful analysis design. In this chapter, we first give an overview of different leakage current mechanisms in present-day short channel MOS devices. In Section 7.2 we show how one can develop efficient leakage current models and estimate leakage power of an SRAM memory block.

7.2   Leakage Current Mechanisms

There are three potentially important contributions to leakage in sub-100 nm MOSFETs: subthreshold leakage, gate tunneling leakage, and junction leakage [1].

7.2.1   Subthreshold leakage

The subthreshold leakage current flows from drain to source, and can be described as

$$I_{ds} = \beta \cdot (1 - n) \cdot (kT/q)^2 \cdot \exp\left(\frac{V_{gs} - V_{th}}{nkT/q}\right) \cdot \left[1 - \exp\left(\frac{-V_{ds}}{kT/q}\right)\right] \quad (7.1)$$

β is the transistor conductance parameter $\beta = \frac{W}{L}\mu C_{ox}$, with μ the carrier mobility, W and L transistor width and length, and C_{ox} the gate capacitance per area. n is a non-ideality factor, $n > 1$. kT/q is the thermal voltage. V_{th} is the threshold voltage of the transistor. V_{ds} is the drain-to-source voltage of the transistor. V_{gs} is the gate-to-source voltage of the transistor. Subthreshold current I_{subth} is typically reported for $V_{ds} = V_{dd}$, hence

$$I_{subth} \propto \exp\left(\frac{V_{gs} - V_{th}}{nkT/q}\right) \propto 10^{\frac{V_{gs} - V_{th}}{S}} \quad (7.2)$$

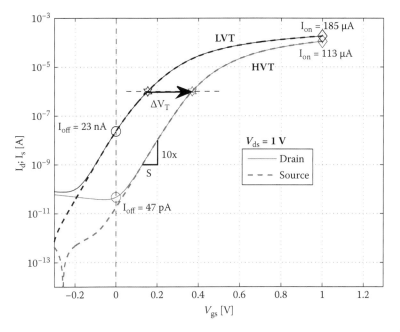

Figure 7.1 Subthreshold leakage for a 90-nm LVT and an HVT NMOS transistor ($W = 200$ nm, $L = Lmin$, $V_{DS} = 1$ V).

S is called the subthreshold slope and is given by $S = n \cdot kT/q \cdot \ln(10)$. Typical values for S are between 60 mV/*decade* and 110 mV/*decade* at room temperature, shifting toward larger values for smaller transistor lengths.

Figure 7.1 shows the current through drain and source terminal of a low threshold (LVT) and a high threshold (HVT) transistor in a 90 nm technology as function of V_{gs} for a fixed $V_{ds} = 1$ V. At low voltages, the drain current becomes larger than the source current. This is due to gate leakage flowing from the drain at 1 V to the gate at a much lower potential. The subthreshold slope S can be readily read from the graph. I_{off} is the current that flows through a transistor with $V_{gs} = 0$. If I_{off} is dominated by the subthreshold current I_{subth}, it is strongly influenced by the selection of the threshold voltage.

$$I_{off} \approx I_{subth} \propto 10^{\frac{-V_{th}}{S}} \qquad (7.3)$$

With $S = 75$ mV /*decade*, a transistor with 200 mV higher threshold voltage has almost 500 times less leakage. A higher threshold voltage, however, also makes the transistor slower because it affects the saturation current of the transistor

$$I_{ds,sat} = \frac{\beta}{2} \cdot (V_{gs} - V_{th})^{\alpha} \qquad (7.4)$$

α is a fit factor. It is 2 for ideal long-channel MOSFETs.

One of the fundamental issues in sub-100 nm MOSFET scaling is V_{th} that must be sufficiently high to limit subthreshold leakage to acceptable values, while the overdrive voltage $V_{dd} - V_{th}$ must be sufficiently large to obtain an acceptable speed, and at the same time, V_{dd} must be as low as possible to reduce the dynamic energy consumption $C{\cdot}V_{dd^2}$. Different application domains will require a different trade-off between dynamic energy, leakage energy and speed, as will be discussed in Section 7.2.4.

Figures 7.2 and 7.3 show drain-induced barrier lowering (DIBL). In the ideal formula for subthreshold current 7.1, V_{ds} has virtually no impact on the current in the region where V_{ds} is larger than 75 mV. In short channel devices, this is no longer true due to DIBL: a larger V_{ds} decreases the threshold voltage. Typical values for DIBL in deep sub-100nm technologies are between 100 mV and 150 mV shift in V_{th} for a 1 V difference in V_{ds}.

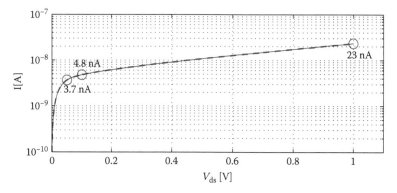

Figure 7.2 Drain and source current for a 90-nm LVT NMOS transistor as function of V_{DS} ($W = 200$ nm, $L = Lmin$). The increase in current for $V_{DS} > 75$ mV is due to DIBL.

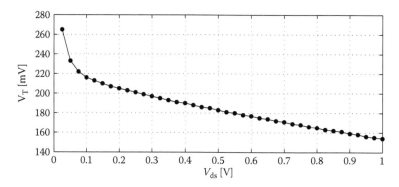

Figure 7.3 Estimated V_{th} for a 90-nm LVT NMOS transistor as function of V_{DS} ($W = 200$ nm, $L = Lmin$). The decrease in V_{th} for $V_{DS} > 75$ mV is due to DIBL.

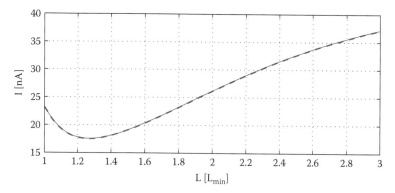

Figure 7.4 Estimated V_{th} for a 90-nm LVT NMOS transistor as function of L ($W = 200$ nm), showing V_{th} roll-off.

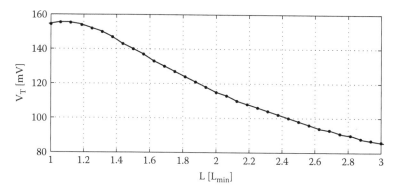

Figure 7.5 Drain and source current for a 90-nm LVT NMOS transistor as function of L ($W = 200$ nm).

In short-channel devices, V_{th} depends on the transistor length L, as shown by Figure 7.4. Figure 7.5 shows that for this specific technology, a slightly longer L ($1.2Lmin$) will decrease subthreshold leakage, while significantly longer values of $L > 1.8Lmin$ will increase the subthreshold leakage.

Threshold voltage of an MOS transistor depends on source-bulk voltage V_{bs} as described by the Equation 7.5. As Figure 7.6 shows that the threshold voltage of a transistor can be adjusted with reverse or forward bias of the bulk-source junction. The body effect is given by [1].

$$V_{th} = V_{th0} + \gamma \cdot [\sqrt{2\phi_b - V_{bs}} - \sqrt{2\phi_b}] \tag{7.5}$$

$$\gamma = \frac{t_{ox}}{\epsilon_{ox}} \cdot \sqrt{2\epsilon_{si}qN_A} \tag{7.6}$$

$$\phi_b = kT/q \ln\left(\frac{N_A}{N_i}\right) \tag{7.7}$$

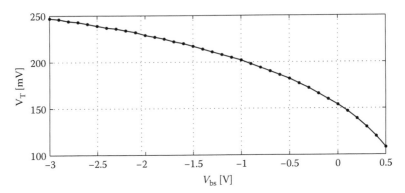

Figure 7.6 Estimated V_{th} for a 90 nm LVT NMOS transistor as function of body bias V_{BS} ($W = 200$ nm, $L = Lmin$).

Because $t_{ox} \cdot \sqrt{N_A}$ reduces with technology scaling [1], the impact of body bias reduces with technology scaling. In this 90 nm technology, the combination of 500 mV forward body bias (FBB) and 1000 mV reverse body bias (RBB) results in a leakage reduction with a factor of 10 compared to a lower threshold transistor with equal speed without body bias. In 22 nm, the reduction would only be a factor of 3.

Temperature and process variability impacts subthreshold leakage severely. Because the threshold voltage reduces with temperature, the subthreshold current is higher at higher temperatures. In this 90 nm technology, the subthreshold leakage increases with a factor of 14 from -20 to $+80$ degrees Celsius.

The threshold voltage also varies stochastically between different transistors. Its distribution is more or less Gaussian. Because of the exponential relation between the subthreshold leakage and the threshold voltage, the leakage current follows a lognormal distribution. Hence there is a significant difference between the average transistor leakage and the nominal transistor leakage. This was discussed in [1] for channel length variations. For threshold voltage variatons, we obtain

$$I_{subth,avg} = I_{subth,nominal} \cdot \exp\left(\frac{1}{2}\left(\frac{\sigma_{V_{th}}}{nkT/q}\right)^2\right) \tag{7.8}$$

$\sigma_{V_{th}}$ is the standard deviation of the stochastic intra-die V_{th} distribution for the transistor under study. In 90 nm, $nkT/q \approx 32$ mV and $\sigma_{V_{th}} \approx 25$ mV for a small transistor, so the difference is a factor 1.36. In 22 nm, $nkT/q \approx 45$ mV and $\sigma_{V_{th}} \approx 60$ mV, so the difference for small transistors is already a factor 2.4.

7.2.2 Gate leakage

Gate leakage is the tunneling current from the gate through the gate oxide to the drain, source, and channel.

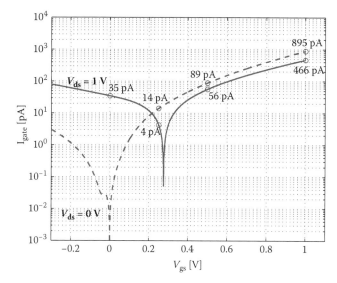

Figure 7.7   Gate leakage for a 90-nm LVT NMOS transistor ($W = 200$ nm, $L = Lmin$).

Figure 7.7 shows the gate tunneling leakage for the example 90 nm technology. Gate leakage is most pronounced for NMOS transistors in strong inversion. In typical circuits, the situation that occurs most in circuits during standby is $V_g = V_{dd}$, $V_d = V_s = 0V$. In this technology, gate leakage for HVT transistors is larger than their subthreshold leakage at room temperature. Gate leakage does not increase significantly with temperature though.

Reducing the supply voltage is a rather effective method for gate leakage reduction. When the supply voltage is halved, gate leakage reduces with a factor 10.

In recent technologies, gate leakage has been controlled by technology design. First, t_{ox} was not scaled down any further and high-k materials were introduced. Recently, high-k metal gate transistors have been introduced that resolve the issue almost entirely [3].

7.2.3   Junction tunneling leakage

A bulk MOSFET has diode junctions between source and bulk and between drain and bulk. It seems as if the leakage from these junctions has been kept under careful technology control and is not an issue in typical designs at this time.

7.2.4   Leakage in different technologies

Different applications differ significantly in their specifications. For a high-performance (HP) desktop processor, speed and dynamic energy consumption

Table 7.1   2009 ITRS roadmap [2] for different technology types

node	Vdd [mV]	Ids, sat [μA/μm]	Isubth [nA/μm]	Igate [nA/μm]
High performance (HP) (Ioff = 100 nA/μm, $V_T \approx 200$ mV) e.g., processor for desktop or server				
90 nm	1200	1110	100	405
65 nm	1100	1210	100	520
45 nm	970	1200	100	374
32 nm	900	1300	100	320
Low operating power (LOP) (Ioff = 5 nA/μm, $V_T \approx 350$ mV) e.g., microprocessor for notebook				
90 nm	1000	589	5	30
65 nm	1000	573	5	51
45 nm	950	746	5	43
32 nm	850	798	5	35
Low standby power (LSTP) (Ioff = 50 pA/μm, $V_T \approx 550$ mV) e.g., low-performance consumer electronics				
90 nm	1200	497	0.05	0.0135
65 nm	1200	519	0.05	0.0143
45 nm	1050	559	0.05	0.0495
32 nm	1000	664	0.05	0.0416

are critical, while for a smart phone processor, dynamic energy consumption and standby power at an acceptable speed are crucial. For many embedded systems such as biomedical implants, the required speed is rather relaxed, but standby power must be extremely low. The ITRS [2] provides a separate technology roadmap for each of these applications. Table 7.1 summarizes the main requirements of these different technologies. High-performance (HP) technologies target a subthreshold leakage of 100 nA/μm and the highest possible speed without exceeding the thermal budget, low operating power (LOP) technologies target a subthreshold leakage of 5 nA/μm at a low dynamic energy consumption (low Vdd), while low standby power (LSTP) technologies target a leakage of 50 pA/μm. According to this table, gate leakage would dominate the leakage power budget in most technologies. However, this does not seem to be the case in typical real manufacturing technologies. Many technologies also provide several transistors types covering a range of 10 to 100 in subthreshold leakage.

Table 7.2 summarizes the specifications of the 90 nm TSMC technology [4] and of the 32 nm TSMC technology with high-k/metal gate [5]. For the 90 nm

Table 7.2 Example technology data for the 90 nm and the 32 nm technology node. [TSMC 90 nm (citeIEDM1) and TSMC 32 nm with high-k/metal gate transistors (citeIEDM2)]. The currents are reported for devices with minimal gate length.

		VT	Vdd [V]	EOT [nm]	Ids,sat [µA/µm]	NMOS Isubth [nA/µm]	Igate [nA/µm]
90 nm	low power	low	1.2	2.2	540	0.400	0.009
		std	1.2	2.2	420	0.015	0.009
		high	1.2	2.2	370	0.004	0.009
	general purpose	low	1	1.6	755	50	0.216
		std	1	1.6	640	5	0.216
		high	1	1.6	520	1	0.216
	high speed	std	1	<1.4	830	300	90
		high	1	<1.4	670	1000	90
32 nm	HS/LP	low	1	<1	1340	100	
		low	0.85		980	100	
		high	1		1020	1	

technology, an up-front selection is made between the low power, the general purpose, or the HP technology. Within each family, devices with different threshold voltages are offered, having a range of 3x in leakage for the HP family and a range in leakage of 100x for the low-power technology. For most devices, gate leakage is small compared to subthreshold leakage. For the 32 nm technology, the gate leakage issue is resolved by the introduction of a high-k dielectric material and a metal gate. This technology offers two devices with different threshold voltage, which provide a 100x difference in leakage current.

Table 7.2 clearly shows that the choice of technology and transistor types is a crucial step to manage the leakage problem.

7.3 Power Breakdown in SoCs

To first analyze leakage it is important to find where the leakage is going. The following sections analyze the power breakdown in a typical SoC platform.

7.3.1 Leakage power breakdown

A typical embedded processor consists of various broad components, its core computational units, a set of register files for intermediate storage, data memories, and instruction memories. Figure 7.8 shows the ADRES architecture [6]. ADRES architecture is a coarse grained reconfigurable architecture that has

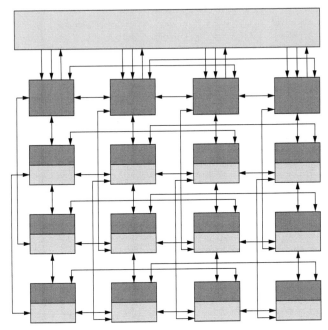

Figure 7.8   ADRES coarse grained reconfigurable architecture.

been tuned for software-defined radio. Figure 7.9 shows the leakage break-down of a processor instance, including the L1 data and instruction memories based on a commercial 40-nm G process.

It is clear from Figure 7.9 that the memories are the most crucial part and leak the most. Since memories are special in both design and their use, they

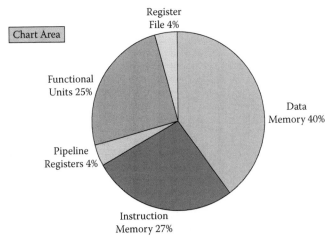

Figure 7.9   Leakage breakdown for a SoC in different components.

are treated separately. More details on system level memory leakage reduction can be found in this chapter, and circuit level techniques to reduce memory leakage can be found in the next chapter. However, logic leakage is also crucial; the treatment of logic is treated later in this chapter.

7.3.2 Use of G versus LP technology

The actual energy lost in leakage depends on how long the core is active and how much the leakage power is. The first part depends on the application profile; however, the leakage power depends on the choice of technology.

Most fabs provide two broad technology flavors:

1. GP or general purpose
2. LP or low power or low leakage

Figure 7.10 shows the total energy in both the LP and GP commercial process for the same architecture at the same frequency of operation. The total power is roughly the same for both instances of the processor in GP and in LP. However, the design in GP can run up to two times faster for twice the dynamic power but the same leakage. While the leakage in GP is worse in few orders of magnitude compared to LP, it is not most evident from the total power perspective, which is the better choice. But in terms of effective area efficiency or computation per sq mm, GP is superior. Therefore the choice of GP versus LP is nontrivial. However, a more rigorous leakage control is needed for GP-based design. The rest of this chapter analyzes modeling methods for leakage. Various system level and EDA methods of leakage control are described in the following chapter.

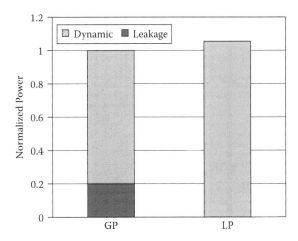

Figure 7.10 Power breakdown for the same architecture on GP versus LP.

7.4   Leakage Current Modeling and Estimation

An efficient and accurate leakage power estimation is an indispensable tool for a circuit designer. It provides a wealth of interesting data for early analysis and optimization. It abstracts underlying technological dependecies and gives the designer the necessary data required for further analysis. Apart from estimating the total leakage power of a design, one can compute leakage power contribution of various subunits of the design. Based on these data, one can choose various circuit implementation styles and architectures to combat leakage power at subblock level and recussively apply it to the entire design. Good leakage estimation also helps in devising leakage power reduction techniques and quantifying their effictiveness in reducing leakage.

In the earlier section, we provided an overview of different leakage current mechanisms in ultra-scaled bulk CMOS devices. These leakage current mechanisms can be modeled and, as a consequence, leakage current equations can be derived. These equations model the dependency of leakage currents on various technological parameters and operating conditions, some of which are design dependent. Technological parameters can be obtained based on accurate measurement or simulation data for a given technology. The other set of parameters are design and operating conditions such as supply voltage, temperature, threshold voltage, circuit topology, etc. Some of these parameters can be chosen to trade off speed for leakage such as supply voltage and threshold voltage, while some others can be used to quantify the leakage power at different operating conditions such as temperature and process corner.

In this section, we first explore the dependency of various parameters on different leakage current mechanisms and then apply these methods to estimate leakage power of an SRAM memory. In the next section, using the estimation methods, we show how choosing different parameters impacts the leakage power of an SRAM memory.

7.4.1   Subthreshold leakage modeling

Subthreshold leakage of an MOS transistor can be written as defined by Equation 7.1 in Section 7.2. As one can see, using this equation to estimate the leakage power of the entire design is not possible due to different gates and components of which the circuit is built having different technology and design-dependent parameters leading to different leakage power values. Butts-Sohi model [7] for subthreshold leakage modeling provides a very simple, yet very effective way to estimate and model subthreshold leakage current for large designs. From [7], subthreshold leakage power can be written as

$$P_{static} = V_{dd} \cdot N \cdot K_{design} \cdot I_{leak,tech} \tag{7.9}$$

where V_{dd} denotes the supply voltage, N denotes the number of transistors in the design, K_{design} is an empirically determined parameter representing the characteistics of an average device, and $I_{leak,tech}$ is a technology parameter describing the subthreshold leakage per device. As we see, the model abstracts subthreshold leakage dependency on various parameters into two main categories, namely design dependent and technology dependent.

7.4.1.1   Design-dependent factor

The subthreshold current in a CMOS circuit depends on two important design-dependent factors. (1) The number of PMOS and NMOS devices switched off. This in turn depends on the input vector applied to the logic circuit in standby mode. (2) Another design-dependent factor is the number of stacked transistors that are switched off. From [8,9], the leakage of a stacked transistor structure can vary significantly with respect to no stack case and hence the leakage depends on number of stacked transistors that are switched off. For example, consider four input NAND gates, we see that there are four transistors that are in series in the pull-down network. Consider the case when the applied input is "1111," then all the NMOS transistors are switched on in the pull-down network and all the PMOS transistors are "off" in the pull-up network. So the leakage of the four parallel PMOS transistors adds up, resulting in maximum leakage of the gate. On the other hand, when the input vector is "0000," then all the NMOS transistors in the pull-down stack are switched "off." Due to negative source-to-gate voltage and reverse body bias effect experienced by transistors on the stack (refer to [9,10] for more details), the leakage is significantly lower. The work in [8] reports up to 70X difference in leakage for the best case and worst case leakage depending on the input applied. Authors in [10] showed that leakage in a stacked transistor case can be written as

$$I_{stack} = I_{nom} \cdot 10^{\frac{1}{S}[\delta V_g + \lambda_d \delta V_d + k_\gamma \delta V_b]} \tag{7.10}$$

where I_{nom} is the leakage at nominal conditions ($V_{gs} = 0$, $V_{ds} = V_{dd}$, and $V_{BS} = 0$). If the gate-drive, body bias, and drain-source voltages reduce by δV_g, δV_b, and δV_{ds}, respectively, then the nominal leakage current I_{nom} changes to I_{stack}. Since it becomes extremely difficult to model leakage of stacked devices on a very large scale using the above equation, [8] proposed a technique for estimating the leakage of stacked devices. The estimation methodology can be detailed as follows:

> *Step 1*: For any given gate in the library, apply all possible input patterns and compute the leakage of the gate. Find the average leakage of the gate based on these data. This step is quite routine in the library characterization phase of a technology library, so it can be easily automated using industry standard tools. Let us denote this leakage as I_{leak}.

Step 2: Find I_{off}, the leakage of a single transistor measured from actual silicon/simulation at a given temperature per unit micron width.

Step 3: Find the average stacking factor X_S for the gate as $I_{off} \cdot W_{tot}/I_{leak}$, where W_{tot} is the total device width of the gate.

Step 4: After estimating the stacking factors for all the gates in the library, one can estimate the leakage of the entire design by

$$I_{tot} = I_{off} \cdot \sum_{i \epsilon cells} \frac{f_i \cdot W_i}{X_{s_i}} \qquad (7.11)$$

The above equation can be written as

$$I_{tot} = I_{off} \cdot N \cdot K_{design} \qquad (7.12)$$

where N is the number of minimum-sized transistors in the design and K_{design} is a co-efficient that captures the subthreshold leakage dependence on stacking and input vector.

7.4.1.2 Technology-dependent factor

From the derivation of the total leakage current of the entire design from Equation 7.12, we have I_{off}, which denotes the leakage current of a unit width of a device. From Equation 7.2, the leakage of unit width device can be written as

$$I_{off} = I_0 \cdot 10^{\frac{V_{gs} - V_{th}}{S}} \qquad (7.13)$$

where I_0 is the drain current with $V_{gs} = V_{th}$ and $V_{ds} = V_{dd}$, and S is the subthreshold slope. By setting V_{gs} to zero, we obtain the off current of the device. From Equation 7.13, the off current depends on two main technology parameters, namely the subthreshold slope S and the threshold voltage of the device V_{th}. Using a different technology or using a different threshold voltage or different operating temperature will alter these parameters, resulting in exponential variation in leakage current. We model these two technological parameter dependences in I_{off} using the Equation 7.13 and denote it as $I_{leak,tech}$. So the total subthreshold leakage power can be written as

$$P_{static} = V_{dd} \cdot N \cdot K_{design} \cdot I_{leak,tech} \qquad (7.14)$$

7.4.2 Gate leakage modeling

The problem of gate leakage modeling was studied in [11]. From [11], we see that the gate leakage is a function of input vector appiled to the circuit. It depends on if a path exits from gate to source/drain to the supply rails. For more details on gate leakage modeling, we refer the interested reader to [11].

Gate leakage current shows exponential dependence on gate-source V_{gs} voltage and shows almost no dependence on the temperature. As in the case of subthreshold leakage current modeling, for gate leakage modeling we apply all possible input patterns for any given gate in the library and find the average gate leakage current. Total gate leakage current can be found by using the following equation:

$$I_{gate} = \sum_{i \in cells} \sum_{j \in states} P(j)I_g(j) \qquad (7.15)$$

where j spans over all possible input vector combinations. $P(j)$ denotes the probability of the gate in state j and $I_g(j)$ denotes the gate leakage current in state $s(j)$. If one assumes an equal probability of all input patterns occuring at the input of a gate, then the I_{gate} denotes the average gate leakage current of the gate. Improvements in process technology such as constant t_{ox}, high-k dielectrics and high-k metal gates have controlled this component of leakage in recent technologies. Given the fact that subtheshold leakage is near exponentially dependent on temperature and reduction in gate leakage current due to technology process improvements, we can fairly assume that at nominal chip operating temperatures, subthreshold leakage dominates and is the main source of leakage power consumption.

7.4.3   Leakage current estimation in SRAM memories

In the previous sections, we studied how to efficiently model and estimate leakage in general CMOS devices. In this section we apply them to estimate leakage power in SRAM memories. Since there have been many different implementation styles of SRAM design, we try to make the estimation methodology as generic as possible so that one can utilize them in estimation based on the relevent architecture. Memories are very regular structures and hence it's easy to quantify their leakage if one can accurately estimate the leakage of each basic unit, which is usually repeated many times in the design. SRAM memory consists of five main subblocks, namely *1. Memory core, 2. Address Decoder, 3. Read column circuit, 4. Write column circuit, 5. Timing circuitry.* Out of these, except for the memory core, the rest can be combined together as *Peripheral circuitry*. The total leakage of an SRAM design can thus be split into leakage of the *Memory core* and that of *Peripheral circuitry*. We observed that the dominant part of the leakage in *Peripheral circuitry* is due to word line drivers, global/local input/output drivers, and address decoder. The drivers are usually large buffers used to drive high capacitive loads to meet memory timing requirements, thus contributing to large leakage power values.

7.4.3.1 Leakage of memory core

As shown in [12] the leakage of the memory core can be approximately written as

$$I_{core} = N_{rows} \cdot N_{columns} \cdot W_{bitcell} \cdot I_{leak,tech} \qquad (7.16)$$

Here N_{rows} and $N_{columns}$ denote the number of rows and columns in the SRAM memory, respectively. $W_{bitcell}$ denotes the effective width of each leaking bit cell. Note that this factor is bit cell implementation dependent. This equation shows that leakage of an SRAM memory core of a given size, technology, and operating conditions is clearly dependent on the implementation of its bit cell.

7.4.3.2 Leakage of drivers

The number of drivers clearly depends on the SRAM architecture. In an architecture that has fully subdivided wordlines, there is one driver per word line, and, on the other hand, where there is no word line subdivision, there is one driver per row. However, note that the load capacitance is different in both cases and hence the size of the drivers. To keep our estimation methodoogy generic, we assume there are N_{driver} number of minimum-sized drivers in the design. So the total leakage of the drivers can be written as

$$I_{driver} = N_{driver} \cdot I_{leak-min} \qquad (7.17)$$

where $I_{leak-min}$ is the leakage of the minimum-sized driver.

7.4.3.3 Leakage of decoder

The address decoder can be implemented in numerous ways, but it is important to note that this block is purely combinational and hence not regular like the memory array. This makes the leakage power estimation of this block not straightforward. One can apply the methodology explained in Section 7.4.1 and use Equation 7.14 to compute the leakage current of this block. So we can write the leakage of the decoder logic as

$$I_{decoder} = N \cdot K_{design} \cdot I_{leak} \qquad (7.18)$$

So the total leakage power of the SRAM memory can be written as

$$P_{SRAM} = (I_{core} + I_{decoder} + I_{driver} + I_{rest}) \cdot V_{dd} \qquad (7.19)$$

Here I_{rest} denotes leakage due to sense amplifiers, timing circuitry, and other circuitry. Figure 7.11 shows split-up of leakage power values of different sections of an SRAM memory. As we see, most of the leakage is in peripheral circuitry and the memory core, and hence our focus in the next chapter is on techniques to reduce leakage in each of these subblocks.

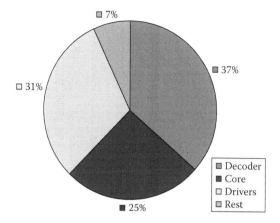

Figure 7.11   Leakage power split in a commercial 90-nm SRAM memory.

References

1. Siva G. Narendra and Anantha P. Chandrakasan. *Leakage in Nanometer CMOS Technologies.* Springer, November 2005.

2. Itrs 2009 edition, http://www.itrs.net/. 2009.

3. K. Mistry, C. Allen, C. Auth, B. Beattie, D. Bergstrom, M. Bost, M. Brazier, M. Buehler, A. Cappellani, R. Chau, et al. A 45nm logic technology with high-k+metal gate transistors, strained silicon, 9 cu interconnect layers, 193nm dry patterning, and 100% pb-free packaging. IEEE International Electron Devices Meeting, IEDM, pages 247–250, 2007.

4. C.C. Wu, Y.K. Leung, C.S. Chang, M.H. Tsai, H.T. Huang, D.W. Lin, Y.M. Sheu, C.H. Hsieh, W.J. Liang, L.K. Han, et al. A 90-nm CMOS device technology with high-speed, general-purpose, and low-leakage transistors for system on chip applications. IEEE International Electron Devices Meeting, IEDM, pages 65–68, 2002.

5. C.H. Diaz, K. Goto, H.T. Huang, Yu. Yasuda, C.P. Tsao, T.T. Chu, W.T. Lu, V. Chang, Y.T. Hou, Y.S. Chao, et al. 32nm gate-first high-k/metal-gate technology for high performance low power applications. IEEE International Electron Devices Meeting, IEDM, pages 1–4, 2008.

6. B. Mei, S. Vernalde, D. Verkest, H. De Man, and R. Lauwereins. ADRES: An architecture with tightly coupled vliw processor and coarse-grained reconfigurable matrix. In *Proc. IEEE Conf. on Field-Programmable Logic and Its Applications (FPL)*, pages 61–70, Lisbon, Portugal, September 2003.

7. J.A. Butts and Gaurinder S Sohi. A static power model for architects. *International Symposium on Microarchitecture*, 2000.

8. W. Jiang, V. Tiwari, E. Iglesia, and A. Sinha. Topological analysis for leakage prediction of digital circuits. *IEEE VLSI Design*, 2002.

9. Z. Chen, M. Johnson, L. Wei, and K. Roy. Estimation of standby leakage power in CMOS circuits considering accurate modeling of transistor stacks. *ACM/IEEE International Conference on Low Power Electronics Design*, 1998.

10. S. Narendra, Vivek De S. Borkar, D.A. Antoniadis, and A.P. Chandrakasan. Full-chip subthreshold leakage power prediction and reduction techniques for sub-0.18-um CMOS. *IEEE Journal of Solid State Circuits*, 39(2):501–510, 2004.

11. R.M. Rao, J.L. Burns, A. Devgan, and R.B. Brown. Efficient techniques for gate leakage estimation. *IEEE/ACM International Symposium on Low Power Electronics Design*, 2003.

12. M. Mamidipaka, K. Khouri, N. Dutt, and M. Abadir. Leakage power estimation in SRAMs. *IEEE International Conference on Hardware Software Codesign*, 2004.

Chapter 8

Leakage Control in SoCs

Praveen Raghavan
IMEC, Heverlee, Belgium

Praveen Raghavan
IMEC, Heverlee, Belgium

Ashoka Sathanur
IMEC, Eindhoven, The Netherlands

Stefan Cosemans
K.U. Leuven, Leuven, Belgium

Wim Dahaene
K.U. Leuven, Leuven, Belgium

Contents

8.1   Leakage Power Reduction Techniques

In the previous chapter, we discussed leakage current mechanisms and how to model leakage currents for SRAMs. In this chapter, we detail different techniques for leakage power reduction at different levels of abstraction and discuss briefly the low-power EDA design flow using common power format. Subthreshold leakage is the dominant leakage mechanism in present-day nanometer CMOS devices, and its impact is more severe at elevated temperatures at which present-day complex SoCs work. In this section we first give an overview of different leakage reduction techniques applied to a general CMOS design. We then focus on applying the leakage reduction techniques on each subblock of an SRAM memory. Leakage reduction techniques can be categorized into device-level, circuit-level, and architectural level techniques.

- *Device-level solutions*: This class comprises various source, drain, and channel engineering techniques such as nonuniform doping, halo doping, retrograde well doping, shallow junctions, raised source-drain, as well as new device architectures such as double-gate CMOS [1];

- *Circuit-level solutions*: This class of techniques rely on a given technology and constructively use it to build leakage-efficient circuit structures. Examples of this class are the use of multiple supply voltages, the exploitation of transistor stacking effect, and the use of multi-threshold CMOS (MTCMOS) technology. MTCMOS technology can be used in various ways to combat leakage currents. Use of dual-V_{th} technique, dynamically varying threshold voltage technique such as body biasing and power shutoff where high-threshold switches are used to shut off supply voltage to the circuit can be used to reduce leakage current.

- *Architectural solutions*: These solutions may either exploit circuit-level solutions at a coarser granularity (e.g., power shutoff), or design explicit structures aimed at the reduction in leakage during idle intervals of the computation (e.g., leakage reduction for memories).

Out of these different solutions at different abstraction levels, circuit-level techniques have gained prominence due to their relative ease of implementation, which can be tailor made for the circuit and the application on hand and their effectiveness in reducing leakage power. One can utilize novel devices and architectural level/system-level policies to complement the circuit-level techniques, thereby achieving very low leakage designs. In this chapter, we focus on circuit-level techniques and choose them to analyze their effectiveness in reducing subthreshold leakage power and the cost of implementing them.

8.1.1 Power shutoff

Power shutoff is one of the variants of MTCMOS technology where high-threshold devices are used as switches to turn off the power supply during the standby state, thus reducing the leakage current of the circuit. One can use either PMOS transistor or NMOS transistor of high-threshold voltage (*sleep transistors*, ST) in series with the transistors of each logic block, thus cutting the path to V_{dd} or VSS. Figure 8.1a shows a part of a logic. As shown in the figure, during standby mode there is subthreshold leakage current flowing from the supply voltage through NAND, AND, and NOR gates to the ground. In Figure 8.1b we show how we can stop this leakage current from flowing by connecting an NMOS high-V_{th} sleep transistor to the source terminals of the gates. In Figure 8.1c, we show how the sleep transistor is ideally modeled as a resistor during the active mode (sleep signal = 0) since it operates in the linear region due to a voltage drop (tens of millivolts) across its drain-source terminals. Note that due to this, the gates experience a degradation in their delay since their gate overdrive voltage (V_{gs}) is being reduced due to the source terminals now at a few millivolts higher than ground. In Figure 8.1d the sleep

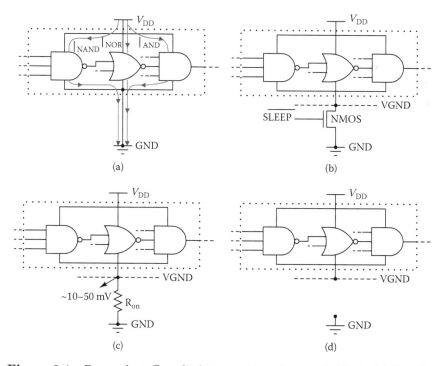

Figure 8.1 Power shutoff applied to a portion of the logic block. (a) Standby leakage current from VDD to VSS without power shutoff. (b) Logic block with NMOS sleep transistor. (c) Sleep transistor in active mode modeled as a resistor. (d) Sleep transistor in standby mode modeled as open circuit.

signal is set at 1, and hence the sleep transistor is turned-off and hence ideally it behaves like an open circuit, thus cutting off the leakage currents flowing from V_{dd} to VSS. In practice, the sleep transistor deviates from the ideal switch characteristics. In the active mode, apart from the resistance, there is also a parallel capacitance contributed by the virtual ground line and the output capacitance of the sleep transistor. In the standby mode since the sleep transistor does not act like an ideal switch, there is a small leakage current contributed by the sleep transistor.

In practice, NMOS sleep transistor is typically used due to its smaller on-resistance for a given size. The sleep transistor has to be carefully sized to guarantee proper functionality. In the active mode since the insertion of sleep transistor results in degradation of delay of all the gates connected to it, one has to carefully size the sleep transistor. A smaller sleep transistor results in higher on-resistance in active mode, thus degrading the speed of the design by a large amount. On the other hand, a large sleep transistor results in very high leakage current in the standby mode since the leakage current of the sleep transistor is proportional to the size of the sleep transistor.

8.1.2   Transistor stacking

Transistor stacking exploits the fact that the subthreshold current that flows through a stack of series transistors is reduced when more than one transistor in the stack is turned off. This is known as *stacking effect*. Stacking effect reduces the subthreshold leakage due to following reasons:

- Negative gate-source voltage V_{gs} experienced by the transistor closer to the output in the stack reduces the subthreshold leakage drastically.

- The source voltage of the transistor closer to the output in the stack being higher than its body voltage, the threshold voltage of this transistor increases due to body effect and hence further reduces the subthreshold current.

- The drain-source voltage of the transistor closer to the output is reduced and hence the effect of drain-induced barrier lowering (DIBL) on this transistor. Due to this, the threshold voltage of this transistor increases and hence suppresses the subthreshold current.

The leakage of a two transistor stacks is an order of magnitude lower than that of a single transistor. This phenomenon can be exploited in several ways. One is to combine it with the proper assignment of circuit inputs so as to maximize the number of gates with NMOS or PMOS stacks with more than one "off" transistor [2,3] (gates with more than two inputs have in fact "natural" NMOS/PMOS stacks).

Another way to exploit transistor stacking is to decouple it from the input assignment problem and use specially modified library cells with "extended" transistor stacks. As in the case of sleep transistors, extra transistors in the NMOS stacks are usually added. Such modified gates incur a delay penalty (similar to replacing a low-V_{th} device with a high-V_{th} one), and must thus be properly designed to minimize the overhead, or, more typically, use these gates on noncritical paths [4,5].

8.1.3 Use of multiple V_{th}

Multiple threshold CMOS technologies (MTCMOS) that provide both high- and low-threshold transistors in a single chip are another common way to reduce leakage. There are several ways to achieve multiple threshold voltages in a design (i.e., through different doping densities, oxide thicknesses, channel lengths, or body bias voltages), which affect the type of circuit techniques that can be used. The latter can be roughly classified into two classes, depending on whether threshold voltage can be varied *dynamically* [Variable threshold CMOS (VTCMOS)], and Dynamic threshold CMOS (DTCMOS), or *statically*, by resorting to a fixed number (typically two) of threshold voltages (dual-V_{th} CMOS). For both classes, the optimization must be coupled with timing analysis in order to identify noncritical gates to which high-V_{th} transistors are assigned.

The general principle in dual-V_{th} technique is that high-V_{th} is assigned to some transistors in noncritical paths to limit the subthreshold current, while low-V_{th} is assigned to transistors in critical paths to achieve high performance. The technique is particularly attractive as its implementation is feasible with existing dual-V_{th} MOSFET process, making it ideally suited to achieve high performance and low power simultaneously. Dual-V_{th} technique helps in reducing active and standby leakage power.

Note that dual-threshold voltage technique is a static technique where only two threshold voltages available in the technology are used; however, one can also use adaptive body bias technique (ABB)/body biasing where the threshold voltages of the gates are varied dynamically by biasing the body of the transistors. By applying reverse body bias during the standby mode, the threshold voltage of the transistors are increased, thereby reducing the leakage power of the design, while in active mode one can still preserve the performance by removing the body bias voltage. Figure 8.2 shows a simple inverter gate to which body bias is applied. By applying a voltage higher than V_{dd} to the body of the PMOS transistor and a voltage lower than the *GND* (negative voltage) to the body of the NMOS transistor, one can raise the threshold voltages of the PMOS and the NMOS transistor, thereby reducing the leakage power. Unlike multi-V_{th} technique, body biasing is used on a coarser granularity (block level and above) due to limitations in implementation on a finer granularity.

8.1.4 Use of multiple V_{dd}

Scaling of supply voltage, normally used to reduce switching power, can indirectly be used as a way to reduce static power. In fact, the subthreshold leakage due to DIBL decreases as the supply voltage is scaled down; experiments

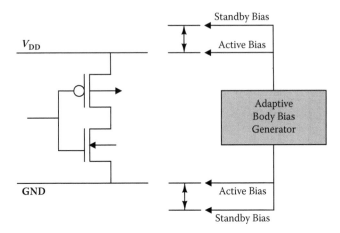

Figure 8.2 Body Biasing technique applied to minimize leakage.

have shown that subthreshold leakage scales as V_{dd}^3 [6], which makes it very attractive as a technique for both static and dynamic power reduction. Since a reduction in supply voltage degrades performance, similarly to multi-V_{th} designs, low V_{dd} must be used on noncritical gates to preserve performance. With respect to the multi-V_{th} case, the use of multiple V_{dd}s has an additional constraint due to the fact that when two different voltage domains have to be interfaced, one needs voltage level shifters for the signals crossing the domains. This usually results in area, timing, and power overhead. For this reason, the voltage domains need to be large to amortize the overhead due to insertion of voltage level shifters.

Table 8.1 shows the comparison of different leakage reduction techniques in terms of leakage reduction, area, and timing overhead in implementing the technique. The data is obtained using a simple 4-bit adder as a test vehicle. As we see, power shutoff incurs the most area and timing overhead and also achieves very high leakage reduction. This makes the technique attractive for standby leakage reduction while making it a very challenging technique for implementation.

Table 8.1 Comparison of different leakage reduction techniques

Technique	Leakage reduction	Area overhead (%)	Timing overhead (%)
power shut-off	67.58X	5–15	5–10
Stacking	1.18X	0	0
Dual V_{dd}	1.2X	<10	0
Dual V_{th}	1.91X	2	0
Body Biasing	2.19X	<10	0

8.2 Leakage Power Reduction Techniques Applied to SRAM Memories

In the previous section, we discussed several leakage reduction techniques that can be applied to reduce subthreshold leakage in general CMOS circuits. Circuit level leakage reduction techniques can be further categorized as static and dynamic techniques.

8.2.1 Static techniques

Static techniques are design time techniques where the designer makes decisions on various parameters that cannot be changed during run-time. The decision on choosing particular value of a parameter at design time involves optimization of one or more cost functions such as timing, power, and area. Multi-V_{th} and stacking are two techniques that can be categorized as static techniques. As an example, in multi-V_{th} optimization, a designer can choose gates with different threshold voltage flavors available in the technology library to map a given circuit. As explained in the previous section, this technique involves timing and leakage power optimization to achieve the best results. In stacking effect, the parameter to choose is the number of transistor stacks of a gate and it involves optimization of timing, area, and power cost functions. In this chapter, we take a deeper look at multi-V_{th} optimization for leakage power control in SRAMs.

8.2.1.1 Multi-V_{th} optimization applied to SRAMs

One of the consequences of increasing importance of subthreshold leakage is that the number of available V_{th}'s in advance process technology nodes is increasing. This is to provide the designer with various options in designing the circuit for performance or for low leakage.

The SRAMs are broken into four components for the purpose of assigning distinct V_{th}'s as described in [7]. The four components are address bus drivers, data bus drivers, decoders, and SRAM arrays with sense amplifiers. From our earlier section on leakage power estimation in previous chapter, we have seen that the leakage power of all these four components can be differentiated into design-dependent and technology-dependent parameters. By assigning different V_{th} values to different components, we change the technology-dependent parameter of the leakage power of these components.

$$P_{leakage} \propto e^{-V_{th}} = K\dot{I}_{leak,V_{th}} \tag{8.1}$$

where $I_{leak,V_{th}}$ is the leakage power per unit width of a transistor of threshold voltage V_{th}. So the objective function of the multi-V_{th} optimization is the total leakage power as given by

$$I_{core,Vth} + I_{decoder,Vth} + I_{driver,address,Vth} + I_{driver,data,Vth} \tag{8.2}$$

The choice of the corresponding V_{th} also affects the timing of each of the four components. One can model the timing of each component as

$$\tau = \frac{C_L \cdot V_{dd}}{(V_{dd} - V_{th})^\alpha} \qquad (8.3)$$

where V_{dd} is the supply voltage, C_L is the load capacitance, V_{th} is the threshold voltage of the gate in the logic block, and α is the velocity saturation index, which can be set at 1 for sub-90 nm technology.

So the problem formulation for multi-V_{th} optimization can thus be written as

Minimize

$$I_{core,Vth} + I_{decoder,Vth} + I_{driver,address,Vth} + I_{driver,data,Vth} \qquad (8.4)$$

Subject to

$$\tau_{decoder,Vth} + \tau_{driver-address,Vth} + \tau_{driver-data,Vth} + \tau_{core,Vth} \leq \tau_{constr} \qquad (8.5)$$

$$V_{th} \in V_{th1}, V_{th2}, V_{th3}, \ldots, V_{thN} \qquad (8.6)$$

The objective function in Equation 8.4 is to minimize the total leakage of the SRAM while adhering to the timing constraints as defined in the constraints in Equation 8.5. Finally, the threshold voltage assignment to the four components can be any of the N available threshold voltages in the technology. So intuitively, the more relaxed the timing constraint is, the chances of a component getting assigned to a right threshold voltage is higher, thus achieving lower leakage power.

8.2.2   Dynamic techniques

Dynamic techniques are run-time techniques where the state of a circuit/block can be changed upon assertion of a control signal. As an example, one can put a memory block in low leakage retention mode through a sleep control signal when it is not being accessed, thus reducing the leakage power of the memory. Designing run-time techniques involves not only optimization of cost functions like timing, area, and power, but also careful consideration on run-time parameters such as wake-up time and energy spent for mode transition. These parameters put constraints on the size and time granularity at which a dynamic technique can be applied to a design. Examples of dynamic techniques are dynamic voltage scaling, ABB, and power shutoff.

In this chapter, we apply two different dynamic techniques to two different components in an SRAM. We apply power shutoff to peripheral circuitry while dynamic voltage scaling is applied to the memory core. The choice is made since power shutoff when applied to the memory core will result in SRAM state being lost.

8.2.2.1 Application of power shut-off in peripheral circuitry

In this section, we analyze the application of power shutoff technique to reduce subthreshold leakage in peripheral circuitry. Two peripheral blocks, which contribute a dominant part of the leakage power in SRAM memories, are decoder logic and drivers. Since they are purely combinational logic, in this section we detail methodology to implement efficient power gating for combinatorial logic blocks. Decoder logic and drivers in an SRAM memory fall on the critical path, and hence when power gating is applied, the timing of the SRAM memory will be impacted. The timing impact depends on the sleep transistor size used. A larger transistor will have lower impact on timing and also results in higher leakage in standby mode. On the contrary, a smaller transistor results in higher impact on timing but lower standby leakage current. So, sleep transistor sizing is a trade-off between active mode timing penalty and standby mode leakage power consumption. In the next section, we present a methodology to size the sleep transistors to keep the timing overhead under a specified value.

8.2.2.2 Sleep transistor sizing with timing constraints

Inserting sleep transistor causes speed degradation, and for timing critical blocks it is very important that we size the sleep transistor such that the timing budget is met. The increase in critical path delay due to the inserted sleep transistor can be computed in the following way. The delay of a gate without power gating can be written as

$$\tau \propto \frac{C_L \cdot V_{dd}}{(V_{dd} - V_{tL})^\alpha} \tag{8.7}$$

where V_{dd} is the supply voltage, C_L is the load capacitance, V_{tL} is the low-threshold voltage of the gate in the logic block, and α is the velocity saturation index, which can be set at 1 for sub-90 nm technology. In the active mode, the sleep transistor behaves as a resistor and hence the low-threshold transistors in the logic block will have their source voltage at V_{VGND} and hence experience a delay degradation. So the delay of the logic gate with sleep transistor inserted can be written as:

$$\tau_{slp} \propto \frac{C_L \cdot V_{dd}}{(V_{dd} - V_{tL} - V_{VGND})^\alpha} \tag{8.8}$$

Since all the gates in the design undergo a degradation in delay due to inserted sleep transistor, the new critical path can be calculated by adding the degraded delay values of the gates. Let us denote the percentage degradation in the overall delay of the design as $\gamma = \frac{D_{new} - D_{old}}{D_{old}} \cdot 100$, where D_{new} (D_{old}) is the critical path delay with (without) sleep transistor. γ is usually a given design constraint. We see that for varying values of V_{VGND}, we can have different values of γ and vice versa. Hence, from the γ specification we can find the required value of V_{VGND}. From Equations 8.7 and 8.8, we can derive γ as

$$\gamma = \frac{V_{VGND}}{V_{dd} - V_{tL}} \tag{8.9}$$

Notice that the maximum value of V_{VGND} one can afford is set by taking into account signal integrity issues. We assume that we are always within the maximum bound set for V_{VGND}.

In order to size the sleep transistor, we need to determine the peak discharge current of the logic block under consideration. Let us denote I_{peak} as the peak discharge current of the decoder block. We can compute the on-resistance of the sleep transistor required as

$$R_{on} = V_{VGND}/I_{peak} = V_{VGND}/I_{sleep}, \quad \text{since} \quad I_{peak} = I_{sleep}.$$

Since the sleep transistor will operate in the resistive (and thus linear) region, we can write the current through the sleep transistor approximately as

$$I_{sleep} \approx \mu_n \cdot C_{ox} \cdot \frac{W_{sleep}}{L} \cdot ((V_{dd} - V_{tH}) \cdot V_{VGND}) \tag{8.10}$$

Here μ_n, C_{ox}, and L are the transistor parameters. V_{tH} is the high-threshold voltage of the sleep transistor and V_{dd} is the supply voltage. From this we can write the sleep transistor on-resistance R_{on} as follows:

$$R_{on} = \frac{V_{VGND}}{I_{sleep}} = \frac{L}{\mu_n \cdot C_{ox} \cdot W_{sleep} \cdot (V_{dd} - V_{tH})} \tag{8.11}$$

From this we can compute the required sleep transistor width as

$$W_{sleep} = \frac{L}{\mu_n \cdot C_{ox} \cdot R_{on} \cdot (V_{dd} - V_{tH})} \tag{8.12}$$

By replacing the R_{on} in Equation 8.12 by V_{VGND}/I_{peak} and deriving V_{VGND} from Equation 8.9, we can establish the link between the size of the sleep transistor and γ as

$$\gamma = \frac{L \cdot I_{sleep}}{\mu_n \cdot C_{ox} \cdot W_{sleep} \cdot (V_{dd} - V_{tH}) \cdot (V_{dd} - V_{tL})} \tag{8.13}$$

By computing the peak discharge current of the decoder logic block, I_{sleep}, and knowing γ, one can use the Equation 8.13 to compute the sleep transistor width required to apply power gating. Similarly, one can apply the above methodology to design power shutoff for drivers.

8.2.2.3 Dynamic voltage scaling applied to the memory core

The leakage of the cell deserves special attention for two reasons: there are more instances of the cell than of any other circuit, and the cells must retain

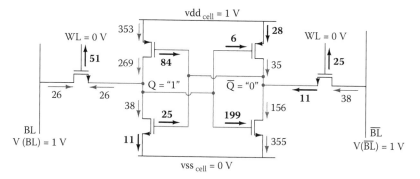

Figure 8.3 Leakage currents in a 90-nm HVT 6T SRAM cell at 27 degrees Celsius. All transistors have $W = 200$ nm, $L = Lmin$. Supply voltage is 1 V.

their state during typical standby periods. This makes it impossible to apply typical power shutoff techniques. Because the cell state is not predefined, state-enforcing leakage reduction techniques such as zigzag super cutoff CMOS [8] cannot be applied either.

Figure 8.3 shows the leakage currents in a 6T cell in the example 90 nm technology at a supply voltage of 1V. Figure 8.4 shows how the leakage current reduces as the supply voltage Vdd reduces. As this reduces the stress over the gate oxide, a lower supply voltage greatly reduces the gate leakage. When Vdd is reduced from 1 V to 0.5 V, the gate leakage reduces with a factor 10. For the same voltage reduction, the subthreshold current only reduces with 36%. This reduction is caused by the combination of reduced DIBL thanks to reduced V_{ds} stress and of the reverse body bias on the PMOS transistors that is obtained by reducing Vdd while keeping $VddBulk$ at 1 V.

If a DC-DC converter is used to generate the lower supply voltage, the leakage power reduces more than the leakage current, as $P_{leak} = I_{leak} \cdot Vdd$.

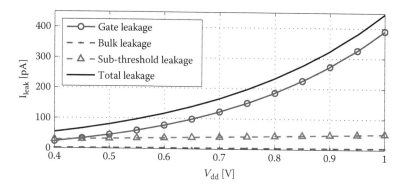

Figure 8.4 Cell leakage current of a 90-nm HVT 6T SRAM cell at 27 degrees Celsius. All transistors have $W = 200$ nm, $L = Lmin$. $VddBulk$ is maintained at 1 V.

Most real-world implementations have used a series regulator though, which results in a leakage power $P_{leak} = I_{leak} \cdot Vdd_{nominal}$.

A lower supply voltage during standby reduces the leakage power more. However, if the supply voltage is reduced too much, the content of specific bit cells will be destroyed. The data retention voltage (DRV) of a cell is the highest voltage at which that specific cell instance loses its contents. This corresponds with the voltage at which the static noise margin [9] (SNM_{hold}) of the cell becomes 0.

The DRV of a matrix is the highest voltage at which a cell in the matrix loses its contents. [10] describes a statistical method to calculate the distribution of the DRV of a single cell, taking only within-die production-time variations into account. From this, it is easy to obtain the distribution of the DRV of a matrix. In [10], the expected value of the DRV for a 1 Gbit matrix in a 90 nm technology was about 210 mV, which is close to our own estimation results. For the choice of the standby voltage for the memory, additional margins should be taken into account to guard against other potential issues. Inter-die variations can cause a shift in strength between NMOS and PMOS transistors, further reducing the nominal SNM_{hold} and increasing the DRV. Transistor aging effects such as negative bias temperature instability (NBTI) and hot carrier injection (HCI) can further degrade SNM_{hold} over time. Ionizing radiation can cause one or more cells to flip state. This is called a single event upset. [11] and [12] provide a good introduction to this topic.

8.2.2.4   Regulating cell supply voltage

The cell supply voltage can be reduced by decreasing the voltage on the power supply rail Vdd or by increasing the voltage on the ground rail VSS, as shown in Figures 8.5 and 8.6. Insertion of a switch in the pull-down path of the cell to increase standby VSS can degrade the NMOS pull-down current and

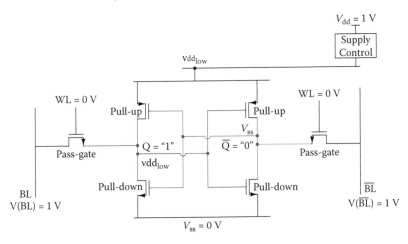

Figure 8.5   Decreasing the cell voltage by decreasing Vdd.

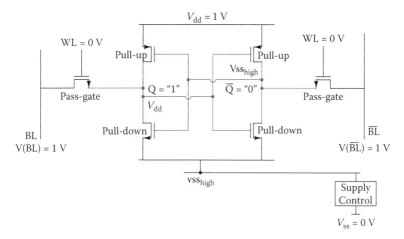

Figure 8.6 Decreasing the cell voltage by increasing VSS.

can reduce the cell read stability. During cell read, hardly any current flows through the PMOS transistors. Hence, a switch in the pull-up path to reduce standby Vdd typically has less impact on active cell performance than increasing VSS, especially for solutions with a small granularity.

Reducing Vdd and increasing VSS don't have the exact same impact on the leakage currents. A lower Vdd reduces the PMOS subthreshold leakage thanks to reverse body bias, while a higher VSS reduces the NMOS pull-down leakage thanks to reverse body bias, and the leakage of the NMOS pass-gate transistors thanks to reverse body bias and—more importantly—thanks to a negative V_{gs}. Reducing Vdd and increasing VSS both reduce the gate leakage of the cross-coupled transistors, but only decreasing Vdd reduces the gate leakage of the pass-gate transistor due to the lower V_{sg}.

For this technology and cell design, the most efficient choice of standby supply voltages to satisfy a given $V_{diff} = Vdd - VSS$ is to increase VSS to approximately 50 mV and lower Vdd to $V_{diff} + 50$ mV. However, the difference with keeping $VSS = 0$ mV and $Vdd = V_{diff}$ is only 2.6%, so it is probably not worth it to insert a switch in the pull-down path of the cell.

Different implementations have been presented to regulate the cell voltage during standby. Methods with a small granularity of control have been proposed in which only the activated part is woken up, sometimes as a step inside the regular access. In [13], each word has its own local series regulator to reduce the cell supply during nonactivity. Before the word line is activated, the series regulator is bypassed and the cell supply is pulled to the full supply voltage. Similarly, the gated-ground cache described in [14] activates a switch in the cell VSS path based on a row decoder output. In [15], a wake-up transistor serves a subblock of 128 Kbit in a 1 Mbit matrix. Many other implementations rely on a single external standby flag, which puts the entire matrix in a sleep mode.

The supply control circuit must reduce the supply voltage as far as possible, but without endangering data stability. Several different approaches have been used to ensure this. Many schemes have relied on simulator- or equation-based sizing of a series regulator transistor, for instance [14]. Sometimes, a predefined external voltage level is applied, for instance 0.5 V in [16] or 0.6 V in [17]. In [15], a programmable sleep transistor strength is provided. This allows it to set the optimal sleep transistor strength during the test after fabrication. A margin for temperature and voltage fluctuations is still needed in this approach. It is, however, the only approach that allows in principle to select the optimal setting for each chip individually, taking the effective inter-die and within-die variations into account. An improved approach was presented in [18]. Here, programming selects a reference voltage from a resistor ladder, and this voltage is copied onto the SRAM supply rail using active feedback control with an integrated op amp. This reduces the sensitivity to temperature variations.

An interesting alternative is to use a closed-loop control system. Here, the available cell margin is continuously monitored. When the available margin becomes too low, the supply voltage is increased. These methods track PVT variations, but must take a safety margin into account for within-die variations. Two interesting monitors have been presented. In [13], SNM_{hold} of the average of a relatively large number of cells is measured with an analog measurement circuit. This is an approximation of the nominal SNM_{hold} on the die with the current supply voltage. When this nominal SNM_{hold} reduces below a predefined margin, the supply voltage is increased. In [19], canary cells are included that consistently toggle their state at a higher supply voltage than the regular cells. The supply voltage is no longer decreased when a predefined number of canary cells have failed. By not reacting to the first failure, the sensitivity to within-die variations in the canary cells is reduced.

8.2.2.5   Other bias voltages

There are three other bias voltages that can be altered to reduce the cell leakage without reducing the available cell stability margins: the bulk voltage, the word line voltage, and the bit line voltage.

Changing the bulk voltage of the transistors allows to apply reverse or forward body bias on the transistors, which significantly affects the subthreshold leakage. As PMOS transistors are created in NWELLs, the bulk voltage of the two PMOS pullup transistors can always be controlled independently of the PMOS bulk voltage in the rest of the chip. In triple well technologies, the bulk voltage of the cell NMOS transistors can also be set independently of the NMOS bulk voltage in the rest of the chip. If triple well is not available or if its use is not desirable, the NMOS transistors are created in the P-substrate, and they should use the same bulk voltage as the rest of the chip. In this case, reverse body bias can still be realized by increasing the cell ground voltage.

Figure 8.7 shows how subthreshold leakage reduces with increasing PMOS reverse body bias, while the leakage flowing into the bulk increases. Notice

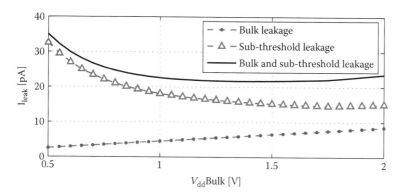

Figure 8.7 Cell leakage as function of PMOS bulk voltage *vddBulk* for a 90 nm HVT cell. *vssCell* = 0 V, *vddCell* = 0.5 V, *vssBulk* = 0 V, all transistors $W = 200$ nm, $L = Lmin$. Gate leakage is omitted from the graph.

that *vddCell* is 0.5 V in this graph; hence 1 V *vddBulk* corresponds to 0.5 V reverse body bias (RBB). Gate leakage was omitted from the figure as it does not depend significantly on the bulk voltage. The total leakage has a relaxed minimum at *vddBulk* = 1.35 V. Figure 8.8 shows similar leakage behavior for NMOS reverse body bias. For this cell at *vssCell* = 0 V and *vddCell* = 0.5 V, the optimal bulk voltage settings are *vssBulk* = −0.5 V and *vddBulk* = 1.35 V. This reverse body bias setting reduces the sum of subthreshold and bulk leakage from 35 pA to 12.7 pA.

The effect of RBB reduces with technology scaling. The range can be extended by also applying forward body bias (FBB) during the active operation. This reduces the threshold voltage during activity, which can increase the read speed and read stability. In [20], forward body bias is applied to the NMOS

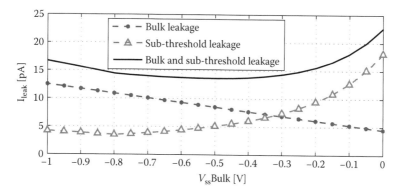

Figure 8.8 Cell leakage as function of NMOS bulk voltage *vssBulk* for a 90 nm HVT cell. *vssCell* = 0 V, *vddCell* = 0.5 V, *vddBulk* = 1 V, all transistors $W = 200$ nm, $L = Lmin$. Gate leakage is omitted from the graph.

transistors of the cells of the accessed sub-matrix. This increases the read speed and improves the write-ability, which allows the cell to be designed with a higher nominal threshold voltage to obtain the same speed and reliability. In [18], FBB is applied on the PMOS transistors of the activated memory part. This increases their strength and improves read stability without increasing standby leakage.

The sub-threshold leakage of the NMOS pass-gate transistors can be reduced by using a negative word line voltage during standby. An interesting implementation is discussed in [16]. This approach is very useful in technologies with little gate and bulk leakage.

Another approach that reduces all the leakage components of the pass-gate transistors is to either reduce the standby precharge voltage of the bit lines, or to make them floating rather than precharged during standby. In [21], the introduction of floating bit lines reduces the leakage power with 18%.

8.3 Leakage Power Reduction Using Low Power EDA Flows

With the explosion of portable devices in the electronics industry coupled with need for more performance in handheld gadgets like smart phones and tablet computers, power consumption has taken a center stage in the past decade as the most important design specification for SoC design in advance technology nodes. An increase in integration of more functionality on a chip and an increase in leakage currents in advance technology nodes has made low-power design an implicit requirement in designing present day SoCs.

Even though there has been considerable effort in the design and research community in developing low-power design techniques in the past decade, seamless integration of these advance low power-techniques in the EDA tool flow and support has been quite slow. Recently, there has been quite some initiative for an industry-wide effort to step up the task of providing a complete and standardized EDA flow and infrastructure that can help designers in the design and verification of advance low-power design techniques from RTL to GDS. The result of this is the emergence of two main low-power formats that can provide the infrastructure for designers to design, analyze, implement, and verify digital circuits with advance low-power design techniques. These two low-power formats are CPF (common power format) and UPF (unified power format). These two low-power infrastructures provide an integrated platform for not only design and analysis of low-power techniques, but also for guiding the tools for correct implementation and verification.

In this chapter, we take a look into CPF-based low-power design flow and see how one can utilize this low-power EDA flow to design advance low-power digital circuits. The requirements for CPF-based flow was created keeping in mind a wide range of view points and applications [22]. Some of the primary requirements for this as described in [22] are **Easy to adopt** to overcome cost, reduce time to market, and risk deployment issues; **Incremental** to existing infrastructure to lower the risk; **Nonintrusive** to existing practices, methodologies, and flows; **Serves IP/reuse** methodologies with a minimal incremental effort **Consolidated** view of the power strategy for a design; **Comprehensive** in capabilities to support the most advanced existing low-power design techniques; and importantly, **Extensible** to new low-power design techniques and to broader design flow scope.

The CPF infrastructure has many advantages and benefits for people using standard design flow. They are

- Enables RTL functional verification of designs with power intent definition.

- Guarantees higher design quality with less functional failures.

- Reduces risk in implementation of advance low-power design techniques.

- Reduces time to market of low-power designs and improves productivity.

- Helps in designing successful low-power chips to drive innovative products.

8.3.1 CPF-based example design flow

Now that we know how CPF-based low-power design is beneficial for designing low-power circuits, in this section we show how to use CPF-based flow by using an example design. CPF allows the designer to define the power intent of the complete system. It facilitates the flow of this power intent information through all the levels of design abstraction. This feature makes sure that at every step of the design and verification stage, the relevant power information is available to smoothly guide the designer and the tools for successful implementation of advance low-power techniques.

The power intent of a system can be defined using CPF by a Tcl (Tool Command Language)-based command file. This file is then used in all stages of implementation (RTL-GDS) to describe the power information of the system. For further information on the CPF , we suggest the readers to this document [22].

Now let's consider an example system as shown in Figure 8.9. Note that this example is far from being complete and acts like a vehicle for understanding of the low-power design flow. This digital system consists of four modules, which are partitioned as four different power domains, namely PD_A, PD_B, PD_C, and PD_D, respectively. While PD_A operates at a lower supply voltage and can be power gated, PD_B and PD_C operate at nominal supply voltage and also can be shut off using power shut off switches. Finally PD_D is an always-on domain. The always-on domain is also the default domain of

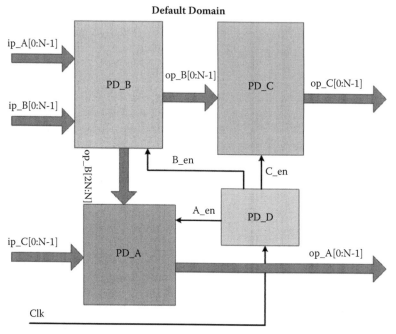

Figure 8.9   Power domains definition in a system.

the system. The Tcl file, which is written to define the power specifications of the system, is as shown in Figure 8.10. In this CPF file, we first define the design libraries that are used to design this system. This includes the standard cell libraries, libraries for memories etc., which are characterized for different PVT corners. This is shown in the *Technology specifications* part of the CPF file. We then define special cells used for designing advance low-power techniques. This includes always-on cells, power switch cells, level shifter cells etc. This is done in the *Library cell specification* part of the CPF file. We then move on to the part where the power intent of the design is defined. This part contains two sections, namely *Design specification* and *Implementation specification*. The *Design specification* part of the CPF file is where the power domains and power modes are defined, and the *Implementation specification* part is the part that guides the tools to implement specific low-power design functionality. For example, the isolation rule creation will define where the isolation cells should be put and what cells to use while the update isolation command will guide the synthesis tools to insert isolation cells in the net-list created after synthesis.

In the *Design specification* section, operating conditions are defined and power domains are created. Note that one has to always define the default power domain where all instances that don't belong to any other domain are placed. Each power domain also has base domains to which they belong.

```
#############################################################
# TECHNOLOGY SPECIFICATION #
#############################################################
define_library_set -name set, _tc -libraries lib1 _tc lib2_lc
define_library_set -name set2_we -libraries lib1_wc lib2_wc


#############################################################
# LIBRARY CELL SPECIFICATION #
#############################################################
# Define always on cells
define_always_on_cell -cells*
# Define isolation cells
define_isolation_cell -cells*
# Define power switch cells
define-power_switch_cell-cells'
# Define state retention cells
define_state_retention_cell-cells
# Define lever shifter cells
define_level_shifter_cell -cells *


#############################################################
# DESIGN SPECIFICATION #
#############################################################
# Define operating condition and associated libraries
create_nominal_condition -name nc1 -voltage #
update_nominal condition -name nc1 -library_set #

# Define power domains
create_power_domain -name PO_default -default
create_power_domain -name PO_A -instances {} -shutoff_condition {en_A} -base_domains PO_default
create_power_domain -name PO_B -instances {} -shutoff_condition {en_B} -base_domains PO_default
create_power_domain -name PO_C -instances {} -shutoff_condition {en_C} -base_domains PO_default

# Define power modes and associated nominal conditions
create_power_mode -name pml -domain_conditions PO_A@ncl, PO_B@nc1,PD_C@nc1-default

# Define isolation rules
create_isolation_rule -name isol -to # -isolation_condition #

# Define level shifter rules
create_level_shifter_rule -name Is1 -from # -to #

# Define power switch rules
create_power_switch_rule -name ps1 -domain # -external_power_net #


#############################################################
# IMPLEMENTATION SPECIFICATION #
#############################################################
# Power supply net definition
create_power_nets -nets # -voltage #
create_ground_nets -nets # -voltage #

# Power domain implementation
update_power_domain -name # -primary_power_net # -primary_ground_net #

# Assign timing constraints to each power mode
update_power_mode -name pm1 -sdc_files

# Implement isolation cell insertion
update_isolation_rules -name iso l -location to -cells #

# Implement level shifter rules
update_level_shifter_rules -name Is1 -from # -to # -cells #

# Implement power switch Insertion
update_power_switch_rule -name -cells #

#End of the cpf creation
end_design
```

Figure 8.10 CPF file definition for the example system.

This allows the designer to hierarchically define power domains. Note that in our example CPF file, we use # and * symbols in a generic sense, and in real implementation one has to correctly specify the right parameters for proper implementation. We also show commands for only one instance while there can be many instances of the same commands to define multiple views. For example, we show only one nominal condition creation while in practice one can have many nominal condition definitions. Similarly for isolation rule, level shifter rule creation, one can have many rules corresponding to many different domains. This is done to reduce the size of the example CPF file and provide clarity. After defining the power domains, we define different power modes in which the system will function: isolation rules that define where to put the isolation cells and control signals for them, level shifter rules that define where to insert the level shifter, and power switch rules that define where to insert power switches. In the example of CPF file, the outputs from power domain PD_A, PD_B, and PD_C need to be isolated since when one of the domains are shut off, the outputs will be floating and hence need to be pulled up or pulled down. Since PD_A will also work at lower supply voltage, the output needs to be shifted back to nominal supply voltage and hence the outputs also need level shifters. The control signals A_en, B_en, and C_en to control the shutoff of different domains are generated by the always-on domain PD_D.

In the *Implementation specification* section, the defined power intent in the *Design specification* is then transferred to the implementation tool to implement the corresponding low-power technique. As we see in the CPF file, first the power supply nets are defined. Then the power domains associated with a pair of nets defining the primary power and ground nets are defined. The primary nets are the nets to which the cells and components inside the power domain are connected. We then assign a timing constraints file for each power mode defined before. Then the update commands help the tools to implement the isolation cells, level shifters, and power switches in corresponding domains with respective control signals.

Note that this is a very simple CPF file that introduces the basics of low-power EDA flow. The advance CPF commands to do power analysis and to implement more advanced low-power design techniques are clearly defined and explained in the document provided at [22].

8.4 Compiler-Driven Leakage Power Reduction

Various compiler-driven techniques can also be applied to reduce the leakage in processor architectures. However, each part of the processor requires a different control method for reducing leakage. Consider the base processor architecture shown in Figure 8.11. It shows a traditional multi-issue VLIW

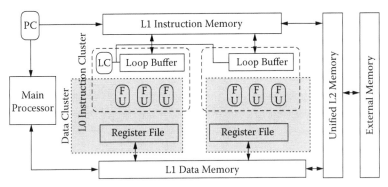

Figure 8.11 Basic processor architecture.

processor with its instruction memory, data memory, clustered register file, and different issue slots. The following subsections show some of the techniques to reduce the energy in such a platform within one core.

8.4.1 Register file

In processors, mostly the register files are clustered or centralized or banked [23]. This is one of the most efficient ways to reduce power consumption in the register file. Such a banking and clustering of register files allows power gating each of the different parts of the register file individually. This is largely application dependent (kernel dependant), as well as the register file occupancy changes from kernel to kernel. [24] shows an approach where the compiler places a role in how the various variables are allocated to ensure that fewer banks are used such that the remaining banks can be power gated.

8.4.2 Functional units/issue logic

Other key parts that also contribute to the leakage in processors are its pipeline and its functional units. Various phases of the same application and more so across different applications, the requirement in terms of computation also changes to a large extent. The evidence of varying workloads across different phases of an application can be found in [25,26]. During these different scenarios or phases of the application, the available ILP or IPC varies; therefore power-gating different issue slots based on this requirement is possible. This would address the 25% of leakage in the functional units of the processor.

8.4.3 Application-driven power reduction in memories

Furthermore, most embedded applications use different memories in different ways. For example in the higher levels of instruction memory in embedded systems, typical use in bursts is between long idle intervals. This allows the higher levels of instruction memory that can be put to sleep with state retention. As discussed before in Section 8.1, one can shut off the pripherals in the memory while the memory core can be brought down to *DRV* voltage to minimize the leakage power, while retaining the state during idle times. Also, based on the memory footprint of the application requirement, the L1 data memory in embedded systems can also be put to sleep with state retention or state loss.

References

1. B. Doyle, et al. Transistor elements for 30nm physical gate lengths and beyond. *Intel Technology Journal*, 6(2):42–54, 2002.

2. V. De Y. Ye, S. Borkar. A new technique for standby leakage reduction in high-performance circuits. *IEEE Symposium on VLSI Circuits*, pages 40–41, 1998.

3. Yuh-Fang Tsai, et al. Characterization and modeling of run-time techniques for leakage power reduction. *IEEE Transactions on VLSI*, 12(11):1221–1233, 2004.

4. S. Narendra, et al. Scaling of stack effect and its application for leakage reduction. *ACM International Symposium on Low Power Electronics Design*, pages 195–200, 2001.

5. S. Narendra, et al. Full-chip subthreshold leakage power prediction and reduction techniques for sub-0.18µm CMOS. *IEEE Journal on Solid State Circuits*, 39(2):501–510, 2004.

6. J.P. Kulkarni, K. Kim, and K. Roy. A 160 mv robust schmitt trigger based subthreshold SRAM. *IEEE Journal of Solid-State Circuits*, 42(10):2303–2313, 2007.

7. D. Blaauw, N.S. Kim, and T. Mudge. Quantitative analysis and optimization techniques for on-chip cache leakage power. *IEEE Transactions on Very Large Scale Integration*, 13(10):1147–1156, 2005.

8. K.S. Min and T. Sakurai. Zigzag super cut-off CMOS (ZSCCMOS) scheme with self-saturated virtual power lines for subthreshold-leakage-suppressed sub-1-VVdd LSI's. In *Proceedings of the 28th European Solid-State Circuits Conference*, pages 679–682, 2002.

9. E. Seevinck, F.J. List, and J. Lohstroh. Static-noise margin analysis of MOS SRAM cells. *IEEE Journal of Solid-State Circuits* 22(5):748–754, October 1987.

10. Wang, A. Singhee, R.A. Rutenbar, and B.H. Calhoun. Statistical modeling for the minimum standby supply voltage of a full SRAM array. Proc. ESSCIRC, 2007, pages 400–403, 2007.

11. T. Karnik and P. Hazucha. Characterization of soft errors caused by single event upsets in CMOS processes. *IEEE Transactions on Dependable and Secure Computing*, 1(2):128–143, 2004.

12. D.F. Heidel, K.P. Rodbell, E.H. Cannon, C. Calabral, M.S. Gordon, P. Oldiges, and H.H.K. Tang. Alpha-particle-induced upsets in advanced CMOS circuits and technology. *IBM Journal of Research and Development*, 52(3):225–232, 2008.

13. P. Geens and W. Dehaene. A dual port dual width 90nm SRAM with guaranteed data retention at minimal standby supply voltage. In *Solid-State Circuits Conference, 2008. ESSCIRC 2008. 34th European*, pages 290–293, 2008.

14. A. Agarwal, Hai Li, and K. Roy. A single-vt low-leakage gated-ground cache for deep submicron. *IEEE Journal of Solid-State Circuits*, 38(2):319–328, 2003.

15. Yih Wang, et al. A 1.1 ghz 12 a/mb-leakage SRAM design in 65 nm ultra-low-power CMOS technology with integrated leakage reduction for mobile applications. *IEEE Journal of Solid-State Circuits*, 43(1):172–179, 2008.

16. Y. Lee, M. Seok, S. Hanson, D. Blaauw, and D. Sylvester. Standby power reduction techniques for ultra-low power processors. In *Solid-State Circuits Conference, 2008. ESSCIRC 2008. 34th European*, pages 186–189, 2008.

17. K. Nii, Y. Tsukamoto, T. Yoshizawa, S. Imaoka, Y. Yamagami, T. Suzuki, A. Shibayama, H. Makino, and S. Iwade. A 90-nm low-power 32-kb embedded SRAM with gate leakage suppression circuit for mobile applications. *IEEE Journal of Solid-State Circuits*, 39(4):684–693, 2004.

18. F. Hamzaoglu, K. Zhang, Y. Wang, H. J. Ahn, U. Bhattacharya, Z. Chen, Y.-G. Ng, A. Pavlov, K. Smits, and M. Bohr. A 3.8 ghz 153 mb SRAM design with dynamic stability enhancement and leakage reduction in 45 nm high-k metal gate cmos technology. *IEEE Journal of Solid-State Circuits*, 44(1):148–154, 2009.

19. Jiajing Wang and B.H. Calhoun. Techniques to extend canary-based standby v_{DD} scaling for SRAMs to 45 nm and beyond. *IEEE Journal of Solid-State Circuits*, 43(11):2514–2523, 2008.

20. C.H. Kim, Jae-Joon Kim, S. Mukhopadhyay, and K. Roy. A forward body-biased low-leakage SRAM cache: device, circuit and architecture considerations. *IEEE Transactions on Very Large Scale Integration (VLSI) Systems*, 13(3):349–357, 2005.

21. Y. Wang, U. Bhattacharya, F. Hamzaoglu, P. Kolar, Y. Ng, L. Wei, Y. Zhang, and M. Bohr. A 4.0 ghz 291mb voltage-scalable SRAM design in 32nm high-k metal-gate CMOS with integrated power management.

In *Solid-State Circuits Conference—Digest of Technical Papers, 2009. ISSCC 2009. IEEE International*, pages 456–457, 457a, 2009.

22. Si2 common power format. *http://www.si2.org*.

23. S. Rixner, W.J. Dally, B. Khailany, P.R. Mattson, U.J. Kapasi, and J.D. Owens. Register organization for media processing. In *HPCA*, pages 375–386, January 2000.

24. J.L. Ayala, G. De Micheli, F. Catthoor, D. Verkest, D. Atienza, P. Raghavan, and Marisa López-Vallejo. Joint hardware-software leakage minimization approach for the register file of VLIW embedded architectures. *Integration*, 41(1):38–48, 2008.

25. F. Vandeputte, L. Eeckhout, and K. De Bosschere. Exploiting program phase behavior for energy reduction on multi-configuration processors. *Journal of Systems Architecture*, 53:489–500, August 2007.

26. S.V. Gheorghita, M. Palkovic, J. Hamers, A. Vandecappelle, S. Mamagkakis, T. Basten, L. Eeckhout, H. Corporaal, F. Catthoor, F. Vandeputte, and K. De Bosschere. System-scenario-based design of dynamic embedded systems. *ACM Transaction on Design Automation of Electronic Systems*, 14:3:1–3:45, January 2009.

Chapter 9

Energy-Efficient Memory Port Assignment

Preeti Ranjan Panda
Indian Institute of Technology Delhi, New Delhi, India

Lakshmikantam Chitturi
Zoran Corp., Sunnyvale, California

Contents

Abstract

The increased bandwidth offered by multiport memories makes them attractive architectural candidates in performance sensitive system designs. When used as storage resources during synthesis, multiport memories can lead to significantly shorter schedules. However, they also have an associated area and energy penalty. We describe a techniques for mapping data accesses to multiport memories during behavioral synthesis that results in significantly better energy characteristics than an unoptimized multiport mapping.

The technique is based on an initial coloring of the array access nodes in the data flow graph based on spatial locality, followed by attempts to consecutively access memory locations with the same color on the same port. Experiments on several applications indicate a significant reduction in address bus switching activity, leading to an overall energy reduction over an unoptimized design, while still maintaining a performance advantage over a single-port solution.

9.1   Introduction

Energy-aware optimization techniques at the system level form an essential feature of modern low-power embedded systems. A substantial amount of research work exists in this domain, ranging from completely new techniques such as predictive algorithms for shutting down the system to modifications of compiler optimizations to handle power and energy as metrics in addition to the traditional performance criterion. Very often, there exists an important trade-off between a performance-optimized and a power-optimized design; the two respective optimal design points are usually different. A good example of this behavior is observed in systems where the clock is slowed down to decrease power dissipation. However, power and energy awareness can also be explicitly built into the system-level synthesis algorithms used to generate design implementations. We study the impact of multiport memories and their associated port allocation strategy on the performance and energy of synthesized designs. Although multiport memories generally lead to better performance, they typically incur a significant area and energy overhead (up to 100% and 75% for area and power respectively for the technology we studied). With an energy-conscious memory port allocation algorithm, it is possible to minimize the energy overhead while still retaining the performance advantage over a single-port memory solution.

Design considerations such as power and energy can be tightly integrated into the inner loop of typical high-level synthesis tasks such as scheduling. A design optimization problem involving the performance and energy coordinates can be phrased in one of two forms:

1. **Optimize for performance**—and while retaining this level of performance, minimize the energy.

2. **Optimize for energy**—and while retaining this level of energy, maximize the performance.

We discuss algorithms for solving both forms of the above optimization problem.

9.2 Background

The memory subsystem has long been recognized as a serious bottleneck in terms of performance, area, and power dissipation in embedded systems. In typical systems, the memory could account for more than 60% of the total area (Chapter 1, Section 1.4) and 30% to 40% of the total power dissipation [1,2]. Memories tend to be significant sources of power dissipation because they are associated with long, high capacitance wires, both inside the memory module (in the form of long word-lines and bit-lines) as well as outside the module (in the form of address and data buses). Consequently, many optimization efforts at reducing memory power have targeted the transition count on the memory address and data buses—specifically, address buses since the pattern of addresses accessed is usually known in advance [3–6].

Sequential addressing of instruction memory in typical programs was exploited by Gray-coding the instruction address in [7]. [3] studied transition count reduction on address buses of a single memory and multiple memory modules connected to a synthesized datapath by rearranging application data. In [5], the observation that programs tend to spend a large fraction of their time in small regions of the memory was used to divide the memory into working zones and encode the address bus so that consecutive accesses within the same zone lead to low transition overhead. The same observation was used to partition the memory architecture itself in [8]. In the T0 encoding scheme [9], a redundant address bit is used to indicate that the current address is consecutive to the previous one. The encoding scheme proposed in [4] assigns codes with a small Hamming distance for data words that are statistically likely to be consecutive.

Several researchers have studied the problem of scheduling instructions in order to reduce power on the instruction memory interface. These approaches include instruction selection, register allocation, and instruction ordering [7,10–12]. Scheduling of operations for low-power synthesis was studied in [13].

Minimizing memory power is also closely related to performance optimizations performed by compilers. Standard optimizations such as induction variable elimination, loop fusion, loop interchange, etc. that result in less memory access also reduce memory power as a direct consequence. In addition, code transformation techniques for reducing memory requirements for an application result in power reduction due to reduced size of the memory modules. Loop optimizations that improve cache performance also reduce power since cache misses impact not only on performance, but also power. Since the actual cache configuration can be customized in embedded systems, a number of research efforts have addressed the problem of determining an application-specific memory hierarchy that optimizes performance and power [14–17].

Multiport memories have been incorporated into traditional behavioral synthesis algorithms by treating the ports as independent schedulable resources [18–20]. These algorithms focus on performance alone and do not study the energy implications. In memory exploration studies such as [3,21], the architectures studied can have multiple memories, but only a single port because multiport memories are area-expensive. In [22], power optimization on multiport memories is applied to a limited set of applications where the data can be divided into tiles. We outline memory port assignment algorithms that can be tightly integrated into the scheduling phase of behavioral synthesis.

9.3 Illustrative Examples

We discuss several examples in this section that illustrate the relationship between performance and energy dissipation in single- and multiport memories. The illustrations use dual-port memories because they are the most commonly used multiport memories, although the algorithms can handle any number of ports. It is also assumed that there is only one memory module used for storing all data in the block being synthesized. In reality, more memory modules could be used, in which case the central analysis used in the techniques would have to be repeated for each individual module and the corresponding arrays stored in it.

Consider the following simple section of code to be synthesized into hardware:

```
int a[100];
⋮
for i = 0 to 99
    a[i] = a[i] + n
```

where a is to be stored in memory; n is a variable; memory reads and writes require one clock cycle; addition and comparison require one cycle. An example schedule of the loop body is shown in Figure 9.1a. The schedule (Schedule A) requires three cycles, and consists of a memory read in cycle 1, addition in cycle 2, and a memory write in cycle 3. A single-port memory is sufficient for this implementation. A loop pipelining transformation shown in Figure 9.1b can optimize the schedule to execute in only two cycles (Schedule B)—here, the addition on the current item proceeds in parallel with reading the next element, leading to a 33% shorter schedule. However, a further improvement is possible if we use a dual-port memory, as shown in Figure 9.1c. Since there are two ports, the computation on the current data can proceed in parallel with writing the previous element and reading the next element, effectively requiring only one cycle, leading to a 66% performance improvement in the steady state (Schedule C).

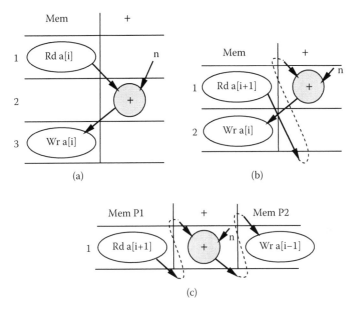

Figure 9.1 (a) Example Schedule A. (b) Pipelined Schedule B. (c) Schedule C using multiport memory.

However, a study of the memory address bus switching activity of the three schedules yields different results. In Schedule A, the sequence of addresses on the memory address bus is

0,0,1,1,2,2,3,3,4,4,...,99,99

Assuming an 8-bit address bus, the bit sequence corresponding to this sequence of addresses is given by the following. The number of bits transitioning on the address bus is given in parentheses, for each new address value.

0: 00000000
0: 00000000 (0 transitions)
1: 00000001 (1 transition)
1: 00000001 (0 transitions)
2: 00000010 (2 transitions)
2: 00000010 (0 transitions)
3: 00000011 (1 transition)
3: 00000011 (0 transitions)
4: 00000100 (3 transitions)
...
98: 01100010
99: 01100011 (1 transition)
99: 01100011 (0 transitions)

For the sequence above, the total number of bits transitioning between successive values is given by

$$0+1+0+2+0+1+0+3+...+1+0=194$$

In Schedule B, the sequence of addresses is:

$$0,\{1,0,2,1,3,2,4,3,5,4,...,98,97,99,98\},99$$

where braces denote accesses in the pipelined loop. The bit sequence corresponding to this address sequence is given by

```
0: 00000000
1: 00000001 (1 transition)
0: 00000000 (1 transition)
2: 00000010 (1 transition)
1: 00000001 (2 transitions)
3: 00000011 (1 transition)
2: 00000010 (1 transition)
4: 00000100 (2 transitions)
3: 00000011 (3 transitions)
...
99: 01100011
98: 01100010 (1 transition)
99: 01100011 (1 transition)
```

For this sequence, the total number of bits transitioning between successive values is given by

$$1+1+1+2+1+1+2+3+...+1+1=386$$

Schedule B is a more expensive alternative in terms of energy dissipation because the additional switching activity on the memory address bus amounts to almost 100%.

Finally, in Schedule C, the sequence of addresses in the two ports are

```
Port 1: 0,1,2,3,...,99
Port 2: 0,1,2,3,...,99
```

Clearly, this leads to twice the number of address bits transitioning compared to Schedule A, since now the address buses of both ports are switching, as opposed to only one.

Of course, address bus switching forms only one part of the total system energy. To determine the impact on the total system energy, we have to measure the dissipation in all the components of the system, including the datapath and the finite state machine. Later in this section we present a detailed analysis of the energy dissipated in the various system components.

When synthesis for low power/energy is an important design objective, the aggressively performance-oriented optimizations may actually result in inferior

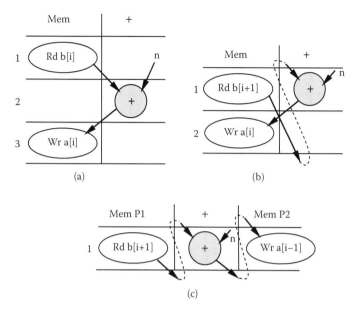

Figure 9.2 (a) Example Schedule A. (b) Pipelined Schedule B. (c) Schedule C using multiport memory.

energy characteristics. However, the unoptimized design from a performance point of view is not always energy-optimal. The following example illustrates this point.

```
for i = 0 to 99
   a[i] = b[i] + n
```

The unoptimized, pipelined, and multiport/pipelined schedules are shown in Figure 9.2. In this case, array *a* is located at addresses 0...99 and *b* occupies 100...199. We have the sequence of addresses as follows:

Schedule A: 100, 0, 101, 1, 102, 2,...,199,99
Schedule B: 100,{101, 0, 102, 1, 103, 2,...,199,98},99
Schedule C: Port 1: {0,1,2,3,...,97},98,99
Schedule C: Port 2: 100,101,{102,103,...,199}

Schedule B results in 9% more switching than Schedule A, but Schedule C results in 57% less switching than Schedule A, possibly making Schedule C a viable candidate from both the performance as well as energy points of view. The dual-port memory configuration actually led to minimum address bus switching because spatial locality could be exploited resulting in sequential accesses. Moreover, addresses being sequential in Schedule C means that we can use appropriate encoding techniques such as Gray code and T0 to further reduce power dissipation both on the memory interface and in the memory module itself.

9.3.1   Analysis

As mentioned earlier, the measured reduction in memory address bus switching does not translate to an equivalent reduction in the total system energy. In order to determine the actual energy dissipation figures for the different designs represented by the three schedules of Figure 9.2a, b, and c, we synthesized them using the commercial synthesis tool Synopsys SystemC Compiler and a 0.18μ IBM ASIC library of components. We used the Synopsys Design Power utility to measure the power dissipation of the resulting circuit.

In order to understand the actual impact of address bus switching, we divided the total system energy into the following components:

1. **Interconnect Energy**—energy dissipated due to switching of the (relatively high capacitance) data and address buses, and other nets in the design.

2. **Memory Internal energy**—energy dissipated inside the memory module during the READ and WRITE accesses. This includes the dissipation in the memory cells as well as at the address decoders, word lines, bit lines, address latches, etc.

3. **Datapath and FSM energy**—energy dissipated in the cells of the datapath and finite state machine generated from synthesizing the application.

The energy dissipated in each of the above components for the three schedules of Figure 9.2a, b, and c, is indicated in Figure 9.3. The difference in the interconnect energy is mainly due to the varying switching activity on address buses reported earlier.

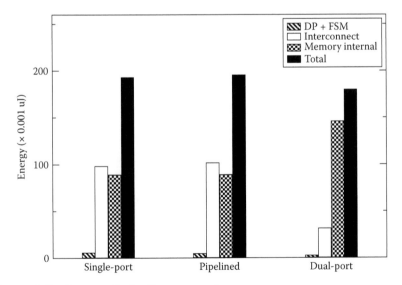

Figure 9.3   Energy dissipation comparison.

The internal memory energy of dual-port memories is, in general, larger than that of a single-port memory because each memory cell drives a larger capacitive load; the address decodes, latches, and other circuitry are duplicated; etc. In the 0.18μ IBM ASIC library, the dual-port memory had a 75% power dissipation overhead compared to a single-port memory of the same size.

The energy dissipated in the datapath and FSM cells is minimum for case A, and is 21% and 31% more in cases B and C, respectively. This is expected since the pipelining leads to slightly more complex control and address generation circuitry.

A few observations are apparent from the above example:

1. The energy dissipated in the datapath and FSM cells is a relatively smaller fraction of the total energy (about 6%). This is obvious in a small example with minimal computation, but the trend of computation-related energy being dwarfed by memory-related energy is also observed in the wider class of data-intensive applications.

2. The interconnect energy and internal memory are the more significant contributors to the total system energy. This is a very important observation and forms the motivation for our research. Dual-port memories incur a higher internal memory energy, but it may be possible to reduce, or even (in some circumstances such as this example) completely negate the overhead by a more efficient addressing mechanism. More importantly, if a dual-port memory has already been chosen for a design due to performance considerations, the techniques presented in the next section help achieve an energy-efficient allocation of memory ports to data so that superior energy characteristics can be obtained for the same or similar performance levels.

9.4 Memory Energy-Aware Synthesis

The memory energy-aware synthesis problem involves the generation of a schedule for a behavioral specification that reduces energy by minimizing switching activity on the memory address bus. The primary optimization criteria may vary depending on the requirements of a specific design, and can be broadly classified into two categories:

1. Optimize the performance and generate a design that operates at this performance level and minimizes memory energy dissipation to the extent possible (i.e., performance is primary; energy is secondary).

2. Optimize the memory energy and maintain the energy-related optimizations while attempting to minimize the schedule length to the extent possible (i.e., energy reduction is the primary objective; performance is secondary).

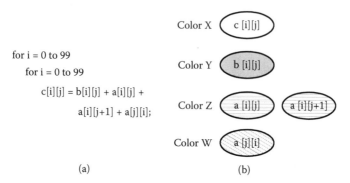

for i = 0 to 99
  for i = 0 to 99
    c[i][j] = b[i][j] + a[i][j] +
      a[i][j+1] + a[j][i];

(a)   (b)

Figure 9.4   (a) Example loop. (b) Coloring.

We address both the above scenarios for generating the appropriate schedules, which require different solution approaches.

9.4.1   Coloring of memory access nodes

The scheduling of memory accesses is preceded by a *coloring* phase where we identify those accesses in a loop body that have spatial locality and are likely to cause only a small number of address bit transitions when accessed consecutively, for instance, $a[i]$ and $a[i + 1]$. Prediction of spatial locality is done by examining whether the array indices differ by a small constant (constrained to be ≤ 4). Standard array index analysis techniques [23] are used for this purpose. An example of coloring memory accesses in a loop is shown in Figure 9.4. $a[i][j]$ and $a[i][j + 1]$ are the same color because they are spatially close, whereas $a[i][j]$ and $a[j][i]$ are colored differently because the two array elements would be located far apart in memory, and consequently, their memory address values would differ by a relatively large number of bits, in general.

9.4.2   Scheduling memory accesses primarily for performance

When the primary objective is performance, we employ energy-optimization as a post-processing step that modifies the generated schedule by assigning the ports to memory accesses in an energy-efficient way without changing the schedule length. This is shown in algorithm REASSIGN_PORTS.

Algorithm REASSIGN_PORTS (G: Scheduled DFG, n: number of ports)
L = Schedule length of G
for ports $i = 1..n$
    PrevColor $[i]$ = NULL
for control steps CS $= 1..L$
    for ports $i = 1..n$
        TempAssign[i] = NULL

```
for ports i = 1..n
    assigned = FALSE
    for ports j = 1..n
        if PrevColor [j] = Color (G, CS, i)
            if (TempAssign [j] ≠ NULL)
                TempAssign [j] = Node (G, CS, i)
                assigned = TRUE
                break
    if (assigned = TRUE) continue;
    for ports m = 1..n
        if TempAssign [m] = NULL
            TempAssign [m] = Node (G, CS, i)
            PrevColor [m] = Color (G, CS, i)
            break
UpdateSchedule (G, CS, TempAssign)
```

The input to REASSIGN_PORTS is a scheduled DFG with memory access nodes already colored, and the number of ports. The color of the previous memory node assigned to a port is stored in PrevColor array, with the array elements initialized to NULL. At every cycle in the schedule, we attempt to assign each memory access to that port whose previous access had the same color. Array TempAssign keeps track of the port assignments made in the current control step. For the ith node in the current control step, given by Node (G, CS, i), we first check whether a node with the same color exists in PrevColor; if yes, then such a color was assigned as the last color on a port, and we attempt to assign this node to the same port so that the same color is retained. If the assignment is not possible at this stage, either because the same color does not exist in PrevColor or there are more nodes of this color in the current control step than in PrevColor, we randomly assign the node to a port. After processing all the ports, the TempAssign array, containing the port assignments for the current control step, is stored into the updated schedule through function UpdateSchedule.

A simple example with an initial schedule is shown in Figure 9.5a. The memory accesses are grouped into two colors: black and white. In the first

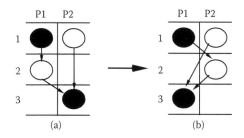

Figure 9.5 (a) Initial schedule. (b) After port reassignment.

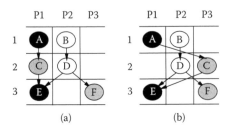

Figure 9.6   (a) Initial schedule. (b) After port reassignment.

iteration of the outer loop (CS $= 1$), the original port assignment is reproduced. When CS $= 2$, the white node is reassigned to port P2 because a white node was accessed in the previous access to P2. Similarly, when CS $= 3$, the black node is reassigned to P1 (Figure 9.5b). This results in an energy-efficient schedule where successive memory accesses on each port have spatial locality and hence results in the minimum number of address bit transitions. The schedule length remains unchanged because the swaps are always *horizontal*, never *vertical*.

A second example is shown in Figure 9.6. A three-port memory is used, along with three colors (BLACK, WHITE, and GREY). The initial scheduled DFG is shown in Figure 9.6a. In the first iteration of the outer loop, the port assignment is copied, resulting in BLACK and WHITE nodes being assigned to ports P1 and P2, respectively. In the second iteration, node C is moved to port P3 so as to avoid the conflict with ports P1 and P2, which have BLACK and WHITE as their previous colors. In the third iteration, node E is retained at port P1 because its color (BLACK) matches the one previously assigned to P1. Node F is also retained at P3 because P3's previous color (GREY) matches F's color. The resulting schedule, after port reassignment (Figure 9.6b), has no color transitions for the three control steps.

9.4.3   Scheduling memory accesses for low energy

The strategy of post-processing a schedule originally optimized for performance is not ideal if the primary objective is energy reduction. This is primarily because of the schedule length constraint. However, if the schedule length is allowed to be modified, that is, performance is allowed to be sacrificed at the expense of energy, then more aggressive energy-optimized schedules are possible. Our strategy, in this case, is to perform the port assignment up front as a *preprocessing* step instead. We first assign the ports to memory accesses using the color information to ensure that each port is assigned memory accesses of the same color as far as possible, resulting in spatial locality being preserved to the maximum extent. Algorithm PREASSIGN_PORTS outlines the strategy.

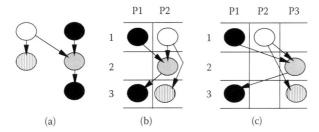

Figure 9.7 (a) DFG. (b) Port assignment and schedule for two-port memory. (c) Port assignment and schedule for three-port memory.

Algorithm PREASSIGN_PORTS (G: DFG, n: #ports)
for all loop bodies L
    Let colors 1..m be used in this loop body
    Sort the colors in decreasing order of their access frequency
        in L into array c[1]..c[m]
    for ports i = 1..n-1
        Assign color c[i] to port i
    Assign remaining colors c[n]..c[m-1] to port n

In algorithm PREASSIGN_PORTS, for each loop, we attempt to assign only one color to each port for all but one port, considering each color in decreasing order of its access frequency in the loop body. We assign the remaining colors to the final port. This configuration is energy-efficient because spatial locality is violated the minimum number of times. The violation takes place when colors change on any port. By assigning one color to each of m-1 ports, we ensure that colors never change for those ports. The assignment is illustrated in Figure 9.7. Since the black-colored memory access occurs more frequently (twice), it is assigned to port P1; others are assigned to P2. This ensures spatial locality (no color change through all loop iterations) in P1, and the violations are restricted to P2 (3 per loop iteration). Note that if, instead, two colors each were to be assigned to P1 and P2 respectively, then there would be at least four violations per iteration. In Figure 9.7c, we show the result of the algorithm when a three-port memory is used. Only one color is assigned to ports P1 and P2, and the remaining colors are confined to port P3. The black nodes are assigned to port P1 because of the higher frequency, as before. The white node is randomly selected for P2, out of the three candidate nodes.

The above assignment strategy may have a negative impact on the schedule length. However, the strategy helps identify important design points on the performance-energy trade-off curve. The final selection decision can be made by the designer.

The port resources are now bound to the memory accesses and scheduling can begin. List scheduling works by invoking a *priority function* to determine the next node to be scheduled among the set of *schedulable nodes* in the current cycle. A common priority function is the *mobility* of operations [24], but this targets a performance-optimized design. Our modification to the priority function that attempts to schedule consecutive nodes of the same color to a port is summarized in function NEXT_NODE. *PrevColor[p]* keeps track of the color of the previous node scheduled on port p. If a schedulable memory access node with the same color as *PrevColor[p]* is found for any p, then we select that node. If no such node is found and there is a schedulable non-memory operation $(Y \neq \phi)$, we use the mobility of that operation to determine the selected node. This serves to defer any unfavorable memory port assignment until absolutely necessary. But if $(Y = \phi)$, we must switch colors on one port (port n, as indicated in PREASSIGN_PORTS). Ideally, we would like to select a node such that future memory access nodes of the same color would get clustered together, but this can be computationally expensive. To prune the search space, we select the node for which the next DFG node with the same color is at a maximum depth, to allow for the possibility of clustering of other colors later on. Function NEXT_NODE omits some details (initial condition, update of *PrevColor*) in the interest of simplification.

Function NEXT_NODE
X = Set of schedulable memory nodes
Y = Set of schedulable non-memory nodes
for all ports p = 1...n
    for all $x \in X$
        if Color(x) = PrevColor[p]
            return x; // matching color found
if $Y \neq \phi$
    return node $y \in Y$ with minimum mobility
else // forced to switch color on port n
    return $x \in X$ for which depth of DFG node with
        Color(x) is maximum

Figure 9.8 shows an example of how NEXT_NODE selects candidate nodes for scheduling. Sets of memory access nodes of the same color are: $\{A, B, C\}$, $\{D, H\}$, and $\{G, I\}$. Suppose the last node scheduled on the single-port memory is A. We have $X = \{B, D, G\}$. B is selected because it is the same color as A. Now, we have $X = \{D, G\}, Y = \{E\}$. Since there is a color mismatch at the port with nodes D and G, we select E. Now we have $X = \{C, D, G\}$. C is chosen to match the color at the port. In the next cycle, we have $X = \{D, G\}$. Note that node I (same color as G) is at a greater depth than H (same color as D). Thus, NEXT_NODE returns node G to allow for the possibility of D and H being consecutive on the port (selecting D here would lead to an additional color switch).

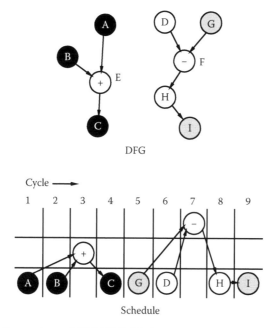

DFG

Schedule

Figure 9.8 Operation of NEXT_NODE.

9.4.4 Loop transformations: unrolling and pipelining

Loop transformations such as unrolling tend to increase the number of consecutive memory accesses of the same color due to the concatenation of array accesses of different iterations. Similarly, loop pipelining improves the throughput by keeping more resources (memory ports) busy in each cycle. The post- and preprocessing steps and the new priority function discussed earlier are directly incorporated into scheduling techniques that involve these loop transformations, since these power optimizations are independent and always applicable.

9.4.5 Alternative architectures: dynamic port assignment

Minimizing transitions across successive accesses to the same memory port can be affected by an alternative architecture that assigns the memory ports dynamically. Instead of statically fixing the port assignment during the scheduling phase, we could have a small circuit that computes the Hamming distances of each new address from the previous address bus values of all the ports and uses the port with minimum distance for the memory transaction. The advantage of such a dynamic port assignment strategy is that it may overcome some of the limitations of the coloring-based approach: if the memory addresses are data-dependent, then the static coloring scheme for grouping memory operations into ports is not effective. However, since the actual addresses are known dynamically, a better port assignment decision can possibly be made.

A dynamic port assignment scheme suffers from some significant overheads. For the set of memory operations scheduled in each cycle, the port assignment circuitry needs to compute the total expected addresses bus switching from assigning each operation on to all the ports, and choose the permutation that leads to minimum overall switching. This increases the complexity (in terms of both area and power) of the datapath. Furthermore, the schedule now does not have an opportunity to optimize the order of accesses targeting low power. In effect, we would have a hardware implementation of algorithm PREASSIGN_PORTS, but would not be able to benefit from the scheduling optimizations of Section 9.4.3. That is, any optimization is restricted to within a cycle, and does not work across cycles.

9.5   Experiments

We studied the effect of our memory port assignment algorithms by performing experiments on several loop- and data-intensive applications involving array accesses and computations. These are ideal candidates because behavioral arrays are stored in memory, and multiport memories can have a significant impact on performance and energy. Since most practical systems use either single-port (SPRAM) or dual-port (DPRAM) memories, we conducted our experiments on these two port types, although the algorithms themselves are general enough to handle a larger port count. We studied the following three cases:

1. **Case A:** Performance-optimized schedule (no power optimization)
2. **Case B:** Performance-optimized schedule with power optimizations applied as a post-processing step (Section 9.4.2)
3. **Case C:** Power optimizations applied as a preprocessing step before and during scheduling (Section 9.4.3)

For each of the above three cases, we performed experiments on both single- and dual-port memories. Thus, there are six configurations corresponding to each example application.

9.5.1   Experimental setup

The experimental setup is shown in Figure 9.9. We used the 0.18μ IBM ASIC library in our experiments. The procedure consisted of the following steps for each design example:

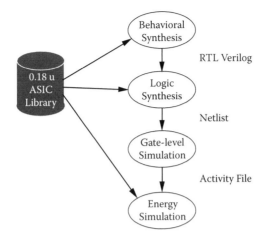

Figure 9.9 Experimental setup.

1. Behavioral synthesis with single- and dual-port memory as target modules using our algorithm in the previous sections. The result of the synthesis was a register transfer level (RTL) Verilog model.

2. Logic synthesis of the RTL Verilog model with Synopsys Design Compiler. The result was a gate level netlist.

3. Simulation of the gate-level netlist with the Cadence NC-Verilog simulator.

4. The resulting *Activity file* from the previous step, parisitics extracted after a placement of the netlist, and the ASIC library were fed into the Synopsys Design Power simulator for measuring the energy dissipation of the example design.

9.5.2 Detailed example

We discuss in detail the experimental results for one important example, the Fast Fourier Transform (*FFT*) algorithm, which is a popular routine used in several digital signal processing applications. We assume that memory accesses, additions, and subtractions require one cycle, and multiplication requires two cycles. The performance-optimized schedules for single- and dual-port memory (case A) are shown in Figure 9.10. The dual-port memory causes a significant reduction in the schedule length, resulting in 30% better performance.

Figure 9.11 shows the schedules for single- and multiport memories for case B, when power optimizations are applied as a post-processing step. The single-port schedule is the same as case A, but the dual-port schedule is modified by interchanging the port assignments of two memory accesses. The memory

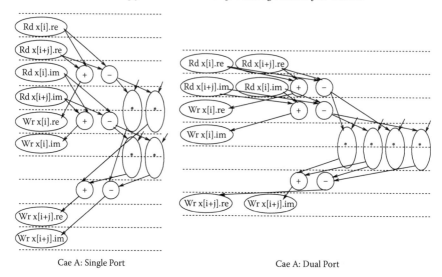

Figure 9.10 Case A: Schedules for single and multiport memory.

access nodes are grouped into two colors, and it is desirable to assign the same color to successive accesses from the same port. $x[i].re$ and $x[i].im$ are grouped into the same color because the fields of the struct have spatial locality.

Figure 9.12 shows the schedules generated by case C. Note that the single-port schedule is longer, but the $x[i]$ elements (colored white) are clustered

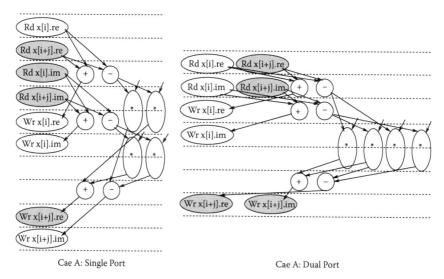

Figure 9.11 Case B: Schedules for single- and multiport memory.

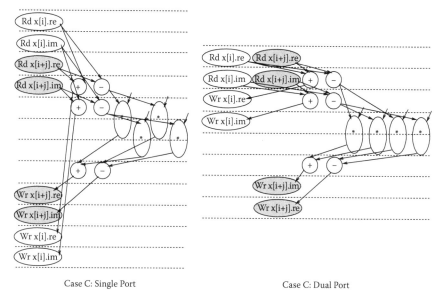

Case C: Single Port   Case C: Dual Port

Figure 9.12 Case C: Schedules for single- and multiport memory.

together, which results in reduced transitions on the address bus. Finally, the case C dual-port schedule is eight cycles long; energy considerations caused our port assignment algorithm to assign each color to a different port. Overall, the power-optimized dual-port memory configuration results in a 20% better performance than the unoptimized SPRAM-based design.

The energy dissipation characteristics of the six schedules discussed previously are summarized in Figure 9.13. For each configuration, we have indicated: (1) the energy spent in datapath and FSM cells; (2) interconnect energy; (3) memory internal energy; and (4) total energy. For the three single-port configurations, the energy dissipation in the various components are almost the same. In fact, the way algorithm REASSIGN_PORTS works, for a SPRAM, cases A and B will always result in an identical schedule (since no port reassignment is possible).

The DPRAM offers an interesting energy comparison. Case A results in higher interconnect energy as well as higher internal memory energy due to the DPRAM. The total energy is 11% higher than the SPRAM-based design. However, our energy optimization, when applied to the DPRAM-based design, results in lower interconnect energy, which offsets the increased internal memory energy of the DPRAM. Cases B and C result in 9% and 14%, respectively, less energy than case A (unoptimized) for the DPRAM. In fact, case C actually results in marginally *lower* energy than even the SPRAM-based design.

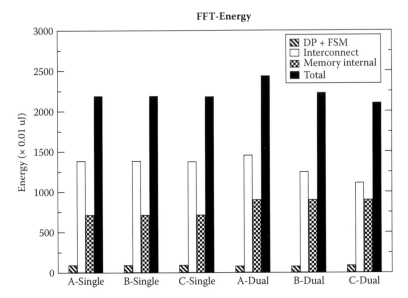

Figure 9.13 Energy dissipation comparison for *FFT*.

9.5.3 Summary of results

We report the performance and energy results on four other examples: *SOR*, *Dprod*, *Planckian*, and *DHRC*. *SOR* is the successive over-relaxation algorithm often used in image processing; *Dprod* is the dot product example from DSP-Stone benchmark set; and *Planckian* is a scientific computing benchmark from the Livermore Loop set. *DHRC* is the Differential Heat Release Computation application.

Figure 9.14 shows the performance comparison of the different configurations on the *SOR* example. The DPRAM-based designs show about 42% over the SPRAM-based ones. The energy dissipation comparisons of Figure 9.15 show that the unoptimized DPRAM-based design (A-Dual) has a 14% total energy overhead over the SPRAM-based one, mainly due to the increased memory internal energy. However, the optimized DPRAM-based design (C-Dual) is able to offset this increase by reducing the interconnect energy. Clearly, C-Dual is a very important configuration in the design space of *SOR*—it results in 42% performance improvement over the SPRAM design, while maintaining comparable energy dissipation.

The performance and energy comparisons for *Planckian* are shown in Figures 9.16 and 9.17. The performance-optimized DPRAM-based designs (A-Dual and B-Dual) are 40% faster than the SPRAM-based ones. In C-Dual, the latency is extended to incorporate the energy optimizations. As Figure 9.17 shows, A-Dual and B-Dual incur about 17% overhead over the SPRAM-based designs due to higher interconnect energy and memory internal energy. However, the energy-optimization in C-Dual reduces the overhead to only 7%.

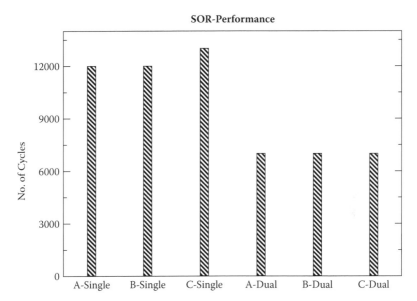

Figure 9.14 Performance comparison for *SOR*.

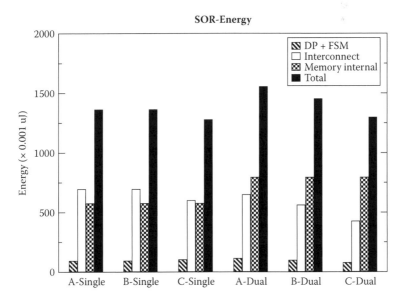

Figure 9.15 Energy dissipation comparison for *SOR*.

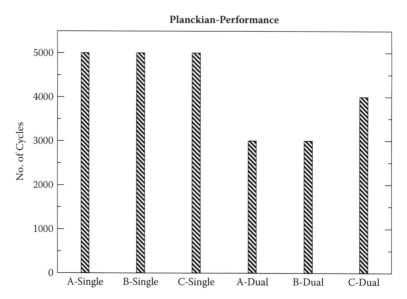

Figure 9.16   Performance comparison for *Planckian*.

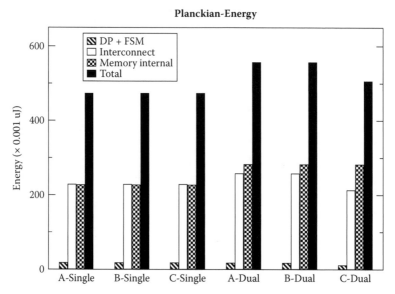

Figure 9.17   Energy dissipation comparison for *Planckian*.

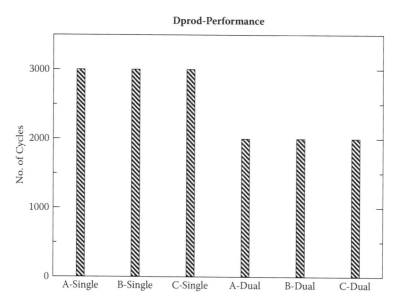

Figure 9.18 Performance comparison for *Dprod*.

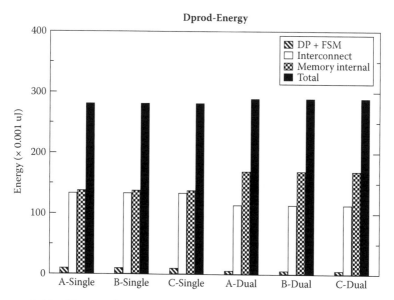

Figure 9.19 Energy dissipation comparison for *Dprod*.

In the *Dprod* example, the schedules for all the SPRAM-based designs were identical, as were all the DPRAM-based designs. The performance-optimal DPRAM-based designs were also energy-optimal. In Figures 9.18 and 9.19 we observe a 33% performance improvement from using DPRAMs, with the total energy overhead being 3% over the SPRAM-based ones. For the *DHRC* application, the DPRAM-based designs showed a performance improvement of about 27% over the SPRAM-based ones, with a 5% energy overhead.

9.5.4   Discussion

1. The most important observation from our experiments is that an efficient assignment of memory ports to data accesses usually ensures a reduction in energy dissipation of synthesized designs based on multiport memories.

2. It is important to note that this work does not attempt to demonstrate that multiport memory-based design can always yield lower overall energy than single-port-based ones. The single-port numbers were presented only as a reference/baseline to perform comparisons. As our experiments show, the energy of the optimized dual-port configurations (C-Dual) was comparable to that of SPRAM-based ones. However, performance considerations might lead designers to choose DPRAMs in system designs. Once this decision is made, our port assignment algorithms help reduce the energy dissipation significantly, as the A-Dual versus C-Dual numbers clearly show.

3. Since energy dissipation and delay can be frequently traded off, it is sometimes useful to observe the energy-delay product when comparing architectural alternatives. In our experiments, we observe that the energy-delay product is smaller for the dual-port memory configurations than those for the single-prt configurations. A comparison of the A-single and C-dual over all the examples shows an average energy-delay product reduction of 27%, thus establishing the dual-port memory as an important design alternative.

   A comparison of A-dual and C-dual (optimized for performance vs. optimized for delay) shows that on an average, the product is roughly unchanged; the energy-delay product is equal in *Dprod*, inferior for C-dual in *FFT* and *Planckian*, and superior for C-dual in *DHRC* and *SOR*. This is because our algorithmic formulation aggressively optimizes either one or the other variable but not both. To minimize the energy-delay product, we would have to modify the basic algorithm by imposing a maximum performance loss constraint in the energy optimization.

4. As manufacturing technology evolves with shrinking geometries, we expect the interconnect-related energy to dominate the total energy dissipation of the complete system. This is already seen in our experiments where the energy consumed in the datapath and FSM cells is a small

fraction of the interconnect-related energy. We verified this trend by runing the experiments on the *Planckian* example for two different technology libraries: 0.25μ and 0.18μ. The results show that the energy dissipated in the datapath and FSM cells was 15% of the interconnect energy in 0.25μ technology, but only 6% in 0.18μ technology.

5. The interconnect energy, as used in our experiments, includes the energy dissipation both due to data/address buses and other high-capacitance wires, as well as due to the computation within the datapath and FSM blocks. Taken together, it makes the interconnect energy comparable to the memory internal energy. As the memory requirement of applications increases, the memory internal energy increases, but since the application complexity also increases, the interconnect energy is likely to still remain comparable.

6. Note that it is not necessary to have multiport memories to benefit from the algorithms we presented in this chapter. The energy optimization strategy can be useful even with single-port memories. In such cases, the problem is not assignment of data to ports, but the appropriate reordering of data accesses to reduce address bus switching (this is incorporated into the NEXT_NODE function in Section 9.4.3). This case is illustrated in Figure 9.15. Case C-Single is more energy-efficient than case A-Single; the interconnect energy is reduced by reordering array accesses.

7. We have restricted our study to the performance and energy implications of single- and multiport memory. The area overhead of the multiport memories will continue to lead to larger overall design area (e.g., area overhead was 37% in *FFT*).

8. The examples chosen for our experiment did not have the memory access patterns that could benefit from the hardware-based dynamic port assignment discussed in Section 9.4.5. As such, including such a scheme would have been an unnecessary datapath overhead. However, it is possible that the dynamic scheme could be useful where it is not possible to make good static inferences, for example, due to run-time dependencies.

9.6 Conclusion

We presented algorithms to reduce memory address bus switching energy by an efficient allocation of memory ports to behavioral array accesses combined with the reordering of data accesses by incorporating energy optimizations into the scheduling phase of behavioral synthesis. The optimizations apply to both single- and multiport memories. Frequently, multiport memories can offer significant performance advantages in system designs because of the increased data throughput, but lead to overheads in area and energy. We have shown that our strategy can help reduce some of this overhead in the overall energy dissipation of the system, while still retaining most of the performance advantages offered by multiport memories.

References

1. P. R. Panda, N. D. Dutt, and A. Nicolau, *Memory Issues in Embedded Systems-On-Chip: Optimizations and Exploration*, Kluwer Academic Publishers, Norwell, MA, 1999.

2. P. R. Panda, A. Shrivastava, B. V. N. Silpa, and K. Gummidipudi, *Power-Efficient System Design*, Springer, New York, NY, 2010.

3. P. R. Panda and N. D. Dutt, "Low-power memory mapping through reducing address bus activity," *IEEE Transactions on VLSI Systems*, vol. 7, no. 3, pp. 309–320, Sept. 1999.

4. L. Benini, G. de Micheli, E. Macii, M. Poncino, and S. Quer, "Power optimization of core-based systems by address bus encoding," *IEEE Transactions on VLSI Systems*, vol. 6, no. 4, pp. 554–562, Dec. 1998.

5. E. Musoll, T. Lang, and J. Cortadella, "Exploiting the locality of memory references to reduce the address bus energy," in *International Symposium on Low Power Electronics and Design*, Monterey, CA, Aug. 1997, pp. 202–207.

6. P. R. Panda, F. Catthoor, N. D. Dutt, K. Danckaert, E. Brockmeyer, C. Kulkarni, A. Vandercappelle, and P. G. Kjeldsberg, "Data and memory optimization techniques for embedded systems," *ACM Transactions on Design Automation of Electronic Systems*, vol. 6, no. 2, pp. 149–206, Apr. 2001.

7. C.-L. Su and A. M. Despain, "Cache design trade-offs for power and performance optimization: a case study," in *International Symposium on Low Power Design*, New York, NY, 1995, pp. 63–68.

8. L. Benini, A. Macii, and M. Poncino, "A recursive algorithm for low-power memory partitioning," in *International Symopsium on Low Power Electronics and Design*, Rapallo, Italy, Aug. 2000.

9. L. Benini and G. De Micheli, "System level power optimization: techniques and tools," *ACM Transactions on Design Automation of Electronic Systems*, vol. 5, no. 2, pp. 115–192, Apr. 2000.

10. V. Tiwari, S. Malik, and A. Wolfe, "Power analysis of embedded software: a first step towards software power minimization," *IEEE Transactions on VLSI Systems*, vol. 2, no. 4, pp. 437–445, Dec. 1994.

11. H. Mehta, R. Owens, M. J. Irwin, R. Chen, and D. Ghosh, "Techniques for low energy software," in *International Symposium on Low Power Electronics and Design*, Monterey, CA, Aug. 1997, pp. 72–75.

12. H. Tomiyama, T. Ishihara, A. Inoue, and H. Yasuura, "Instruction scheduling for power reduction in processor-based system design," in *Design, Automation, and Test in Europe*, Paris, France, Feb. 1998.

13. E. Musoll and J. Cortadella, "High-level synthesis techniques for reducing the activity of functional units," in *International Symposium on Low Power Design*, Dana Point, CA, Apr. 1995, pp. 99–104.

14. P. R. Panda, N. D. Dutt, and A. Nicolau, "Local memory exploration and optimization in embedded systems," *IEEE Transactions on Computer Aided Design*, vol. 18, no. 1, pp. 3–13, Jan. 1999.

15. W.-T. Shiue and C. Chakrabarti, "Memory exploration for low power embedded systems," in *Design Automation Conference*, June 1999, pp. 140–145.

16. Y. Li and J. Henkel, "A framework for estimating and minimizing energy dissipation of embedded hw/sw systems," in *Design Automation Conference*, San Francisco, CA, June 1998, pp. 188–193.

17. D. Kirovski, C. Lee, M. Potkonjak, and W. Mangione-Smith, "Application-driven synthesis of core-based systems," in *Proceedings of the IEEE/ACM International Conference on Computer Aided Design*, San Jose, CA, Nov. 1997, pp. 104–107.

18. M. Balakrishnan, A. K. Majumdar, D. K. Banerji, J. G. Linders, and J. C. Majithia, "Allocation of multiport memories in data path synthesis," *IEEE Transactions on Computer-Aided Design of Integrated Circuits and Systems*, vol. 7, no. 4, pp. 536–540, Apr. 1988.

19. T. Kim and C. L. Liu, "Utilization of multiport memories in data path synthesis," in *Design Automation Conference*, Dallas, TX, June 1993, pp. 298–302.

20. J. Seo, T. Kim, and P. R. Panda, "Memory allocation and mapping in high-level synthesis: an integrated approach," *IEEE Transactions on VLSI Systems*, vol. 11, no. 5, pp. 928–938, Oct. 2003.

21. S. Wuytack, F. Catthoor, G. De Jong, and H. De Man, "Minimizing the required memory bandwidth in VLSI system realizations," *IEEE Transactions on VLSI Systems*, vol. 7, no. 4, pp. 433–441, Dec. 1999.

22. P. R. Panda and N. D. Dutt, "Behavioral array mapping into multiport memories targeting low power," in *Proceedings of the 10th International Conference on VLSI Design*, Hyderabad, India, Jan. 1997, pp. 268–272.

23. M. E. Wolf and M. Lam, "A data locality optimizing algorithm," in *Proceedings of the SIGPLAN'91 Conference on Programming Language Design and Implementation*, Toronto, Canada, June 1991, pp. 30–44.

24. D. Gajski, N. Dutt, S. Lin, and A. Wu, *High Level Synthesis: Introduction to Chip and System Design*, Kluwer Academic Publishers, Norwell, MA, 1992.

Chapter 10

Energy-Efficient Address-Generation Units and Their Design Methodology

Ittetsu Taniguchi
Ritsumeikan University, Shiga, Japan

Guillermo Talavera
Universitat Autònoma de Barcelona, Spain

Francky Catthoor
IMEC/Katholieke Universiteit, Leuven, Belgium

Contents

10.1   Introduction

Modern embedded systems, especially nomadic embedded systems, are becoming more and more complex, and they often require both high performance and low energy consumption. To achieve both requirements, exploiting ILP (instruction level parallelism) and DLP (data level parallelism) of embedded applications are well known to be good solutions. High-level compiler optimizations like data flow and loop transformations targeted toward data memory sub-systems have been shown to achieve high performance and low energy consumption. Such techniques transform C code to improve spatial and temporal locality of data and reduce unnecessary memory access. By improving data locality in the application loops, this data can be stored within a smaller memory footprint so that accesses to larger memories like level 1 (or higher) memories are drastically reduced. These transformations decrease the quantity of memory accesses but increase considerably the complexity of address generation. Such complex address calculations initially result in expensive operations such as multiplications or modulo operations.

For the purposes of our investigation, we have used a cavity detector application as our major test-vehicle. This image processing application is a data-intensive algorithm composed of nested loops where each loop kernel includes memory access to calculate input data. As memory is accessed for each iteration, this image processing application requires a large amount of memory access for its execution. We have counted the number of address calculations and data calculations manually, and have calculated a ratio. We have observed that after data-flow/loop transformations and memory hierarchy optimization that 70% of all compute operations are address computation operations that now form the new performance bottleneck.

To alleviate this problem, modern architectures often include a dedicated unit for address calculation that works in parallel with the main computing elements: the address-generation unit (AGU).

In this chapter, we propose an energy-efficient address generation unit (AGU) and outline its supporting design methodology. The proposed AGU is based on a coarse grained reconfigurable architecture. To use the reconfigurable AGU effectively, it is important to specify and differentiate its architecture from the many potential architectural candidates. Systematic architecture exploration from a vast solution space represents a demanding challenge because it is very difficult to identify the best architecture both quickly and accurately: indeed, an accurate evaluation usually consumes a long time. We therefore propose a fast and systematic architecture exploration method for the proposed reconfigurable AGU.

The rest of the chapter is organized as follows: Section 10.2 highlights the importance of address generation units and provides a summary of address generation methods. Section 10.3 describes a proposed AGU model and its cost model including area model and energy model. Section 10.4 defines the architecture exploration problem for the proposed reconfigurable AGU model. Section 10.5 describes a performance-evaluation method, and Section 10.6 proposes an effective architecture exploration based on a proposed metric used to identify promising solutions. Section 10.7 demonstrates the efficiency of the proposed AGU exploration framework. Finally, Section 10.8 provides a conclusion to the chapter.

10.2   Motivation behind Exploration of AGUs

Programmable architectures oriented to exploit parallelism, digital signal processors (DSPs), and multimedia processors (a mixture of RISC and DSP processors) usually follow the VLIW paradigm: a number of functional units (FUs) running in parallel, following a schedule generated by a compiler [1]. These architectures focus on real-time performance and often deal with infinite, continuous streams of data.

In many architectures, an address-generator unit works in parallel with the main data calculation units to ensure efficient feed and storage of the data from/to the datapath. Leaving aside the access time and the parallel access constraints, the main problem is the efficient generation of the address sequences for a given application.

The generation of an address sequence is done from an address equation (AE), which is a function extracted from the software description of the algorithm where the parameters are the indexes (I_n) of nested loops or range addresses (r_m): the bounding box where addresses are generated.

$$AE = f(I_1, I_2, \ldots, I_n, r_1, r_2, \ldots, r_m), \qquad (10.1)$$

where I_i represents the i-th loop index and r_j represents the j-th loop range.

In a broad sense, an address-generation unit is the unit that uses the address equation (AE) to generate an address sequence (AS). The resulting Address Sequence contains "what" address to access and "when" to access it.

10.2.1   Types of address equations

The address equation is a function extracted from the software description of the algorithm. The different AEs can be categorized in terms of their regularity or their flexibility.

10.2.1.1   Regularity

The regularity of the AE is correlated with the complexity of the index expressions.

```
for (i=0; i<=N1-N2; i++){
    y[i] = 0.0;
    for (j=0; j<N2; j++){
        y[i] += w[j]*x[i+j];
    }
}
```

Figure 10.1   Example of an affine address equation: a FIR filter code.

```
for (i=0; i<8; i++){
    for (j=0; j<8; j++){
        tmp=0.0;
        for (k=0; k<8; k++)
            tmp+=c[k][j]*block[8*i+k];
        res[8*i+j]=tmp;
    }
}
```

Figure 10.2   Example of an affine address equation: piece of code from the MPEG2 decoder kernel.

Affine AE: An AE is affine when the address equation is a linear expression of the indexes I_n and constants C_n as shown in the following equation:

$$AE = C_0 + C_1 \cdot I_1 + C_2 \cdot I_2 + \cdots + C_n \cdot I_n \qquad (10.2)$$

This is the typical case for addresses generated by a number of manifest nested loops. In Figure 10.1, we can recognize three AE. For arrays y and w, the AEs are a direct function of the loop indexes i and j respectively, and for array x, it is a function of i and j with the coefficients $C_0 = 0$ and $C_1 = C_2 = 1$.

A more complex example is shown in Figure 10.2. In this case, the address equation calculates the address indexes for the arrays c, *block*, and *res*. Assuming c is a 4-byte integer, the address equation of c can be written in the following form $AE_c([k][j]) = c[k * 4 + j * 4 * 8]$, and under this form we can clearly see that the AE of c is affine with parameters $C_0 = 0, C_1 = 4$, and $C_2 = 32(= 8 * 4)$.

Piece-wise affine AE: An AE is called a piece-wise affine AE when parts of the AE can be written as linear expression of the indexes and constants. This is the case of AE in a nested loop with conditional statements based on the iterators (manifest conditions).

In Figure 10.3, a piece of code from the MPEG-4 video decoder core is shown. In this case, the conditional expressions limit the possible optimizations in the search space.

```
int *DCstore;
DCstore = (int*)malloc(LB*6*15*sizeof(int));
...
initialize (Xtab, Ytab, Ztab, Xpos, Ypos);
loop{
   if(comp==1||comp==3){
       blockA=DCstore[0+15*Xtab[comp]+90*(((mbnum/MB)*MB+(mbnum%MB+
          Xpos[comp]))%LB)];
   }else{
       block A=mg*8;
   }
   if(comp==3){
       blockB=DCstore[0+15*Ztab[comp]+90*(((mbnum/MB+Ypos[comp])*MB
          +(mbnum%MB+Xpos[comp]))%LB)];
   } else{
       block B = mg*8;
   }
   if(comp==2||comp==3){
       blockC=DCstore[0+15*Ytab[comp]+90*(((mbnum/MB+Ypos[comp])*MB+
          mbnum%MB)%LB)];
   }else{
       block C = mg*8;
   }
   mbnum++;
}
```

Figure 10.3 Example of a piece-wise affine AE extracted from the MPEG-4 video decoder core. Conditional expressions limit the search space for address arithmetic optimization.

Non-linear AE: An AE is called non-linear when there is no linear relation between the AE and the address indexes. This is the most general case of AE. Figure 10.4 shows an example of real code where we can see square expressions ($j*j$ and $i*i$). The non-linearity of those expressions highly constrain the search space for the optimization of address generation and related hardware, but this type of AE does not occur very often in real life application codes.

10.2.1.2 Flexibility

The flexibility of the address equation gives the idea of the range of the flexibility needed by the AGU to create the address sequence.

Manifest or predefined with constants: This is a special case that occurs for architectures that target a specific algorithm, or a small number of algorithms. In these cases, the AE is manifest, and then address sequence patterns are predefined before the hardware design and the knowledge of the program can be efficiently exploited by constructing optimized AGUs.

Parameterizable or predefined with parameters: This is the case for AS controlled by some input parameters. *E.g.*, this is the case for domain

```
for (j=0; j<l_code; j++)
for (i=0; i<l_code; i++){
    if(i<=c1)
        if(i==c1)
            rdm=h2[l_code-1-i];
        else rdm=rr[i*(i-1)/2];
    else rdm=rr[j*(j-1)/2];
    ...mul(rdm, ...);
    if(i<=c2)
        if(i==c2)
            rdm=h2[l_code-1-i];
        else rdm=rr[j*(j-1)/2];
    else rdm=rr[i*(i-1)/2];
    ...mul(rdm, ...);
```

Figure 10.4   Example of a non-linear AE: fragment for the GSM codebook algorithm.

specific designs where the AGU can be parametrized to the application since it has to support a limited number of algorithms.

Dynamic address equation: This is the most general case of address sequence where the AE depends on the AGU's external events and results (*e.g.*, data-dependent addresses). Here, the AE and thus the AS can be modified during run-time execution. This is the general case for address sequences in processors or DSPs. In the embedded systems domain, based on application scenarios [2] that occur most frequently, the majority of dynamic address equations can be modeled as manifest or parameterizable address equations [3–5]. For the rest of the cases, a fully run-time backup scenario without compiler support must be used.

10.2.1.3   AE design space exploration

Figure 10.5 summarizes Section 10.2.1 and shows the various possible combinations of regularity and flexibility of the AE. The complexity of building an efficient AGU increases as we move away from the origin of the graph. For each point in the design space, an optimal AGU can be found to meet the design metrics targeted by the design.

10.2.2   AGU classification

In the literature, we can find different names for Address Generator Units (AGUs), for example, Address Calculation Unit (ACU), Address Arithmetic

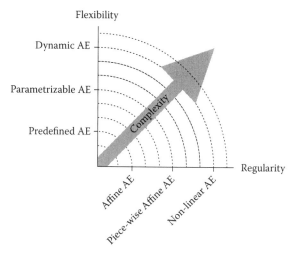

Figure 10.5 Design space of AGUs.

Unit (AAU), Data Address Generator (DAG), Memory Management Unit (MMU), Direct Memory Access (DMA), etc. In this chapter we will generally refer to all these address generators as AGUs, and in this section we will explain the different possible hardware implementations. We can distinguish two big types of AGU architectures: the ones based on tables and the ones based on a datapath.

Let us take the following address sequence as an example:

$$AE : \&[2 * i + 1] = 1, 3, 5, 7, 9, 11, 13, 15 \qquad (10.3)$$

10.2.2.1 Table-based AGUs

Table-based AGUs can be used to implement short address sequences with a few indexes because the increase of indexes leads to a large area penalty to store the addresses for all combinations. This approach needs a controller to complete any deterministic address sequences. In the simplest case, the controller can be a simple increment/decrement counter, but often a finite state machine (FSM) is needed. When the AE is more complicated and a LUT (lookup table) solution becomes too big, the logic (that can be custom, programmable, or configurable) to complete the AE can be implemented using a programmable logic array (PLA) (Figure 10.6), as those devices have a smaller size compared to the same implementation in a LUT.

Until the beginning of the 1990's, little research had been done on address-generation optimizations, and the first optimizations were oriented to reduce area on table-based AGUs. Once the possibility of integrating SRAM memories onto chips was available, area savings became less important and energy

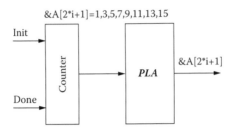

Figure 10.6 Example of a table-based AGU.

became the main problem. Even if area is no longer a limiting factor, some of the techniques can be reused for energy reduction.

Table-based AGUs implement manifest address sequences and although they cannot deal with dynamic applications, the knowledge of the specific address equation enables hardware design time optimizations. In [6,7] the authors assumed that every AE is mapped to a separate address unit. Miranda *et al.* presented in [8] an area-optimization technique for application-specific address-generation units for large memories in real-time signal-processing systems. Their techniques rely on a system level analysis of the trade-offs involved in algorithmic specifications: when the signal-processing application is being defined, savings are exploited by sharing index expressions among several individual AEs.

Miranda *et al.* present in [8–11] the ADress OPTimization (ADOPT) environment. The framework gives a formalized methodology and an automated technique to support address arithmetic optimizations in flow-graph expressions [12–15] for distributed memory architectures. The ADOPT environment targets architecture and system level optimizations at the same time and the methodology relies on two stages: one independent of the target architectural style and a second one stage specific to the selected AGU.

The trade-offs between area and performance for known address sequences have been discussed in [16]. Their work studies the impact on area and performance of memory access related circuitry in eliminating row and column address decoders from the memory and incorporating the necessary hardware decoding in the address-generation circuit. They show how circuit delay can be nearly halved at the expense of increased area.

10.2.2.2   Datapath-based AGUs

Datapath based AGUs are more suited to complex or long address sequences when there is a dependency on loop iterators, pointers, or flags coming from other units (*e.g.*, the datapath) or for address generation that needs some control or programmability so that a wider address space can be covered. Those AGUs have an arithmetic unit inside, and they calculate addresses directly.

For complex and long calculations, an expensive arithmetic unit may be required. Datapath-based AGUs are classified into custom AGU and programmable AGU based on the flexibility they provide.

Custom AGUs are an optimal architectural style for irregular memory accesses that consume too much mapping logic if implemented in table-based AGUs or for long array-based address sequences (Figure 10.7). These cases are usually based on a set of nested loops, where the address parameters are combined to give addresses at each iteration of the loop. Those units have fixed arithmetic units and interconnections, and these functionalities cannot be changed. Miranda *et al.* [9] proposed counter-based AGU, which is an AGU whose input indexes are given by a modulo counter, architecture exploration method. The same authors in [11] also proposed a custom AGU architecture exploration method. The implementation of non-linear calculations is often expensive, thus the exploration of the trade-off between custom based-AGUs and table-based AGUs is necessary.

Predefined AEs can also be implemented using bit level sequence generators. In the most straightforward case, this will lead to the direct mapping of each bit level AE onto a dedicated hardware block. This strategy could lead to a large area overhead, and bit level optimizations are mainly oriented to the reuse of hardware between bit level address generators. The work of Grant [17] and Lippens [18] shows, for very regular AE, optimizations at bit level for counter/table-based architectures. In [19] the same authors present the address-generation techniques of the ZIPPO toolbox, which relies on hardware-based multiplexing optimization at word level and bit level address among several AEs. The authors show that address-generation optimizations can result in area savings between 10% and 60%.

Programmable AGUs are targeted for an optimal calculation of loop array indexes. These AGUs have dedicated registers available that can be programmed, flexible arithmetic units and interconnections, and their functionality can be modified at run-time. A very general programmable AGU is shown in Figure 10.8. This flavor of AGU is the most suitable architectural style for parameterizable or dynamic address equations providing flexibility

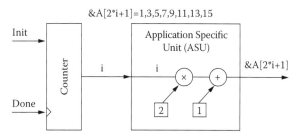

Figure 10.7 Example of a custom datapath-based AGU.

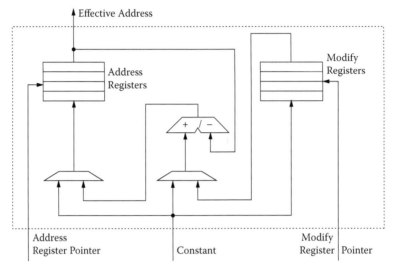

Figure 10.8   Example of a general programmable AGU.

at the price of increasing area and energy. This flexibility of programmable AGUs pays an area overhead depending on the following parameters:

- number of memories to address
- size of the arithmetic blocks (adders, multipliers, etc.)
- address register files
- program memory and instruction decoders

Programmable AGUs are normally used in digital signal processors such as TI TMS320C2x/5x/6x, AT&T DSP 16xx, and other VLIW processors. In those processors, the AGU is dedicated and can be used to reduce the number of address calculation operations with auto-increment and auto-decrement capabilities. Some research topics related to AGUs on DSPs can be found in [20–22]. Leupers [23] has defined a generic AGU model that covers major DSPs and proposed some design methodologies. Mathew and Davis [24] have proposed a distributed address generation and loop acceleration architecture for VLIW processors that have multiple SRAMs.

Programmable AGUs benefit from flexibility but have an important area penalty due to the arithmetic units, registers, and counters needed to give enough versatility, and must mainly rely on software and compiler optimizations.

Traditional compiler approaches aim at address pre-calculation and post-calculation [25]. Address pre-calculation consists of an addition of a base address with a variable. In this method, the address must be calculated before the reference. For innermost loop kernels, this approach presents a significant decline in the performance of the algorithm. Address post-calculation reduces an array reference to an address pointer reference and incrementing/decrementing the resulting address for the next reference of the array. This method allows the calculation of the next address in parallel with a main data operation. Optimizations on address post-calculation were first studied by Liem *et al.* in [26,27]. In their work, the authors show code transformations that produce code with optimized index array references for fast post-modification combining address variables with the same addresses offset among loop iterations. These optimizations improve the efficiency of programmable AGUs.

For data-dependent address generation (Section 10.2.1.2) using a programmable AGU, address optimizations can be achieved via data ordering and address register allocation. Data ordering determines the order of data stored in the memory. Address register allocation assigns an address register to each data access for address generation. The goal of this approach to optimization is to maximize the usage of auto-increment/decrement and hence reduce the number of address loading instructions. Many research papers and industrial compiler groups work in auto-increment/decrement optimizations but they are mainly oriented toward timing optimization, not power. In [28] Cheng and Lyn present an address optimization technique for loop execution for DSPs with auto-increment/decrement architecture. They propose a new graph model that takes care of constraints on memory allocation and data ordering. In [29], other optimization techniques targeting DSPs are explained in detail.

In [30], the authors present arithmetic and address computation optimization for a set of typical very regular kernels that appear frequently in several kinds of applications. Their optimization resides in the observation that, for those kernels, the elements accessed are usually stored close to one another in memory. Their work studies scalar conversion and common sub-expression optimization between successive iterations of loop bodies. Actual compilers do not carry out this kind of analysis, and for successive iterations the value of common sub-expressions is computed at each iteration.

Programmable AGUs are a key element nowadays in most designs. As an example, multimedia-handheld devices require high-performance specific computation, usually accompanied by real-time and Quality of Service (QoS) constraints, which ensure the device can run at low energy levels and have a long battery life. Having a flexible architecture is also needed to meet short time-to-market restrictions and to facilitate the update to new versions of the applications and/or to add a new application to the device with similar requirements. Hence, the ideal multimedia device will present high-quality multimedia content and will be networked, portable, inexpensive, and easy to use. Moreover, in order to cope with the dynamism of current and future multimedia applications, modern embedded systems demand a programmable and high-performance solution running at low energy consumption to deal with all these requirements—this is even more relevant when it comes to the devices that will be designed in the future.

```
...
for(i=0;i<N;i++) {
for(j=0;j<M;j++) {
...
tmp_val[i]= myImages[TypeMacroblock*(ImageMCRef-1)
         *(IMG_ROWS+IMG_ROWS/2)*IMG_COLS + IMG_ROWS*IMG_COLS
         + (v*M + j )*IMG_COLS/2 + (h*N + i)];
...
```

Figure 10.9   Example of complex address calculation.

In this context, having a high-performance and flexible AGU is crucial to meet all the requirements. Based on our research we propose a reconfigurable AGU intended for VLIW-like processors or "regular super-scalar processors." This AGU is a high-performance and highly-flexible reconfigurable AGU and belongs to the category of programmable AGUs.

The most relevant existing work in this research area is outlined in [24] by Mathew and Davis. Mathew's work targets the speeding up of two-dimensional affine address equations so that $A[i * P + Q][j * R + S]$. The two-dimensional affine address equation is a subset of all practically relevant address equations, but our reconfigurable AGU can handle all address equations that can occur in valid C code. For two-dimensional affine address equations, Mathew's method can calculate them in one cycle because the unit is coupled with counters and is optimized for the targeted calculations. Our proposed reconfigurable AGU cannot achieve higher performance for the calculation of this $A[i * P + Q][j * R + S]$ expression. However, Figure 10.9 shows an example of the address calculation from a loop kernel extracted from a real-life application after significant optimization. These calculations cannot be effectively performed by the proposed architecture by Mathew and Davis [24].

In contrast to these approaches, the objective of our research is to produce a reconfigurable AGU template accompanied by mapping techniques inspired by the reconfigurable computing domain. We also include an exploration of special low power custom instructions like *shift-add, add-shifts*, and *shift-shift-adds*. In addition, our aim is to explore different possible AGU configurations and to analyze the various trade-offs rather than focusing on one specific instance for a particular metric (*e.g.*, performance) as is the case in most of the available related work in this field.

10.3   Reconfigurable AGU: What Do We Execute the Calculations on?

In order to generate appropriate address sequences, various architectures have been proposed as we previously outlined. In this chapter, we propose a reconfigurable AGU model as a general address-generation unit. The proposed

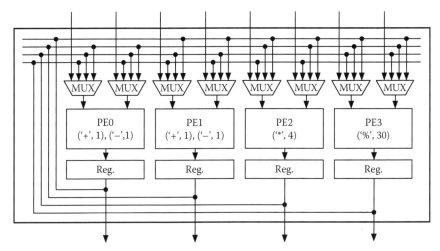

Figure 10.10 Reconfigurable AGU model.

reconfigurable AGU model is based on a highly abstracted reconfigurable architecture model proposed by Taniguchi *et al.* [31]. It consists of several processing elements (PEs) and the corresponding pipeline register. Each PE is composed of dedicated functional units of heterogeneous instructions that have different latencies. Each PE is fully connected to the other PEs.

Figure 10.10 shows one instance of the reconfigurable AGU template, which has four PEs. PE0 and PE1 have one cycle *add* and *sub* instructions indicated by "+" and "−", and PE2 has 4 cycles *multiply* instruction indicated by "*". PE3 has 30 cycles *modulo* instruction indicated by "%". Because of full connection, the order in which the PEs are organized is not important. For example, the array would have the same functionality if PE0 was to be swapped with PE3. Thus we can look at the AGU model in which any PE can be swapped to any other in an AGU model, for example, the one shown in Figure 10.10.

Please note that the reconfigurable AGU is only a model, and the architecture shown in Figure 10.10 is only a single instance. This means the reconfigurable AGU has a variety of architectural configurations such as a different number of PE or a different instruction assignment. In this chapter, we regard address-generation unit design as a decision based on the following reconfigurable points of the AGU model:

- the number of PE
- the instruction assignment for each PE

10.3.1 Variation of the reconfigurable AGU model

What we want to do, ultimately, is to formulate a decision regarding the optimum number of PE and the instruction assignment for each PE. At that point, we can review the reconfigurable AGU model again in more formal way.

Let I, I^A, P, and S be a set of instructions that could be used, a set of instructions used in a given application, a set of PE implementation patterns, and a set of solutions (i.e., instances of reconfigurable AGU model), respectively. If $I(p)$ and $P(s)$ represent a set of instructions executed in PE p and a set of PEs in solution s, then $I(p)$ and $P(s)$ satisfy the following condition: $I(p) \subseteq I$ and $P(s) \subseteq P$. Now, in order for a given solution s to execute a given application, the following Relation 10.4 must hold true.

$$I^A \subseteq \bigcup_{\forall p \in P(s)} I(p) \tag{10.4}$$

Figure 10.11 shows an example set of I, P, and S. Solution $s0$, which is one instance of the reconfigurable AGU, is composed of three PEs, and three PE implementation patterns, $p0$, $p2$, and $p4$, are assigned to solution $s0$. Then $I(p0) = \{add\}$, $I(p2) = \{add, sub, sft\}$, and $I(p4) = \{mul, mod\}$. When we assume $I^A = \{add, sub, sft, mul\}$, solution $s0$ can execute I^A. However, when we assume $I^A = \{add, sub, sft, mul, sp0\}$, solution $s0$ cannot execute I^A because $s0$ does not include instruction $sp0$.

The number of architecture candidates in the solution space S equals $|P|^m$, where m is the number of assumed PEs shown in Figure 10.11, because each PE has $|P|$ implementation patterns.

However, because of the full connectivity of the reconfigurable AGU template, the proposed AGU can achieve the same level of functionality as the AGU whose PEs are swapped. Therefore, the number of architecture candidates with m PEs can shrink to $_{|P|}H_m$, which means repeated m-combinations from $|P|$ elements. Finally, the number of architecture candidate $|S|$ is described as follows:

$$|S| = \sum_{m=1}^{max_{PE}} {}_{|P|}H_m = \sum_{m=1}^{max_{PE}} \frac{(|P| + m - 1)!}{m!(|P| - 1)!}, \tag{10.5}$$

where max_{PE} is maximum limit on the number of PEs.

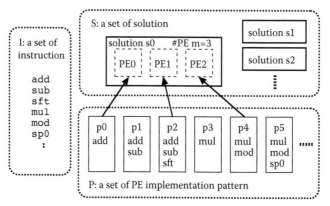

Figure 10.11   PE implementation pattern and its assignment.

Usually, the number of functional units in embedded processors is limited, and the maximum number of PEs max_{PE} cannot readily be increased significantly in the real world. The number of PE implementation patterns $|P|$ is expected to increase when we consider special instructions for more effective address calculation. Then, $|S|$ may explode because $|S|$ increases in the factorial order of $|P|$. It is through the formal description of the reconfigurable AGU model that we notice the variation.

10.3.2 Cost model for the reconfigurable AGU model

The previous section explained the variations of the reconfigurable AGU model, and how its variations increase in the factorial order of the number of patterns.

We often use the term "design." However, it is worth reflecting on the meaning of "design" within this context. We define the "design" as the act of choosing the optimal solution from the solution space. The potential variations of the solution is called the "solution space," and the size of the solution space is calculated by the Equation 10.5. Therefore, the size of the solution space impacts on the difficulty of the design.

In order to choose the optimal solution from the solution space, we have to evaluate each AGU model. To evaluate an instance of the reconfigurable AGU model, cost models including area and energy estimation should be defined. In this chapter, a simple cost model is used as follows.

Let $a(i)$, $e^L(i)$, $e^D(i)$ be area $[mm^2]$, leakage energy $[J/cycle]$, dynamic energy $[J]$ that means energy per activation of instruction i, respectively. Area $A(s)$ of solution s is calculated as follows:

$$A(s) = \sum_{\forall p \in P(s)} \sum_{\forall i \in I(p)} a(i). \tag{10.6}$$

Let $C(s)$ represent the execution cycle counts, obtained by AGU mapping, required to execute a given application on solution s, and $E(s)$ represents the energy consumption required to execute all instructions in I^A on solution s. $E(s)$ is calculated as follows:

$$E(s) = C(s) \cdot E^L(s) + E^D, \tag{10.7}$$

where $E^L(s)$ and E^D are the leakage energy of the solution s, and the dynamic energy to calculate I^A, respectively. $E^L(s)$ and E^D are defined as follows:

$$E^L(s) = \sum_{\forall p \in P(s)} \sum_{\forall i \in I(p)} e^L(i), \tag{10.8}$$

$$E^D = \sum_{\forall i \in I^A} n(i, I^A) \cdot e^D(i), \tag{10.9}$$

where $n(i, I^A)$ is the number of instructions i contained in a given application I^A.

Please note that a modification of cost models would be needed for more precise evaluation including clock and power gating schemes, dynamic voltage and frequency scaling, etc. because the currently used models lack complexity. They do, however, suffice for illustrative purposes. Equation 10.9 also means each instruction consumes the same amount of energy per activation in all PE implementation patterns that include the instruction. For a more precise evaluation, the definition of E^D should be modified when more precise data are available given that each instruction consumes different energy amounts for each implementation pattern, etc.

10.4   Architecture Exploration Problem: What Is the Optimal Solution?

In the previous section, we explained that the term "design" represents the choosing of the optimal solution from the solution space. The solution space can become very large as many design parameters exist. However, the big questions are as follows: What is optimal? Is the optimal solution the highest performance architecture? Is the optimal solution the lowest energy architecture? This is quite a challenging issue in general terms, but before discussing it in more detail, it is worth considering a better solution in a formal way.

Let S be a solution space, a set of solutions. Then $C(s)$, $A(s)$, and $E(s)$ represent cycle counts, area, and energy for solution $s \in S$, respectively. Let $BS_{CA}(s)$ and $BS_{CE}(s)$ be *better solutions* of solution s under cycle versus area/energy trade-off, respectively. $BS_{CA}(s)$ and $BS_{CE}(s)$ are defined as follows:

$$BS_{CA}(s) = \{x | x \in S, C(x) < C(s), A(x) < A(s)\}, \qquad (10.10)$$

$$BS_{CE}(s) = \{x | x \in S, C(x) < C(s), E(x) < E(s)\}. \qquad (10.11)$$

This is quite a standard definition of the better solution, and we can use this definition to address in more detail the identification of the optimal solution. Actually, the optimal solution is dependent on factors such as the designer, target products, etc. So it is impossible to designate one solution as the optimal solution. Therefore, we introduce the concept of Pareto optimal in this discussion. Pareto solutions of cycle versus area (PS_{CA}) and Pareto solutions of cycle versus energy (PS_{CE}) are defined as follows:

$$PS_{CA} = \{x | x \in S, BS_{CA}(x) = \phi\}, \qquad (10.12)$$

$$PS_{CE} = \{x | x \in S, BS_{CE}(x) = \phi\}. \qquad (10.13)$$

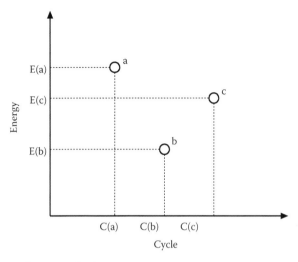

Figure 10.12 Better solution.

Pareto solutions are usually regarded as the solution on the trade-off curve, and the "design" can be classified into two layers as follows:

- Searching for Pareto solutions from within the solution space.
- Choosing the optimal solution from Pareto solutions.

We call this approach the "Searching Pareto solutions from solution space" type of architecture exploration. Once the Pareto solutions can be obtained, all the designer has to do is choose the appropriate Pareto solution because by definition, for any Pareto solution there is no better solution. Then the architecture exploration problem is defined as follows.

– ARCHITECTURE EXPLORATION PROBLEM –

For given application I^A, to find Pareto solutions PS_{CA} and PS_{CE} from the available solution space S.

$\square$

For example, Figure 10.12 shows an example of better solutions and Pareto solutions under cycle versus energy trade-off. The *better solution* $BS_{CE}(\cdot)$ for each solution in solution space $S = \{a, b, c\}$ can be obtained as follows:

- $BS_{CE}(a) = \phi,$
- $BS_{CE}(b) = \phi,$
- $BS_{CE}(c) = \{b\}.$

Then Pareto solution set PS_{CE} includes two solutions: solution a and b, and designers then have to select the best solution considering their specific constraints.

10.5 AGU Mapping Framework: How Is the Address Calculation Mapped on the AGU Model?

Address calculation is recognized as a significant performance bottleneck and conventional approaches do not provide any silver-bullet solution to the problem. This section explains the performance-evaluation method for a given AGU model and application via the use of a practical example.

In the following section, we take the example from Figure 10.3, and pick up the piece of code in case *comp* is 1 or 3 in Figure 10.3. Figure 10.13 shows a piece of sample code and the extracted address calculation data-flow graph.

Address calculation can be defined as a calculation between brackets of an array. Therefore, the address calculation can be extracted as shown in Figure 10.13; the data-flow graph (DFG) is extracted as well. This DFG is

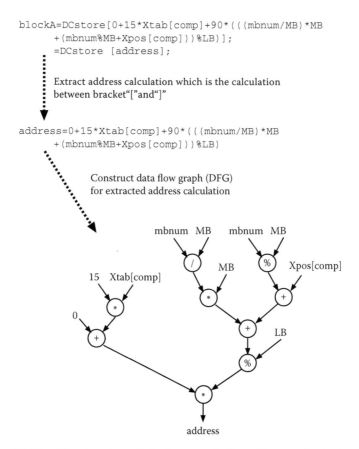

Figure 10.13 Constructing data-flow graph for address calculation.

now our target calculation, and we answer the two questions in the following section: "What do we perform the target calculation on?" and "How do we design the suitable address generation unit?" Notice that the address calculation does not include any control flow, and this enables a straight forward evaluation.

In order to evaluate performance, we have to take two inputs: reconfigurable AGU spec and address calculation DFG. The AGU mapping framework maps the given DFG to the given reconfigurable AGU model, and at this point the performance can be evaluated.

Figure 10.14 shows an example of performance evaluation. Now we consider the reconfigurable AGU, which has 4 PEs, and the input DFG is from the piece of code. As the reconfigurable AGU has 4 PEs, the given DFG from the piece of code has to be scheduled for each step, and to be mapped for each PE.

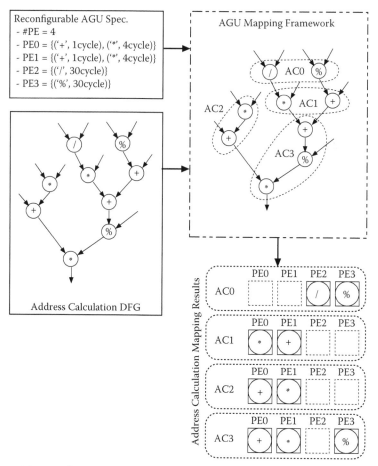

Figure 10.14 Overview of AGU mapping framework.

Once this information is obtained, the total execution cycle is also calculated. For example, AC0 takes 30 cycles because "/" and "%" can be executed in parallel. In the same way, AC1, AC2, and AC3 take 4 cycles, 5 cycles, and 35 cycles respectively. Therefore, the total execution cycle is 74 cycles, and now we obtain the performance to execute the piece of code on the given AGU model. This is performance evaluation using the AGU mapping framework.

An important challenge for AGU mapping framework is how to obtain good scheduling and good mapping in practical time. Normally, this problem is NP-hard, and various heuristics have been proposed. We created the AGU mapping framework using the graph partitioning algorithm based on SA (simulated annealing). Graph partitioning by SA is a common challenge, and a more detailed explanation can be found in the reference [31].

Once the AGU mapping framework works well, changing the AGU specification, which is the input to the AGU mapping framework, we can evaluate various architectures in practical time. The AGU mapping framework now becomes a performance-evaluation engine for the AGU exploration framework explained in the next section.

10.6   AGU Exploration Framework: How Are Pareto Solutions Obtained from the Solution Space?

Exhaustive architecture exploration makes automated architecture exploration possible, but it is very difficult to identify the best architecture from the vast solution space because our architecture template has many design parameters, and the available hybrid combinations are enormous. This makes architecture exploration in practical time impossible as exhaustive architecture exploration evaluation iterates evaluations for all solutions in the available solution space.

Systematic architecture exploration from this vast solution space is a demanding challenge and represents a significant dilemma. Fast and accurate evaluation, which is a crucial step during architecture exploration, consumes for each solution a potentially overly-large elapsed time. Therefore, the best approach to enable fast and accurate architecture exploration is to limit the evaluation to promising solutions, that is, those that might become a Pareto solution. This approach will bring significant exploration time reduction; the next challenge to be addressed is, therefore, how to identify these promising solutions. The objective of architecture exploration is to find trade-off curves of cycle versus area and cycle versus energy. This means, a method to find promising solutions for each trade-off point is necessary. The fastest way to find these is based on using estimations for each metric. This enables crude trade-off curves to be obtained. As such, this section proposes the concept of

"Optimistic Cycle (OC)" as a new metric for cycle estimation. Additionally, this section also shows that a set of cycle versus energy Pareto solutions becomes a subset of cycle versus area Pareto solutions based on certain practical assumptions. Combining the new metric "Optimistic Cycle" and the specific feature of our Pareto solutions, we can achieve fast and accurate architecture exploration.

10.6.1 Specific features of the Pareto solution

To solve the previously discussed architecture exploration problem, it is necessary to consider both PS_{CA} and PS_{CE}. However, an interesting feature of Pareto solutions concerning cycle versus area and cycle versus energy should be noted, that is: a set of cycle versus energy Pareto solutions becomes a subset of cycle versus area Pareto solutions under certain defined assumptions.

The energy model in question is constructed based on the following assumptions and restrictions:

- Clock and power gating schemes are not applied.
- Voltage and frequency remain constant.
- Each instruction consumes the same energy per activation in all PE implementation patterns that include the instruction.

Empirically, it is known that leakage energy increases proportionally to area [32]. When $E^L(s)$ refers to leakage energy (Equation 10.8) for solution s, then $E^L(s) < E^L(t)$ is assumed for two solutions $s, t \in S$ that satisfy $A(s) < A(t)$ under the assumption that clock and power gating schemes are not applied. Then the following lemma can be introduced.

Lemma 10.1 *When $A(s) < A(t)$ and $C(s) < C(t)$ is satisfied for two solution $s, t \in S$, $E(s) < E(t)$ can also be satisfied.*

Proof Dynamic energy E^D as defined by Equation 10.9 remains constant because of the characteristic of the energy model definition. With the assumption of leakage energy above mentioned, $E^L(s) < E^L(t)$ is true when $A(s) < A(t)$. Therefore, $E(s) < E(t)$ is clearly satisfied by the definition of the energy consumption in Equation 10.7. $\Box$

Based on the lemma, the following theorem can be introduced.

Theorem 10.1 *For a given application, PS_{CA} and PS_{CE} satisfy the following equation under the predefined assumptions:*

$$PS_{CE} \subseteq PS_{CA} \subseteq S. \tag{10.14}$$

Proof Consider solution $s \in S$ is such that $s \in PS_{CE}$ and $s \notin PS_{CA}$. By definition of PS_{CE} and PS_{CA}, $PS_{CE} \subseteq S$ and $PS_{CA} \subseteq S$ are obviously satisfied and the following equations are also satisfied for solution s:

$$BS_{CE}(s) = \phi, \tag{10.15}$$

$$BS_{CA}(s) \neq \phi. \tag{10.16}$$

Then conditions $C(v) < C(s)$ and $A(v) < A(s)$ are true, where $v \in BS_{CA}(s)$. From Lemma 10.1, $E(v) < E(s)$ is also satisfied for solution $s, v \in S$. Since $C(v) < C(s)$ and $E(v) < E(s)$ are satisfied for solution $s, v \in S$, Equation 10.17 is also satisfied.

$$\exists v \in BS_{CE}(s) \tag{10.17}$$

Now, we see that Equations 10.15 and 10.17 contradict each other. Therefore, solution s does not exist and Equation 10.14 is satisfied. ☐

Based on Theorem 10.1, we can focus on obtaining PS_{CA} to get both PS_{CA} and PS_{CE}. The energy consumption can be calculated easily once the execution cycle counts are obtained. Therefore, according to Equation 10.14, PS_{CE} can be obtained by calculating energy consumption for all solutions in PS_{CA} and plotting the cycle versus energy trade-off curve.

10.6.2 Systematic architecture exploration method

In this section, the systematic architecture exploration algorithm is described that is used to draw up fast cycle versus area tradeoff curve [32]. To obtain a fast cycle versus area trade-off curve, the following approach is taken: (1) find out the promising architectures expected to fall in the cycle versus area Pareto solutions space, and (2) apply AGU mapping only to these promising architectures.

To identify promising architectures under cycle versus area trade-off, a new metric called *optimistic cycle (OC)* is proposed. OC provides a crude cycle estimation based on the following equation:

$$OC = \sum_{\forall i \in I^A} \frac{n(i, I^A) \cdot latency(i)}{para(i)}, \tag{10.18}$$

where $latency(i)$ and $para(i)$ mean a latency of instruction i and the number of PEs, which can execute instruction i simultaneously on the given processor architecture, respectively. A solution that achieves a lower OC will achieve a lower cycle count, and promising architectures may be obtained by exploring OC versus area trade-off for all architectures.

Algorithm 10.1 shows the proposed fast architecture exploration algorithm that can be used to obtain PS_{CA} and PS_{CE}. The proposed algorithm first calculates the OC metric for all architectures, and explores OC versus area trade-off to obtain promising architectures under cycle versus area trade-off. Then AGU mapping is applied for the promising architectures, namely Pareto solutions of OC versus area, and cycle versus area/energy Pareto curves are then obtained.

Under the aforementioned assumptions and based on the proof in Theorem 1, focusing on cycle versus area trade-off will also lead to cycle versus energy trade-off solution points.

Algorithm 10.1 Fast Architecture Exploration Algorithm

Calculate OC for all solutions in S.

Calculate area for all solutions in S.

Find Pareto solutions from OC versus area trade-off.

Apply AGU mapping for all OC versus area Pareto solutions.

Plot the OC versus area Pareto solutions on cycle versus area field.

Plot the OC versus area Pareto solutions on cycle versus energy field.

Obtain PS_{CA} and PS_{CE}.

10.7 Experimental Results

To demonstrate the proposed systematic architecture exploration method, the proposed algorithm is implemented with a Perl script and executed on a PentiumD 2.8 GHz with 2 GB memory. The AGU mapping was performed by a simulated annealing (SA) based algorithm as defined in [31]. We prepare four benchmarks: Handmade, Cavity, Motion, and QSDPCM. Handmade is created artificially. Cavity, Motion, and QSDPCM are all real application kernels, where the address calculation DFGs have been separated manually. Cavity is part of an image-processing application to detect cavities in brain tomographies. Motion and QSDPCM are from video motion compensation and quad-tree structured difference pulse code modulation, which are used in multimedia codecs.

Figure 10.15 shows a piece of Cavity benchmark. Variables _race_* are inserted by compiler optimization, and calculations related to the variables are regarded as "address calculation." Obviously the address calculations are dominant in the application.

Table 10.1 shows the exploration time and the number of evaluated solutions. "Time" and "#Evaluated" refer to exploration time and the number of evaluated solutions that are applied in the AGU mapping, respectively. "Exhaustive" and "OC" refer to the exhaustive search [31] by applying AGU mapping for all architecture candidates and the proposed algorithm using OC, respectively.

```
...
for ( ;_race_91 = _race_90 ,_race_91 < 400; )
{
  int _race_82;
  int _race_106;
  int _race_102;
  int _race_92;
  int _race_96;
  int _race_76;
  int _race_94;

  _race_92 = 0;
  _race_4 = _race_3;
  _race_96 = _race_4;
  _race_57 = _race_56;
  _race_82 = _race_57;
  _race_75 = _race_74;
  _race_76 = _race_75;
  _race_81 = _race_80;
  _race_94 = _race_81;
  _race_101 = _race_100;
  _race_102 = _race_101;
  _race_105 = _race_104;
  _race_106 = _race_105;
  _race_112 = _race_91 % 2;
  _race_119 = _race_91 * 636;
  _race_120 = _race_119 - 4;
  _race_130 = _race_91 >= 2;
  _race_122 = _race_91 >= 4;
  _race_115 = _race_91 >= 6;
  _race_123 = _race_119 - 640;
  _race_126 = _race_91 <= 398;
  _race_128 = _race_115 ? 6 : 640;
  _race_127 = _race_122 && _race_126;
  _race_139 = _race_130 && _race_126;

  for ( ; _race_93 = _race_92 , _race_93 < _race_128; )
  {
    in_image_0 = in_image_1;
    in_image_1 = in_image_2;
    in_image_2 = input();
    _race_95 = _race_94;
    _race_129 = _race_93 >= 2;
    _race_118 = &_race_8[_race_95];

  if (_race_129)
  {
    g_acc_x = 0;
    g_acc_x += gauss_filt(in_image_0, -1);
    g_acc_x += gauss_filt(in_image_1, 0);
    g_acc_x += gauss_filt(in_image_2, 1);
    *_race_118 = g_acc_x / _race_10;
  }
  ...
```

Figure 10.15 Benchmark cavity code.

Table 10.1 Exploration Time and Number of Evaluated Solutions

	Time [sec]	#Evaluated
	Exhaustive / OC	Exhaustive / OC
Handmade	12,384 / 236	1,909 / 38
Cavity	10,341 / 157	2,387 / 35
Motion	14,790 / 136	1,120 / 12
QSDPCM	46,356 / 282	3,586 / 23

We can see the exploration time is drastically decreased by the proposed algorithm using OC because AGU mapping is applied only to a limited set of solutions. To obtain a Pareto curve for QSDPCM, it took about 13 hours by means of exhaustive search. However, applying the proposed algorithm, the exploration time of QSDPCM is decreased to about 5 minutes. By applying the proposed algorithm for QSDPCM, architecture exploration can be done up to 164 times faster than exhaustive search.

Figures 10.16 and 10.17 show cycle versus area Pareto curves obtained by exhaustive search and by the proposed algorithm, respectively. The black dot means one solution evaluated by AGU mapping, and the white point means a Pareto solution. The proposed algorithm obviously applies AGU mapping only to the solutions that are located near the Pareto curves.

Figures 10.18 and 10.19 show the comparison of cycle versus area/energy Pareto curves of QSDPCM, and Figures 10.20 and 10.21 show the comparison

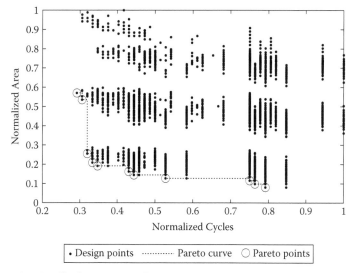

Figure 10.16 Cycle vs. area Pareto curve obtained by exhaustive search (QSDPCM).

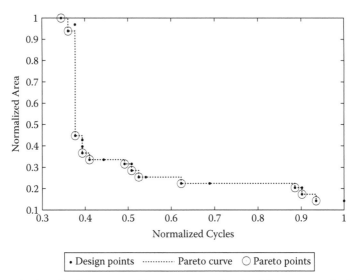

Figure 10.17   Cycle vs. area Pareto curve obtained by proposed algorithm (QSDPCM).

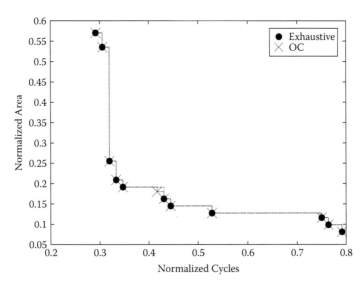

Figure 10.18   Comparison of cycle vs. area Pareto curves (QSDPCM).

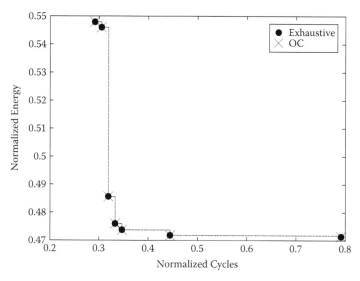

Figure 10.19 Comparison of cycle vs. energy Pareto curves (QSDPCM).

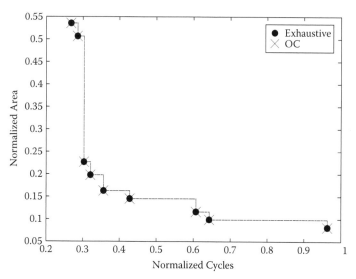

Figure 10.20 Comparison of cycle vs. area Pareto curves (motion).

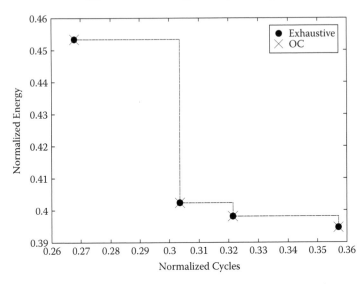

Figure 10.21 Comparison of cycle vs. energy Pareto curves (motion).

of cycle versus area/energy Pareto curves of Motion. The black point and the X-mark correspond to Pareto solutions obtained by exhaustive search and the proposed algorithm, respectively. Both Pareto curves in each figure are normalized to the same scale. We can see that the obtained Pareto solutions are completely overlapped and the proposed algorithm can obtain the same Pareto curves as the exhaustive search for QSDPCM and Motion for both cycle versus area and cycle versus energy. The proposed algorithm obtains almost the same Pareto curves, and this not only for QSDPCM and Motion but also for the other benchmarks. Therefore, focusing on only the cycle versus area trade-off, the proposed systematic architecture exploration method can draw up the same Pareto curves for both cycle versus area and cycle versus energy in short execution times.

10.8 Conclusion and Future Work

In this chapter, we have proposed an energy-efficient reconfigurable address-generation unit and its associated design methodology. First, we have summarized the state of the art address-generation units and have outlined the motivation for exploration in this field. We have also proposed an energy-efficient flexible AGU template and its corresponding architecture exploration framework.

Various techniques to achieve even lower energy consumption are being proposed in ongoing research, and most of the future architectures and related tools will include these techniques. Our current energy model is quite crude, and some of our assumptions will not necessarily remain true when some of these techniques are taken into account. For example, by applying effective power gating, which is one of the more promising techniques, the proof in Lemma 1 will not be true anymore. To be applicable in these future low-energy oriented designs, our exploration technique needs a more elaborate energy estimation model. This is part of our future work.

The current technique can effectively handle loop kernels. As another avenue of future work, additional pruning heuristics for even faster architecture exploration are needed for when we want to address even larger applications.

10.9 Exercises

1. Describe the reasons why address is becoming more and more important.

2. Discuss the advantages in separating the data calculation and address calculation.

3. Regarding the following address equation:

$$AE : \delta A[3 * i + j] \tag{10.19}$$

   What do you think will be the best implementation option for small ranges (*e.g.*, from 0 to 3) of i and j? And for medium ranges (*e.g.*, from 0 to 50)? And large ranges (*e.g.*, from 0 to 10,000)?

4. Regarding the following address equation:

$$for(i = 0; i < N; i + +)res[k + i] = tmp; \tag{10.20}$$

   What do you think will be the best implementation style if N is small ($N < 10$) and k is a known parameter? And if k depends on the application?

5. Summarize the address optimization techniques at the different levels. Among the previous optimizations:

   (a) Which ones are best suited for computing intense and regular applications?

   (b) Which ones are best suited for computing intense and irregular applications?

6. When we assume 10 PE implementation patterns, calculate the solution space where the number of PE is 5.

7. Describe the DFG of address calculation from Figure 10.4, and design the appropriate AGU.

References

1. Rainer Leupers. Code generation for embedded processors. In *ISSS '00: Proceedings of the 13th International Symposium on System Synthesis*, pages 173–178, Washington, DC, USA, 2000. IEEE Computer Society.

2. Stefan Valentin Gheorghita, Martin Palkovic, Juan Hamers, Arnout Vandecappelle, Stelios Mamagkakis, Twan Basten, Lieven Eeckhout, Henk Corporaal, Francky Catthoor, Frederik Vandeputte, and Koen De Bosschere. System-scenario-based design of dynamic embedded systems. *ACM Trans. Des. Autom. Electron. Syst.*, 14(1):1–45, 2009.

3. Martin Palkovic, Erik Brockmeyer, Peter Vanbroekhoven, Henk Corporaal, and Francky Catthoor. Systematic preprocessing of data dependent constructs for embedded systems. In *Proceedings of PATMOS*, pages 89–98, 2005.

4. Martin Palkovic, Henk Corporaal, and Francky Catthoor. Global memory optimisation for embedded systems allowed by code duplication. In *SCOPES '05: Proceedings of the 2005 Workshop on Software and Compilers for Embedded Systems*, pages 72–79, New York, NY, USA, 2005. ACM Press.

5. Stefan Valentin Gheorghita, Sander Stuijk, Twan Basten, and Henk Corporaal. Automatic scenario detection for improved WCET estimation. In *DAC '05: Proceedings of the 42nd Annual Conference on Design Automation*, pages 101–104, New York, NY, USA, 2005. ACM Press.

6. Jan Vanhoof, Ivo Bolsens, Karl Van Rompaey, Gert Goossens, and Hugo De Man. *High-Level Synthesis for Real-Time Digital Signal Processing*. Kluwer Academic Publishers, Norwell, MA, USA, 1993.

7. D. Grant, P.B. Denyer, and I. Finlay. Synthesis of address generators. In *ICCAD-98: IEEE International Conference on Computer-Aided Design*, pages 116–119, 1989.

8. M. Miranda, F. Catthoor, and H. De Man. Address equation multiplexing for realtime signal processing applications. In *VLSI Signal Processing VII*, pages 188–197, La Jolla, California, USA, 1994.

9. M. Miranda, M. Kaspar, F. Catthoor, and H. de Man. Architectural exploration and optimization for counter based hardware address generation. In *EDTC '97: Proceedings of the 1997 European Conference on Design and Test*, page 293, Washington, DC, USA, 1997. IEEE Computer Society.

10. Miguel Miranda, Francky Catthoor, Martin Janssen, and Hugo de Man. Adopt: Efficient hardware address generation in distributed memory architectures. *ISSS*, 00:20, 1996.

11. Miguel A. Miranda, Francky Catthoor, Martin Janssen, and Hugo J. De Man. High-level address optimization and synthesis techniques for data-transfer-intensive applications. *IEEE Trans. Very Large Scale Integr. Syst.*, 6(4):677–686, 1998.

12. Heiko Falk, Cdric Ghez, Miguel Miranda, and Rainer Leupers. High-level control flow transformations for performance improvement of address-dominated multimedia applications. *Proceedings of 11th Workshop on Synthesis and System Integration of Mixed Information Technologies (SASIMI)*, pages 135–150. Hall, 2003.

13. Heiko Falk and Peter Marwedel. Control flow driven splitting of loop nests at the source code level. In *DATE '03: Proceedings of the Conference on Design, Automation and Test in Europe*, page 10410, Washington, DC, USA, 2003. IEEE Computer Society.

14. Sumit Gupta, Miguel Miranda, Francky Catthoor, and Rajesh Gupta. Analysis of high-level address code transformations for programmable processors. In *DATE '00: Proceedings of the Conference on Design, Automation and Test in Europe*, pages 9–13, New York, NY, USA, 2000. ACM Press.

15. C. Ghez, M. Miranda, F. Catthoor, and D. Verkest. Systematic high-level address code transformations for piece-wise linear indexing: illustration on a medical imaging algorithm. In *Proc. IEEE Workshop on Signal Processing Systems (SIPS2000)*. IEEE Press, 2000.

16. S. Hettiaratchi, P. Cheung, and T. Clarke. Performance-area trade-off of address generators for address decoder-decoupled memory. In *DATE '02: Proceedings of the Conference on Design, Automation and Test in Europe*, page 902, Washington, DC, USA, 2002. IEEE Computer Society.

17. D.M. Grant and P. B. Denyer. Address generation for array access based on modulus m couters. In *EDAC '91: Proceedings of the 2nd ACM/IEEE European Conference on Design Automation (EDAC)*, European Design Automation Conference, pages 118–123, Feb. 1991.

18. P. Lippens, J. V. Meerbergan, A. V. der Werf, and W. Verhaegh. Phideo: a silicon compiler for high speed algorithms. *Proceedings of the European Conference on Design Automation*, European Design Automation Conference, pages 436–441, 1991.

19. D.M. Grant, J. V. Meerbergen, and P. Lippens. Optimization of address generator hardware. *DATE '94: Proceedings of the 5th ACM/IEEE European Design and Test Conference*, European Design Automation Conference, pages 325–329, 1994.

20. A. Basu, R. Leupers, and P. Marwedel. Register-constrained address computation in dsp programs. In *DATE '98: Proceedings of the Conference on Design, Automation and Test in Europe*, pages 929–930, Washington, DC, USA, 1998. IEEE Computer Society.

21. Ashok Sudarsanam, Stan Liao, and Srinivas Devadas. Analysis and evaluation of address arithmetic capabilities in custom dsp architectures. In *DAC '97: Proceedings of the 34th Annual Conference on Design Automation*, pages 287–292, New York, NY, USA, 1997. ACM Press.

22. Rainer Leupers and Peter Marwedel. Algorithms for address assignment in DSP code generation. In *ICCAD*, pages 109–112, 1996.

23. Rainer Leupers. *Retargetable Code Generation for Digital Signal Processros*. Kluwer Academic Publishers, 1997.

24. Binu Mathew and Al Davis. A loop accelerator for low power embedded vliw processors. In *Proc of CODES and ISSS*, Stockholm, Sweden, September 2004.

25. Alfred V. Aho, Ravi Sethi, and Jeffrey D. Ullman. *Compilers: Principles, Techniques, and Tools*. Addison-Wesley Longman Publishing Co., Inc., Boston, MA, USA, 1986.

26. Clifford Liem, Pierre Paulin, and Ahmed Jerraya. Address calculation for retargetable compilation and exploration of instruction-set architectures. In *DAC '96: Proceedings of the 33rd Annual Conference on Design Automation*, pages 597–600, New York, NY, USA, 1996. ACM Press.

27. C. Liem, P. Paulin, and A. Jerraya. Compilation methods for the address calculation units of embedded processor systems. In *Proceedings of the Design Automation for Embedded Systems*, pages 61–77, Netherlands, 1997. Springer.

28. Wei-Kai Cheng and Youn-Long Lin. Addressing optimization for loop execution targeting dsp with auto-increment/decrement architecture. In *ISSS '98: Proceedings of the 11th International Symposium on System Synthesis*, pages 15–20, Washington, DC, USA, 1998. IEEE Computer Society.

29. Rainer Leupers. *Code Optimization Techniques for Embedded Processors Methods, Algorithms, and Tools*. Kluwer, 2000.

30. J. Ramanujam, Satish Krishnamurthy, Jinpyo Hong, and Mahmut Kandemir. Address code and arithmetic optimizations for embedded systems. In *ASP-DAC '02: Proceedings of the 2002 Conference on Asia South Pacific Design Automation/VLSI Design*, page 619, Washington, DC, USA, 2002. IEEE Computer Society.

31. Ittetsu Taniguchi, Praveen Raghavan, Murali Jayapala, Francky Catthoor, Yoshinori Takeuchi, and Masaharu Imai. Reconfigurable agu:

An address generation unit based on address calculation pattern for low energy and high performance embedded processors. *IEICE Transactions on Fundamentals of Electronics, Communications and Computer Sciences*, E92.A(4):1161–1173, 2009.

32. Ittetsu Taniguchi, Murali Jayapala, Praveen Raghavan, Francky Catthoor, Keishi Sakanushi, Yoshinori Takeuchi, and Masaharu Imai. Systematic architecture exploration based on optimistic cycle estimation for low energy embedded processors. In *Proc. of ASP-DAC09*, pages 449–454, 2009.

Index

A

Address-generation units, 309–341
  architecture exploration, 324–325,
    330–331
  benchmark cavity code, 332
  classification, 314–320
  datapath-based AGUs, 316–320
  exploration framework, 328–331
  mapping framework, 326–328
  Pareto solutions, 328–331
    features of, 329–330
  reconfigurable AGU, 320–324
    cost model, 323–324
    variation, 321–323
  table-based AGUs, 315–316
  types of address equations, 311–314
    AE design space exploration, 314
    flexibility, 313–314
    regularity, 311–313
Affine dependences, 22, 53–64
  affine recurrence equations, 53
  algorithm, 58–63
    broadcast-reductions, 62
    Hermite normal forms, 58–60
    null space computing, 58–60
    pipelining vectors, 62–63
    redundant iteration spaces,
      60–62
  broadcast, 57
  classification, dependences, 55–58
    affine hull, 56
    broadcast, 55
    dependence graphs, dependences, 55
    local, 55
    non-local, 55
    null spaces, 56–58
    reduction, 55
    redundant iteration space, 58, 60
  dependences, 54–55
  reduction, 57
  regular array processors, 53
  systolic arrays, 53
  variable, 54

Affine specifications, data storage, 67–116
  decomposition, array references,
    88–92
  lattice size, 92–98
  linearly bounded lattices, 82–88
    difference, linearly bounded lattices,
      83–88
    intersection, linearly bounded
      lattices, 82–83
  loop fusion, storage requirement, 69
  loop interchange, storage
    requirement, 69
  memory size computation, 71–74
  minimum data storage, 74–83, 98–105
    concepts, 76–78
    definitions, 76–78
    delays, array references with,
      103–104
    estimation operating mode, 75
    flow of algorithm, 98–103
    index space, array reference,
      78–82
    inter-array memory sharing, 75–76,
      195, 208, 211, 232
    maximum iterator vector of array
      element, 104–105

B

Back-substitution function, 23
Barvinok, 124
Benchmark cavity code, address-
  generation units, 332
Binary occupied address-time domain,
  169–170
BOAT. *See* Binary occupied address-time
  domain
Broadcast-reductions, 18, 57, 62
Broadcasts, 53, 55–57, 60, 62–63

C

CLooG, 123
Coloring memory access node, 290

343

Printed and bound by CPI Group (UK) Ltd, Croydon, CR0 4YY

18/10/2024

01776243-0012